T I M E

GARLAND REFERENCE LIBRARY
OF THE HUMANITIES
(VOL. 1506)

John Harrison's no. 1 Marine Timekeeper, 1737, and in center foreground his prize winning no. 4, 1759; on the right Larcum Kendall's copy, K.1, used by Cook and Vancouver; on the left Kendall's K.2, once carried by Bligh in the *Bounty*. (*By courtesy of the National Maritime Museum—on loan from Ministry of Defence—Navy.*)

TIME
A Bibliographic Guide

Samuel L. Macey

GARLAND PUBLISHING, INC. • NEW YORK & LONDON
1991

Library of Congress Cataloging-in-Publication Data

Macey, Samuel L.
Time : a bibliographic guidc / Samuel L. Macey.
p. cm. — (Garland reference library of the humanities ; vol. 1506)
Includes indexes.
ISBN 0–8153–0646–6 (acid-free paper)
1. Time—Bibliography. I. Title. II. Series.
Z7876.M33 1991
[QB209] 91–25934
016.529—dc20 CIP

Printed on acid-free, 250-year-life paper
Manufactured in the United States of America

For J. T. Fraser

CONTENTS

INTRODUCTION

In varying degrees, concern with time as an experience, its role in collective and individual life, and the nature of time as a subject of enquiry has penetrated our daily lives no less than it has penetrated the formal domains of learning. This bibliography of selected time-related books and articles—divided into twenty-five existing academic disciplines and about one hundred subdisciplines—confirms the validity of that claim, particularly insofar as the domains of learning are concerned. Even those of us who have been closely connected with the study of time over many years have found it extremely difficult to acquire a sense of the subject as a whole. During the course of writing three books on time—dealing respectively with literature and philosophy, religion and the history of ideas, and social and economic history—I became acutely aware of the need for a multidisciplinary guide in the form of a good—though necessarily incomplete—bibliography. Indeed, during the past few years I could sense and am presently able to demonstrate that there has been a greatly accelerating growth in the study of time. Having now completed such a bibliography, I can even better appreciate the extent to which the nature and dimensions of the subject it reflects are a widely ranging theme of concern for the academic community.

FORMAT

The interdisciplinary nature of the study of time has made it essential to use a bibliographic format readily understandable by readers in every discipline. As a result, virtually all abbreviations—such as the acronyms for journals in English literature and the equally enigmatic shortened titles used in the sciences—have been avoided. In the main, abbreviations have been restricted to *J* for *Journal* and *UP* for *University Press*.

This bibliography adopts a format that has been widely used by North American scholars whose work in time studies cuts across the disciplines. The format employed respectively for books, for articles in journals, and for articles in collections of essays is represented by the following examples:

> Landes, David S. *Revolution in Time: Clocks and the Making of the Modern World*. Cambridge: Belknap P of Harvard UP, 1983.

Thompson, E. P. "Time, Work-Discipline and Industrial Capitalism." *Past and Present* 38 (December 1967): 61-84.

Davies, P. C. W. "Space-Time Singularities in Cosmology and Black Hole Evaporation." *The Study of Time III*. Ed. J. T. Fraser, N. Lawrence, and D. Park. New York: Springer, 1978. 74-91.

Since the above format comes very close to Basic Style *A* in *The Chicago Manual of Style* 13th ed., Section 16.5, and to that recommended in *The MLA Style Manual*, I have followed those publications in other appropriate areas. But with respect to the preference for the use of first names or initials, which varies both between disciplines and among scholars, no attempt has been made to impose a consistent form.

SOURCES

This bibliography has been developed by supplementing my existing collection of books, articles, and bibliographic references on time with the judicious use of references drawn from a number of sources that are readily available in most university libraries. Some readers may wish to use these sources for broadening appropriate sections of the bibliography or, in due course, for including more recent works in their particular areas of interest. Conveniently, the subject headings that can be searched in the sources indicated below are similar to the disciplines and subdisciplines as set out in this bibliography.

For bibliographic information on books the most comprehensive source is the *Library of Congress Catalog* and the related *Library of Congress Subject Catalog*. But the reader should not overlook the subject headings in his or her own library's card or computerized catalog as well as in the current issue of *Books in Print*. In addition, *Time's News*, distributed to members of the International Society for the Study of Time, contains useful reviews of recent publications. Though I have consciously omitted dissertations, the appropriate bibliographic information can be readily obtained by looking up entries under time-related words in the "Keyword Title Index" of *Dissertation Abstracts International*.

Articles present a special problem because no single index of journals can cover all of them. Readers will need to accommodate themselves to the particular abbreviations used in disciplines other than their own. But reference librarians are uniformly helpful; they will usually let one use such books as *Periodical Title Abbreviations*, which list the shortened forms employed for the many thousands of journals currently being published. Among the most useful periodical indexes—generally to be found in the reference section of university libraries—are the following:

Cumulated Index Medicus (subject headings include many aspects of science), 1960 to date

The Social Sciences Index, 1974 to date

The Humanities Index, 1974 to date

(the above two indexes were previously *International Index to Periodicals*, 1907-65, and *Social Sciences and Humanities Index*, 1965-74)

The Readers' Guide to Periodical Literature (for articles in magazines), 1900 to date

Art Index

The Music Index

Cumulative Subject Index to Psychological Abstracts

Index to Legal Periodicals

In addition, those who wish to supplement the entries selected for this bibliographic guide can use the range of specialized bibliographies indicated within their respective disciplines. For the requirements of most readers, however, this book will offer sufficiently detailed bibliographies. And for all readers it should provide an exciting new entry into the very wide range of disciplines that have been inspired and energized through an active concern with time and time measurement.

In some subjects, the time-oriented nature of the discipline as a whole is immediately evident. These include aging, archaeology, history, horology, navigation, time management, and time measurement. In other subjects, important subdisciplines have developed in which the time component is equally obvious. Examples are film and TV in art; chronobiology and evolution in biology; demography in human geography; paleontology and geochronology in geology and geophysics; limitation of actions in law; utopian, futuristic, and science fiction—as well as prophecy—in literature; dynamics and time series analysis in mathematics; chronopharmacology in medicine; rhythm, meter, and tempo in music; astronomy and cosmology, quantum and relativity theory, space-time, and thermodynamics and entropy in physics; memory and time perception in psychology; and the argument from design, creation, and eschatology in religion.

As can be readily ascertained from the Table of Contents to this book, the disciplines and subdisciplines indicated above account for slightly more than half of the subdivisions. Since they are also recognized components of their own disciplines, the subjects that relate directly to time have been given preference as the headings or subheadings for this bibliography. The use of so many time-related headings and subheadings has proved invaluable for structuring the bibliography and for extracting references from comparable subject headings in the periodical indexes and the *Library of Congress Subject Catalog*.

ESTIMATING THE TOTAL NUMBER OF TIME-RELATED ARTICLES

The high percentage of the headings and subheadings in this bibliography that refer directly to time has also helped in the difficult problem of attempting a numerical estimate of time-related literature at large. Only one previous attempt seems to have been made, though it involved a somewhat different approach.

J. T. Fraser, one of the very few scholars who has acquired an overview of this subject, published in 1981 a bibliography of some 800 books on time (see under "Multidisciplinary Studies: Bibliographies"). In an introduction, he examines the academic abstracts of periodical literature and, based on the number of citations for each, suggests the approximate proportions of some of the major subjects—like aging, biological rhythm, philosophy, and time measurement—to the number of time studies as a whole. Of the subject as a whole, he says, "We may...settle on the hypothetical figure of 65,000 as the number of periodical articles in this century potentially related to a systematic study of time."

Fraser's figures were obtained through the DIALOG Information Retrieval System, but there is no record of the descriptors and identifiers used at the time. In explaining his procedure, Fraser notes that "Fifteen databases of an educational information retrieval system were searched covering, more or less, the period 1965-80." The number of entries related to time were a few short of 18,500. These were reduced to 12,000 to allow for multiple listings, but then increased to 40,000 because the indexers "must be regarded as untutored in the study of time." Finally, the 40,000 was increased to 65,000 in order to allow for articles published during the first sixty-five years of this century. The small proportion that Fraser allows for articles published prior to 1965 shows that he judged the literature to have been rapidly increasing during the period 1965-80. These assumptions are fully corroborated by my own figures, which include the most recent decade and which, for the first time, have the advantage of being coordinated with an extensive multidisciplinary bibliography.

My own method has been to use a physical rather than a computer count from which to estimate the number of articles published on time during the past century. For the years 1980-89—and using the appropriate time-related subject headings in the main periodical indexes comparable to those in this book's Table of Contents—the total count is as follows:

Applied Science and Technology	2,127
Art Index	218
Chemical Abstracts	792
The Humanities Index	7,658
Cumulated Index Medicus	12,146
Index to Legal Periodicals	1,538
The Music Index	812
Psychological Index (Cumulative Subject)	14,919
The Social Sciences Index	10,083
	50,293

A comparative sampling of the entries suggests that the figure of 50,293 time-related articles during the period 1980-89 should be adjusted in two directions. For the purpose of arriving at a final estimate, this figure has therefore been reduced by approximately 40 percent to 30,000. The adjustment is to allow for multiple listings and any items included that are not significantly concerned with time. In the other direction, the figure of 30,000 has been increased by a factor of three to some 90,000. This adjustment—also based on a comparative sampling—makes appropriate allowance for entries under subject headings not directly related to time and also for what Fraser calls indexers untutored in the study of time.

Two methods suggest themselves for extrapolating from the figure of 90,000 for 1980-89 to the ninety years since the beginning of the century. As the entries in *Time: A Bibliographic Guide* were originally chosen on a basis that was relatively random insofar as dates were concerned, the figure of 90,000 can be adjusted by the proportion that the articles in the bibliography for the past decade bears to the articles in the bibliography for the period 1900-90. This produces a figure of approximately 205,000 time-related articles since the beginning of the century. Alternatively, if we add to our figure of 90,000 for the last decade, the 65,000 articles that Fraser estimates for the first eight decades of the century, we have a total figure of approximately 155,000 time-related articles since the beginning of the century. The two figures of 205,000 and 155,000 provide a broad band within which it is reasonable to assume that the true figure must lie. Since the modern mind craves numerical precision, we might suggest that about 180,000 time-related articles have been published since the beginning of the century.

Anybody who has not been closely involved with the study of time during the past twenty-five years, might well be expected to question the rapidly accelerating increase in the number of articles here estimated to have been published during that period. But the growth in publication is graphically corroborated when one considers the publication of time-related books during the same period. Neither Fraser nor anyone else seems to have attempted such an estimate for books.

ESTIMATING THE TOTAL NUMBER OF TIME-RELATED BOOKS

The system that I have used for estimating the number of time-related books published since 1900 grows directly out of the method used for estimating the number of articles published. Though this involves counting and tabulating the entries for individual books, it is greatly simplified by the fact that there is no need to employ a number of disparate periodical indexes. By a happy coincidence, the *Library of Congress Subject Catalog*, which has been published continuously since 1950, comprises some four million books in English and many other languages even during its first twenty-five years of publication. This has facilitated counting and tabulating time-related books under just over thirty main subject headings—in addition to the related subsidiary headings—from Age, Archaeology, Biological Rhythm, and Calendar through to Sundials, Tempo, Thermodynamics, and Time Study.

While the older volumes of the subject catalog have been cumulated in books for five-year periods, the material for the most recent years is available on fiches. The fiches for 1983-84 are cumulated, but those for 1985-89 are separated into individual years. In order to reduce the labor involved, only one five-year period has been counted for each decade up to and including the seventies. The annual averages are as follows: 1950-54 346, 1960-64 553, 1970-74 1,385, 1983-84 1,775, 1985 2,547, 1986 3,498, 1987 4,048, 1988 4,311, and 1989 4,310. When these figures are charted and the curve on the chart is extrapolated back to 1900, they add up to a total of some 63,000 books catalogued under time-related headings since the turn of this century.

The figure of 22,263 time-related books for the last seven years alone fully confirms the rapidly accelerating increase that we discovered when dealing with articles. Though we cannot develop comparable statistics for the years prior to 1950, a surprisingly early publication suggests that the figures before 1900 would have been very small. The *Library of Congress Catalogue of Subjects* in 2 volumes (Washington: Government Printing Office, 1869)—which curiously appears to have had no immediate successors—lists books under subject headings from the earliest time until the date of publication. Many of the time-related subjects with which we have thus far been dealing did not then, of course, exist. But the seven such subjects which do parallel those in the *Library of Congress Subject Catalog* (1950-89) yielded only 417 titles in all.

A sampling suggests that to allow for multiple entries of the books in the *Library of Congress Subject Catalog* a similar reduction is needed to that appropriate in the case of articles. The 63,000 has therefore been reduced by 40 percent to 38,000. In the other direction, the figure of 38,000 has been increased by a factor of some two and a half times. This adjustment has been made to allow for entries under subject headings not directly related to time, for indexers untutored in the study of time, and for books not included in the *Library of Congress Catalog*. We may therefore estimate that, since the beginning of this century, about 95,000 books have been published on time-related subjects.

SIZE, STRUCTURE, AND CONTENT OF THE BIBLIOGRAPHIC GUIDE

Clearly a multidisciplinary bibliography of all the books and articles on time can no longer be envisaged in book form. It would currently need about fifty volumes equal in size to this bibliographic guide. If and when such a work does appear, it will have to do so as an on-line computer bibliography, but even that would not eliminate the usefulness of introducing readers to the subject through this more traditional form of bibliography. Insofar as constructing a future computerized bibliography is concerned, the present work may be considered as a preparatory step for an open database under the name TIMELINE first mentioned in print in Fraser's article of 1981. There are surely many other and perhaps better ways than mine for classifying the material. But the construction of an improved framework depends on the extent of the material at hand, and on the desirable condition that many people have given thought to the classification. This book, therefore, is also a contribution to the development of a classification scheme for the study of time, one that need not copy those of the established disciplines but obtain its guidance from the study of time itself.

As far as possible, the guiding principle for the construction of this book has been to reflect the proportions and nature of the subject as a whole. To begin with, the size of the bibliography has been limited to the largest number of bibliographic entries that can conveniently be included within a single volume. The internal structure is equally pragmatic. The main headings have been derived from the disciplines now being taught in university departments. To these have been added subjects like Aging, Horology, Navigation, Time Management, and Time's Measurements, which have a wide application in relation to time studies. Approximately half of the subheadings, as we have already noted, are provided by time-related subject headings derived from the periodical indexes and the *Library of Congress Subject Catalog*. Where time-related subheadings are not available, preference is given to chronological subdivisions (as in Philosophy) or geographical subdivisions (as in Horology).

The size of individual divisions and subdivisions has been similarly controlled by pragmatic considerations. Large subjects—like Aging,

Chronobiology, and Psychology—cannot be allowed to crowd out completely the smaller subdisciplines, and the smaller entries must be at least large enough to give the reader a sense of the subject. Though there are exceptions, the preferred length of sections has been held to 80-450 entries and that of subsections to 12-150 entries. The intention has been to compile a multidisciplinary bibliography on time with at least a general sense of the appropriate weight that should be given to sections and subsections if they are to reflect the subject as a whole.

By compiling a bibliography of approximately six thousand entries with at least a sense of the proportion that should be given to sections and subsections, one acquires a real feeling for the full spectrum of this subject. In the same spirit, an attempt has been made to reflect the rapidly accelerating growth of time studies that has earlier been indicated. For example, the books listed here for 1983-89 represent almost exactly one third of the total number of books for the current century. This mirrors the proportions of 22,000 to 63,000 in the initial data. Approximately 45 percent of the articles in the bibliography were published during 1980-90, and this figure also reflects very closely the real growth in the publication of time-related articles which occurred in that period.

While it has been virtually impossible to assess with any confidence, the percentage of works that should be allocated to languages other than English, it was felt that publications in German, French, and Italian should be adequately represented. Titles in languages other than English account for almost 11 percent of the bibliography, and a further 2 percent are English translations of foreign titles. Languages other than English are represented generously in the *Library of Congress Catalog* and the *Index Medicus*; representation in the other periodical indexes varies considerably. Perhaps 20 percent of the total bibliography should be allocated to titles originating in languages other than English. An educated guess has also been exercised in restricting the total number of entries before 1900 to rather less than 2 percent of the whole.

In a bibliography restricted to one volume, no section, however important the reader may consider it, can possibly be exhaustive. But fortunately a bibliography of some six thousand entries is of sufficient length to ensure that almost all of the disciplines and subdisciplines include a body of entries at least comparable in number to the specialized bibliographies, where these exist. The ultimate goal is to provide a reasonably well-balanced indication of what the study of time has to offer. In this connection I have attempted to apply critical insights in choosing publications for interest, quality, and relevance to the study of time. Merely having the word *time* in the title has by no means been a sufficient criterion for inclusion in this work. What is required rather is that the primary focus of a publication should incorporate such questions as the human experience of time's passage, the idea of time or the timeless, or time as an ordering principle of physical change.

As a result, in addition to the works that I have myself read, I have been obliged to use my judgment—based on a long involvement with the interdisciplinary study of time—when including works which have been classified by others as relating to the study of time. As far as possible, I have done so by consulting reviews, by including publications used or accepted by my peers, and by assessing the reputation of the publisher, or the goals and reputation of the classifier. And beyond this, as indicated in my acknowledgements, I have, in respect of the greater part of the manuscript, arranged for specialists in particular fields to suggest corrections, deletions, and additions. Nevertheless, any factual or judgmental fallacies that remain—including the errors of transcription which plague all bibliographers—are my responsibility.

The limitation imposed by the length of this work as a whole means also that an article or book can only be entered once, and in a multidisciplinary bibliography the decision of where to enter it is not always easy. The Subject Index, however, will help to overcome a number of the problems related to the choice of classification in marginal cases. It will also guide readers seeking information on major figures or on subjects not included in the Table of Contents. I have tried to restrict the classification of "Multidisciplinary Studies" to those works which—although related to time, like every other entry in the bibliography—fall outside the category of a marginal case. Generally, they are too much concerned with a range of disciplines to be included comfortably under a single discipline or subdiscipline. I am, however, all too well aware of the potential for disagreement inherent in the use of this classification. Because of the essentially multidisciplinary and interdisciplinary nature of time studies and the inclusion of multiple authors—normally when these do not exceed three per publication—an Author Index has also been provided.

Insofar as I know, no previous bibliography has dealt with the multidisciplinary study of time using criteria in any way comparable to those as set out in this Introduction. Such few multidisciplinary bibliographies as presently exist have been listed separately under "Multidisciplinary Studies," but their limitations will be immediately apparent. However immodest it may seem, I must confess that working on the project has given me something like the sense of awe felt by Keats's Cortez as he looked out on the Pacific for the first time. And it has seemed important to let the reader share that vision by controlling the ratio of reduction. In this way I have attempted to make the model a reflection of what is now more clearly revealed as a large, wide-ranging, and dynamic discipline.

For those of us who have experienced the intellectual exhilaration and fruitfulness of the International Society for the Study of Time's interdisciplinary conferences, the multiplicity of academic endeavor reflected in this bibliography can only confirm what we have long suspected. Most people will use the bibliography as a tool for specialized research. But for many of us it will also represent an opportunity to savor and learn from a refreshing insight into the wide spectrum of human interests inherent in our subject as a whole.

I am grateful to the University of Victoria and to the Canada Council for financial aid related to the study of time; to the Institute of Management Services, of which I have been a member and a fellow for over thirty years; to Corpus Christi College, Cambridge, which facilitated my early research into matters temporal by electing me a visiting fellow in 1972; and to The International Society for the Study of Time, of which I am currently the president. My indebtedness to that society—which has members in thirty-three countries and over twenty disciplines—and to its founder, J. T. Fraser, is considerable. I am also indebted, to the staff of the McPherson Library, to my colleagues at the University of Victoria, to Brenda Gerth for advice on computers, and to two research assistants—Jolene Demers and Janelle Jenstad—who have given me invaluable help with this work. In addition, I owe particular thanks to Barbara Adam, Jacob Arlow, Mark Aultman, Richard Block, J. T. Fraser, Leonie Kellahar, Jonathan Kramer, Gaston Labrecque, David Park, Charles Sherover, and Marlene Soulsby. These colleagues have generously interrupted their own research on the study of time to read and correct those parts of this work in which they specialize. The credit is theirs, but any faults that may remain are my own. Finally, and above all, I am grateful to a surprisingly cheerful wife, whose remarkable competence and quiet encouragement have sustained me for more years than either of us cares to remember.

Time

1. MULTIDISCIPLINARY STUDIES

A. BIBLIOGRAPHIES

Das, T. K. *The Time Dimension: An Interdisciplinary Guide*. New York: Praeger, 1990. Of the 2,975 entries two disciplines take up more than sixty percent of the work: "Philosophy and Physics" (709), and Psychology (1,105). Geology receives 4 entries, Political Science 6, Geography 12, and Biology 89. There are no foreign language titles, the nineteen disciplines are not subdivided and there is no Subject Index.

Fraser, J. T. "A Report on the Literature of Time, 1900-1980." *The Study of Time IV*. Ed. J. T. Fraser, N. Lawrence, and D. Park. New York: Springer, 1981. 234-70. Bibliography of about 800 books grouped under 12 disciplinary areas.

Krudy, Elmer S., Bernadine T. Bacon, and Rollo Turner, eds. *Time: A Bibliography*. London: Information Retrieval, 1976. Of the 3,005 entries all but 512 are in Biology (1,200), Physics (615), and Psychology (673). Religion receives 57 entries, History 27, and Arts (including Literature, Language and Music) 50. Includes Subject Index.

Patrides, C. A. "Bibliography." *Aspects of Time*. Ed. C. A. Patrides. Manchester: Manchester UP, 1976. 250-70. Of the 562 books and articles in this bibliography 307 are in Literature.

Zelkind, Irving, and Joseph Sprung. *Time Research: 1172 Studies*. Metuchen: Scarecrow P, 1974. Largely concerned with Psychology and related subjects. Includes Subject Index, but no foreign language titles.

B. COLLECTIONS OF ESSAYS

Butler, S. T., and H. Messel, eds. *Time: Selected Lectures on Time and Relativity, the Arrow of Time and the Relation of Geological and Biological Time, and on Men of Science*. Oxford: Pergamon, 1965.

Campbell, Joseph, ed. *Man and Time: Papers from the Eranos Yearbooks*. Trans. Ralph Manheim, and R. F. C. Hall. New York: Pantheon, 1957.

Čapek, Milič, ed. *The Concepts of Space and Time: Their Structure and Their Development*. Dordrecht: Reidel, 1976.

Fischer, Roland, ed. "Interdisciplinary Perspectives of Time." *Annals of the New York Academy of Sciences* 138.2 (1967): 367-915.

Forman, F. J., and Caoran Sowton, eds. *Taking Our Time: Feminist Perspectives on Temporality*. Oxford: Pergamon, 1989.

Fraser, J. T., ed. *The Voices of Time: A Cooperative Survey of Man's Views of Time as Expressed by the Sciences and the Humanities*. 2nd ed. Amherst: U of Massachusetts P, 1981.

Fraser, J. T., F. C. Haber, and G. H. Müller, eds. *The Study of Time I*. New York: Springer, 1972.

Fraser, J. T., and N. Lawrence, eds. *The Study of Time II*. New York: Springer, 1975.

Fraser, J. T., N. Lawrence, and D. Park, eds. *The Study of Time III*. New York: Springer, 1978.

——, eds. *The Study of Time IV*. New York: Springer, 1981.

Fraser, J. T., N. Lawrence, and F. C. Haber, eds. *Time, Science, and Society in China and the West: The Study of Time V*. Amherst: U of Massachusetts P, 1986.

Fraser, J. T., ed. *Time and Mind: The Study of Time VI*. Madison, Connecticut: International Universities P, 1989.

Freeman, Eugene, and Wilfrid Sellars, eds. *Basic Issues in the Philosophy of Time*. La Salle: Open Court, 1971.

Gale, Richard M., ed. *The Philosophy of Time: A Collection of Essays*. London: Macmillan, 1968.

Gardet, L., et al., eds. *Cultures and Time*. 3 vols. Paris: Unesco P, 1976-79.

Gold, Thomas, and D. L. Shumacher, eds. *The Nature of Time*. Ithaca: Cornell UP, 1967.

Grant, John, ed. *The Book of Time*. North Pomfret: David and Charles, 1980.

Healy, R., ed. *Reduction, Time and Reality: Studies in the Philosophy of the Natural Sciences*. Cambridge: Cambridge UP, 1981.

Heinemann, G., ed. *Zeitbegriffe*. Freiburg: Karl Alber, 1986.

Hersch, Jeanne, and René Poirier, eds. *Entretiens sur le temps*. Paris: Mouton, 1967.

Kamper, C., and C. Wulf, eds. *Die sterbende Zeit*. Darmstadt: Luchterhand, 1987.

Kirsch, G., P. Nijkamp, and K. Zimmermann, eds. *Time Preferences: An Interdisciplinary Theoretical and Empirical Approach*. Berlin: Science Center, 1985.

Klibansky, R., and H. J. Paton, eds. *Philosophy and History: Essays Presented to E. Cassirer*. Oxford: Oxford UP, 1936.

Machamer, Peter K., and Robert G. Turnbull, eds. *Motion and Time, Space and Matter: Interrelations in the History of Philosophy and Science*. Columbus: Ohio State UP, 1976.

Meyer, R. W., ed. *Das Zeitproblem im 20. Jahrhundert*. Bern: Francke, 1964.

Pacault, Adolphe, et al. *À chacun son temps*. Paris: Flammarion, 1975.

Paflik, H., ed. *Das Phänomen der Zeit in Kunst und Wissenschaft*. Weinheim: Acta Humaniora, 1987.

Peisl, A., and A. Mohler, eds. *Die Zeit*. Munich: Oldenburg Verlag, 1983.

Petri, H., and I. Zepf, eds. *Geht uns die Zeit Verloren?: Beitrage zum Zeitbewusstsein*. Bochum: Studienverlage Dr. N. Brockmeyer, 1982.

Pratt, Sally, ed. *Man vs. Time*. Minneapolis: Graduate School Research Center, U of Minnesota, 1966.

Smart, J. J. C., ed. *Problems of Space and Time*. New York: Macmillan, 1964.

Swinburne, R., ed. *Space, Time and Causality*. Dordrecht: Reidel, 1983.

Wendorff, Rudolf, ed. *Im Netz der Zeit: Menschliches Zeiterleben interdisciplinär*. Stuttgart: S. Hirzel, 1989.

Zeman, Jiri, ed. *Philosophische Probleme der Zeit*. Prague: Czechoslovak Academy of Sciences, 1987.

——, ed. *Time in Science and Philosophy: An International Study of Some Current Problems*. New York: Elsevier, 1971.

Zoll, E., ed. *Zerstörung und Wiederaneignung von Zeit*. Frankfurt: Suhrkamp, 1988.

C. INDIVIDUAL WORKS

Alkon, Paul K. "Historical Development of the Concept of Time." *Biorhythms and Human Reproduction*. Ed. Michael Ferin, et al. New York: Wiley, 1974. 3-22.

Blumenberg, Hans. *Lebenszeit und Weltzeit*. Frankfurt: Suhrkamp, 1986.

Boorstin, Daniel J. "Time" in his *The Discoverers*. New York: Random House, 1983. Book 1, 4-78.

Bronowski, Jacob. *A Sense of the Future: Essays in Natural Philosophy*. Ed. Piero E. Ariotti, and Rita Bronowski. Cambridge: MIT P, 1977.

Burgelin, P. *L'homme et le temps*. Paris: Aubier, 1945.

Caponigri, A. Robert. *Time and Idea*. London: Routledge and Kegan Paul, 1953.

Cassirer, E. *An Essay on Man*. New Haven: Yale UP, 1945.

Denbigh, Kenneth G. *Three Concepts of Time*. New York: Springer, 1981.

Eisenhardt, P., D. Kurth, and H. Stiehl, eds. *Du steigst nie zweimal in denselben Fluss: Die Grenzen der wissenschaftlichen Erkenntnis*. Reinbek bei Hamburg: Rowolt, 1988.

Elton, L. R. B., and H. Messel. *Time and Man*. New York: Pergamon, 1978.

Fraser, J. T. "The Interdisciplinary Study of Time." *Annals of the New York Academy of Sciences* 138.2 (1966): 822-47. Reprinted in *Studium Generale* 19 (1966): 705-20.

——. "The Many Dimensions of Time and Mind: An Epistemic Jigsaw Puzzle Game." *Time and Mind: The Study of Time VI*. Ed. J. T. Fraser. Madison: International Universities P, 1989. 1-14.

——. *Of Time, Passion, and Knowledge: Reflections on the Strategy of Existence*. 2nd ed. Princeton: Princeton UP, 1990.

——. "Out of Plato's Cave: The Natural History of Time." *The Kenyon Review* n.s., 2 (Winter 1980): 143-62. Reprinted in *Temps Libre* 7 (1983): 5-21 (in French), and *Revista de Occidente* 76 (September 1987): 5-34 (in Italian).

——. "Reflections on Time, Science, and Man." *Britannica Yearbook of Science and the Future*. Chicago: Encyclopedia Britannica, 1970. 94-105.

——. "The Study of Time." *The Study of Time I*. Ed. J. T. Fraser, F. C. Haber, and G. H. Müller. New York: Springer, 1972. 479-502.

——. "The Study of Time." *The Voices of Time*. Ed. J. T. Fraser. 2nd ed. Amherst: U of Massachusetts P, 1981. 582-92.

——. "Temporal Levels: Sociobiological Aspects of a Fundamental Synthesis." *J of Social and Biological Structures* 1 (1978): 339-55.

——. "Time as a Hierarchy of Creative Conflicts." *Studium Generale* 23 (1970): 597-689.

——. *Time as Conflict: A Scientific and Humanistic Study*. Basel: Birkhäuser, 1978.

——. "Time in Western Philosophy, Natural Science and History." *Main Currents in Modern Thought* 28 (1972): 115-30.

——. "Time: Removing the Degeneracies." *Fantasy, Myth, and Reality*. Ed. H. P. Blum, et al. Madison, Connecticut: International Universities P, 1988. 467-85.

——. *Time, The Familiar Stranger*. Amherst: U of Massachusetts P, 1987. Redmond: Tempus Books, 1988 (paperback). Stuart: Triformation Braille Service, 1989 (Braille). Basel: Birkhäuser, 1988 (German translation). Milan: Feltrinelli, 1990 (Italian translation).

——. "Von der Suche nach Zeitlosigkeit." *Im Netz der Zeit*. Ed. Rudolf Wendorff. Stuttgart: S. Hirzel, 1989. 61-67.

Gebser, J. *Ursprung und Gegenwart*. 2nd ed. 3 vols. Munich: Deutsches Taschenbuch, 1986.

Glasser, Richard. *Time in French Life and Thought*. Trans. C. G. Pearson. Manchester: Manchester UP, 1972.

Granier, Jean. *La mesure du temps*. Paris: Presses Universitaires de France, 1943.

Gribbin, John R. *Timewarps*. London: Sphere, 1979.

Gunn, J. Alexander. *The Problem of Time*. London: Allen and Unwin, 1929.

Hammerschmidt, William W. *Whitehead's Philosophy of Time*. New York: King's Crown P, 1947.

Harrison, Edward. *The Masks of the Universe*. New York: Macmillan, 1985.

Heiss, R. "Die Zeit des Menschen und die Zeit überhaupt." *Freiburger Dies Universitatis* (1960-61): 117-31.

Jantsch, E. *The Self-Organizing Universe: Scientific and Human Implications of the Emerging Paradigm of Evolution*. 1980; reprint, Oxford: Pergamon, 1985.

Jaques, Elliott. *The Form of Time*. New York: Crane, Russok, 1982.

Knoll, Max. "Transformations of Science in Our Age." *Man and Time: Papers from the Eranos Yearbooks*. Trans. Ralph Manheim. Ed. Joseph Campbell. New York: Pantheon, 1957. 264-307.

Lawrence, Nathaniel. "My Time Is Your Time." *The Study of Time IV*. Ed. J. T. Fraser, N. Lawrence, and D. Park. New York: Springer, 1981. 1-12.

Lestienne, Rémy. *Les fils du temps: causalité, entropie, devenir*. Paris: Presses du CNRS, 1990.

Lumsden, C. J., and E. O. Wilson. *Genes, Mind and Culture: The Coevolutionary Process*. Cambridge: Harvard UP, 1981.

Macey, Samuel L. *The Dynamics of Progress: Time, Method, and Measure*. Athens: U of Georgia P, 1989.

Mellor, D. H. *Real Time*. Cambridge: Cambridge UP, 1981.

Palmer, L. S. *Man's Journey Through Time*. London: Hutchinson, 1957.

Pomian, K. *L'ordre du temps*. Paris: Gallimard, 1984.

Rawlence, C., ed. *About Time*. London: Cape, 1985.

Rifkin, Jeremy. *Time Wars*. New York: Henry Holt, 1987.

Rossi, P. *The Dark Abyss of Time*. Chicago: U of Chicago P, 1984.

Schlesinger, George N. *Aspects of Time*. Indianapolis: Halkett, 1980.

Schwarz, Brita, and Uno Svedin. *Methods in Future Studies: Problems and Applications*. Boulder: Westview, 1982.

Shallis, M. *On Time: An Investigation into Scientific Knowledge and*

Human Experience. Harmondsworth: Penguin, 1983.

Shoham, Yoav. *Reasoning About Change: Time and Causation from the Standpoint of Artificial Intelligence*. Cambridge: MIT P, 1988.

Treder, H. J. "Kosmische und subjective Zeit." *Forschungen und Fortschritte* 39 (1965): 325-27.

Voisé, W., et al. "The Study of Time in Poland, Czechoslovakia, and the Soviet Union." *The Study of Time III*. Ed. J. T. Fraser, N. Lawrence, and D. Park. New York: Springer, 1978. 471-99.

Von Foerster, H. "On Self-Organizing Systems and Their Environment." *Self-Organizing Systems*. Ed. M. Yavits, and S. Cameron. Oxford: Pergamon, 1960.

Watanabe, Michael S. "Creative Time." *The Study of Time I*. Ed. J. T. Fraser, F. C. Haber, and G. H. Müller. New York: Springer, 1972. 159-89.

——. *Time* (in Japanese). Tokyo: Kawade Shoboh Shinsha, 1974.

Wendorff, Rudolf. *Zeit und Kultur: Geschichte des Zeitbewusstseins in Europa*. 2nd ed. Wiesbaden: Westdeutscher Verlag, 1981.

Whitrow, Gerald J. "Man and Time: Some Historical and Critical Reflections." *Time and Mind: The Study of Time VI*. Ed. J. T. Fraser. Madison: International Universities P, 1989. 295-304.

——. *The Natural Philosophy of Time*. 2nd ed. Oxford: Clarendon P, 1980.

——. *The Nature of Time*. New York: Holt, Rinehart, 1973.

——. "Reflections on the History of the Concept of Time." *The Study of Time I*. Ed. J. T. Fraser, F. C. Haber, and G. H. Müller. New York: Springer, 1972. 1-11.

Whittaker, E. T. *The Beginning and End of the World*. Oxford: Oxford UP, 1942.

Wright, Lawrence. *Clockwork Man*. London: Elek Books, 1968.

2. AGING

Though both spellings are acceptable in either country, "Aging" is the preferred spelling in the United States while "Ageing" is preferred in Britain. In the interest of conformity, "Aging" is the preferred spelling for this text.

A. GENERAL

Abrams, M. "Transitions in Middle and Later Life." *Transitions in Middle and Later Life*. Ed. M. L. Johnson. London: British Society for Gerontology, 1980.

Adelman, Richard C., and George S. Roth, eds. *Testing the Theories of Aging*. Chicago: CRC P, 1982.

Andrews, G. R., et al. *Aging in the Western Pacific: A Four-Country Study*. Albany: World Health Organization, 1987.

Ashley, Montagu. *Growing Young*. 2nd ed. Westport: Bergin and Garvey, 1989.

Atchley, Robert C. *Aging: Continuity and Change*. 2nd ed. Belmont: Wadsworth, 1987.

Auerbach, Doris N., and Richard L. Levenson. "Second Impressions: Attitude Change in College Students toward the Elderly." *The Gerontologist* 17.4 (1977): 362-66.

Avery, Anne. *Successful Aging: A Sourcebook for Older People and Their Families*. New York: Ballantine, 1987.

Becker, Gaylene. *Growing Old in Silence*. Berkeley: U of California P, 1980.

Bee, Helen. *The Journey of Adulthood*. New York: Macmillan, 1986.

Bergener, Manfred, et al. *Aging in the Eighties and Beyond*. New York: Springer, 1985.

——, eds. *Dimensions in Aging*. San Diego: Academic P, 1987.

——, eds. *Thresholds in Aging*. San Diego: Academic P, 1985.

Binstock, Robert H., and Linda K. George, eds. *Handbook of Aging and the Social Sciences*. 3rd ed. San Diego: Academic P, 1990.

Birren, James E., ed. *Aging: A Challenge to Science and Society, Vol. 3: Behavioral Sciences and Conclusions*. New York: Oxford UP, 1983.

Birren, James E., and Vern L. Bengtson, eds. *Emergent Theories of Aging*. New York: Springer, 1988.

Blau, Zena S., ed. *Current Perspectives on Aging and the Life Cycle*. Vol. 3. Greenwich: Jai P, 1988.

Bond, John, and Peter Coleman. "Aging into the Twenty-First Century." *Aging in Society*. Ed. John Bond, and Peter Coleman. London: Sage, 1990. 276-90.

Bond, John, Roger Briggs, and Peter Coleman. "The Study of Aging." *Aging in Society*. Ed. John Bond, and Peter Coleman. London: Sage, 1990. 17-47.

Borgatta, Edgar R., and Rhonda J. Montgomery, eds. *Critical Issues in Aging Policy: Linking Research and Values*. Newbury Park: Sage, 1987.

Botwinick, Jack. *Aging and Behavior: A Comprehensive Integration of Research Findings*. 3rd ed. New York: Springer, 1984.

Bould, Sally, et al. *Eighty-Five Plus: The Oldest Old*. Belmont: Wadsworth, 1989.

Bruboker, Timothy H., and Edward A. Powers. "The Stereotype of 'Old': A Review and Alternative Approach." *J of Gerontology* 31.4 (1976): 441-47.

Burnet, F. Macfarlane. *Endurance of Life*. New York: Cambridge UP, 1978.

Busse, Ewald W., and George L. Maddox. *The Duke Longitudinal Studies of Normal Aging 1955-1980: Overview of History, Design and Findings*. New York: Springer, 1985.

Busse, Ewald W., and Eric Pfeiffer, eds. *Behavior and Adaptation in Late Life*. Boston: Little, Brown, 1977.

Butler, Robert N., and Herbert P. Gleason. *Productive Aging: Enhancing Vitality in Later Life*. New York: Springer, 1985.

Carmichael, Carl W., et al. *Human Communication and the Aging Process*. Prospect Heights: Waveland P, 1988.

Charness, Neil, ed. *Aging and Human Performance*. New York: Wiley, 1985.

Chen, Linda H. *Nutritional Aspects of Aging*. 2 vols. Chicago: CRC P, 1986.

Cheyne, George. *An Essay of Health and Long Life*. Ed. Robert Kastenbaum. 1724; reprint, Salem: Ayer, 1979.

Cohen, Lily, and Marie Oppedisano-Reich. *Funding in Aging: Public, Private, and Voluntary*. Garden City: Adelphi UP, 1979.

Coleman, Peter, and John Bond. "Aging in the Twentieth Century." *Aging in Society*. Ed. John Bond, and Peter Coleman. London: Sage, 1990. 1-16.

Cosby, Bill. *Time Flies*. New York: Doubleday, 1987.

Courtois, Y., et al. *Modern Trends in Aging Research*. New York: Scientific and Medical Publications of France, 1987.

Cowgill, Donald O. *Aging Around the World*. Belmont: Wadsworth,

1986.

Crockett, Walter H., Allen N. Press, and Marilyn Osterkamp. "The Effect of Deviations from Stereotyped Expectations upon Attitudes toward Older Persons." *J of Gerontology* 34.3 (1979): 368-74.

Danon, D., ed. *Aging: A Challenge to Science and Society, Vol. 1, Biology*. New York: Oxford UP, 1981.

Datan, Nancy, ed. *Transitions of Aging*. San Diego: Academic P, 1980.

Davis, Roger T., and Charles W. Leathers. *Behavior and Pathology of Aging in Rhesus Monkeys*. New York: Liss, 1985.

Davitz, Joel R., and Lois L. Davitz. *Making It: Forty and Beyond, Surviving the Mid-Life Crisis*. New York: Harper and Row, 1979.

Dowd, James J. "The Problems of Generations: A Generational Analysis." *International J of Aging and Human Development* 10.3 (1979-80): 213-30.

Dulin, Rachel Z. *A Crown of Glory: A Biblical View of Aging*. Mahwah: Paulist P, 1988.

Eckert, J. Kevin. *The Unseen Elderly*. San Diego: San Diego State UP, 1981.

Ellison, Jerome. *Life's Second Half*. Greenwich: Devin-Adair, 1978.

Featherstone, Mike, and Mike Hepworth. "Images of Aging." *Aging in Society*. Ed. John Bond, and Peter Coleman. London: Sage, 1990. 250-75.

——. *Surviving Middle Age*. Oxford: Blackwell, 1989.

Foner, Anne, and Karen Schwab. *Aging and Retirement*. Pacific Grove: Brooks Cole Publishing, 1981.

Ford, J., and R. Sinclair. *Sixty Years On: Women Talk about Old Age*. London: The Women's P, 1987.

Freeman, Joseph T. *Aging: Its History and Literature*. New York: Human Sciences P, 1979.

Fry, Christine L., ed. *Dimensions: Age, Culture and Health*. New York: Praeger, 1981.

Fry, Christine L., et al. *New Methods for Old-Age Research*. Westport: Bergin and Garvey, 1986.

Gelfand, Donald E., and Charles Barresi, eds. *Ethnic Dimensions of Aging*. New York: Springer, 1987.

Gelfand, Donald E., and Donald V. Fandetti. "Suburban and Urban White Ethics: Attitudes toward Care of the Aged." *The Gerontologist* 20.5 (1980): 588-94.

Gonzalez-Crussi, F. *Three Sudden Forms of Death, and Other Reflections on the Grandeur and Misery of the Body*. London: Picador, 1987.

Gruman, Gerald J., and Robert Kastenbaum, eds. *The "Fixed Period"*

Controversy: Prelude to Modern Ageism, an Original Anthology. Salem: Ayer, 1979.

Harris, Diana K. *The Sociology of Aging: An Annotated Bibliography and Sourcebook*. New York: Garland, 1985.

Hazan, Haim. *The Limbo People: A Study of the Constitution of the Time Universe Among the Aged*. London: Routledge and Kegan Paul, 1980.

Hendricks, Jon, and C. Davis Hendricks. "The Age Old Question of Old Age: Was It Really So Much Better Back When?" *International J of Aging and Human Development* 8.2 (1977-78): 139-54.

Hendricks, Jon, and M. M. Seltzer, eds. *Aging and Time*. Beverly Hills: Sage, 1986.

Henig, Robin M. *How a Woman Ages*. New York: Ballantine, 1985.

Hess, Beth B., ed. *Growing Old in America*. New Brunswick: Transition Books, 1976.

Holden, Karen C. "Comparability of the Measured Labor Force of Older Women in Japan and the United States." *J of Gerontology* 33.3 (May 1978): 422-26.

Hultsch, David, and Francine Deutsch. *Adult Development and Aging*. New York: McGraw-Hill, 1980.

Hutchinson, Martha L., and Hamish N. Munro. *Nutrition and Aging*. San Diego: Academic P, 1986.

Jones, Maxwell. *Growing Old: The Ultimate Freedom*. New York: Human Sciences P, 1988.

Kafer, Rudolph A., et al. "Aging Opinion Survey: A Report on Instrument Development." *International J of Aging and Human Development* 11.4 (1980): 319-33.

Kent, Barbara, and Robert Butler, eds. *Human Aging Research: Concepts and Techniques*. New York: Raven P, 1988.

Kimmel, Douglas. *Adulthood and Aging: An Interdisciplinary, Developmental View*. 2nd ed. New York: Wiley, 1980.

Laslett, P. *A Fresh Map of Life: The Emergence of the Third Age*. London: Weidenfeld and Nicholson, 1989.

Lawton, M. Powell, Miriam Moss, and Mark Fulcomer. "Objective and Subjective Uses of Time by Older People." *International J of Aging and Human Development* 24.3 (1986): 171-88.

Lessius, Leonard, and Luigi Cornaro. *A Treatise of Health and Long Life with the Future Means of Attaining It*. Ed. Robert Kastenbaum. 1743; reprint, Salem: Ayer, 1979.

Levinson, Daniel, et al. *The Seasons of a Man's Life*. New York: Ballantine, 1986.

Low, Jaclyn F. "Time Perception and Rehabilitation of the Elderly."

Physical and Occupational Therapy in Geriatrics 5.4 (1987): 17-30.

McFarland, David D. "The Aged in the 21st Century: A Demographer's View." *Aging into the 21st Century: Middle-Agers Today*. Ed. Jarvik F. Lissy. New York: Gardner P, 1978. 5-22.

McKee, Patrick L., ed. *Philosophical Foundations of Gerontology*. New York: Human Sciences P, 1982.

Maddox, George L., ed. *The Encyclopedia of Aging*. New York: Springer, 1986.

Maddox, George L., and E. W. Busse, eds. *Aging: The Universal Human Experience*. New York: Springer, 1987.

Maitland, David J. *Aging: A Time for New Learning*. Louisville: Westminster John Knox, 1987.

Marshall, Victor W. "No Exit: A Symbolic Interactionist Perspective on Aging." *International J of Aging and Human Development* 9.4 (1978-79): 345-48.

Marshall, Victor W., and Joseph A. Tindale. "Notes for a Radical Gerontology." *International J of Aging and Human Development* 9.2 (1978-79): 163-75.

Midwinter, Eric. *Re-defining Old Age*. London: Centre for Policy on Aging, 1987.

Minirth, Frank B., et al. *Beating the Clock: A Guide to Maturing Successfully*. Grand Rapids: Baker Book House, 1986.

Minot, Charles S. *The Problem of Age, Growth, and Death*. Ed. Robert Kastenbaum. 1908; reprint, Salem: Ayer, 1979.

Moment, Gairdner B. *Nutritional Approaches to Aging Research*. Chicago: CRC P, 1982.

Monroe, Margaret E., and Rhea J. Rubin. *The Challenge of Aging: A Bibliography*. Englewood: Libraries Unlimited, 1983.

Moody, Harry R. *Abundance of Life: Human Development Policies for an Aging Society*. Ed. Abraham Monk. New York: Columbia UP, 1988.

Morse, Donald R., and Robert L. Pollack. *Nutrition, Stress and Aging: An Holistic Approach*. New York: AMS P, 1988.

Newitt, Jane. *Older Americans: Nineteen Eighty to Two Thousand: An Exercise in Demographics-Based Forecasting*. Indianapolis: Hudson Institute, 1980.

Osterbind, Carter C., et al., eds. *Data Based Planning in the Field of Aging*. Gainesville: U Presses of Florida, 1983.

Palmore, Erdman, ed. *International Handbook on Aging: Contemporary Developments and Research*. Westport: Greenwood, 1980.

——. "When Can Age, Period and Cohort Be Separated?" *Social Forces* 57.1 (1978): 282-95.

Piper, M. *Erwachsenalter und Lebenslauf: Zur Sociologie der Lebens-*

stufen. Munich: Kösel, 1978.

Pippin, Roland N. "Assessing the Needs of the Elderly with Existing Data." *The Gerontologist* 20.1 (1980): 65-70.

Ridley, Jeanne C., Christine A. Backroch, and Deborah A. Dawson. "Recall and Reliability of Interview Data from Older Women." *The J of Gerontology* 34.1 (January 1979): 99-105.

Riley, Matilda W. *Aging from Birth to Death: Interdisciplinary Perspectives*. Boulder: Westview P, 1979.

Robinson, Pauline K., et al., eds. *Aging and Technological Advances*. New York: Plenum, 1984.

Ross, Robert F., Jr., and Carl B. Freitag. "Comparison of Adolescent and Adult Attitudes toward the Aged." *Educational Gerontology* 1 (1976): 291-95.

Rowles, G. "Towards a Geography of Growing Old!" *The Human Experience of Space and Place*. Ed. A. Buttimer, and D. Seaman. London: Croom Helm, 1980. 55-72.

Saxon, Sue V., and Mary J. Etten. *Physical Change and Aging: A Guide for the Helping Professions*. 2nd ed. New York: Tiresias P, 1987.

Schaie, K. Warner, et al., eds. *Methodological Issues in Aging Research*. New York: Springer, 1988.

Schulz, James. *The Economics of Aging*. 4th ed. Dover: Auburn House, 1988.

Schwartz, Arthur N., and James A. Peterson. *Introduction to Gerontology*. New York: Holt, Rinehart, 1979.

Seefeldt, Carol, et al. "Children's Attitudes toward the Elderly: Educational Implications." *Educational Gerontology* 2 (1977): 301-10.

Segali, Paul, and Carol Kahn. *Living Longer, Growing Younger: Remarkable Breakthroughs in Life Extension*. New York: Times Books, 1989.

Shepherd, Roy J. *Physical Activity and Aging*. Rockville: Aspen, 1987.

Smith, T. Lynn, ed. *Problems of America's Aging Population*. Gainesville: U Presses of Florida, 1951.

Stonecypher, D. D., Jr. *Getting Older and Staying Young*. New York: Norton, 1980.

Strumpf, N. E. "Probing the Temporal World of the Elderly." *International J of Nursing Studies* 24.3 (1987): 201-14.

Sweetser, William. *Human Life Considered in Its Present Condition and Future Developments*. Ed. Robert Kastenbaum. 1867; reprint, Salem: Ayer, 1979.

Tedrick, Ted., ed. *Aging: Issues and Policies for the 1980's*. New York: Praeger, 1985.

Thomas, L. Eugene, ed. *Research on Adulthood and Aging: The Human*

Science Approach. Albany: SUNY P, 1989.

Treas, Judith. "Socialist Organization and Economic Development in China: Latent Consequences for the Aged." *The Gerontologist* 19.1 (February 1979): 34-43.

Van Es, J. C., and Michael Bowling. "A Model for Analyzing the Aging of Local Populations: Illinois Counties between 1950 and 1970." *International J of Aging and Human Development* 9.4 (1978-79): 377-87.

Watson, Wilbur. *Stress and Old Age*. New Brunswick: Transaction Publications, 1980.

Weinberger, Arthur. "Stereotyping of the Elderly: Elementary School Children's Response." *Research on Aging* 1 (1979): 113-36.

Weindruch, Richard, and Roy L. Walford. *The Retardation of Aging and Disease by Dietary Restriction*. Springfield: Charles C. Thomas, 1988.

Weininger, Ben, and Eva L. Menkin. *Aging Is a Lifelong Affair*. Santa Barbara: Capra P, 1978.

Wolff, Kurt. *The Biological, Sociological, and Psychological Aspects of Aging*. 1959; reprint, Westport: Greenwood, 1978.

Wood, W. Gibson, and Merrill F. Elias, eds. *Alcoholism and Aging: Advances in Research*. Chicago: CRC P, 1982.

Woodruff, Diana S., and James E. Birren. *Aging: Scientific Perspectives and Social Issues*. Pacific Grove: Brooks Cole Publishing, 1983.

Young, Eleanor A. *Nutrition, Aging and Health*. New York: Alan R. Liss, 1986.

Zarit, Steven H., ed. *Readings in Aging and Death: Contemporary Perspectives*. New York: Harper and Row, 1977.

Zito, Dorothea R., and George V. Zito. *A Guide to Research in Gerontology: Strategies and Resources*. Westport: Greenwood, 1988.

B. BIOLOGY AND MEDICINE

Adelman, Richard C., and Eugene E. Dekker. *Modification of Proteins During Aging*. New York: Alan R. Liss, 1985.

Adelman, Richard C., and George S. Roth., eds. *Altered Proteins and Aging*. Chicago: CRC P, 1983.

——. *Endocrine and Neuroendocrine Mechanisms of Aging*. Chicago: CRC P, 1982.

Adelman, Richard C., et al. *Enzyme Induction in Aging and Protein Synthesis*. New York: Irvington, 1974.

——, eds. *Neural Regulatory Mechanisms During Aging*. New York: Alan R. Liss, 1980.

Age Words: A Glossary on Health and Aging. Washington: U.S. Government Printing Office, 1986.

Alder, William H., and Albert A. Nordin. *Immunological Techniques Applied to Aging Research*. Chicago: CRC P, 1981.

Alvi, Effraim. *The Amador Study on Aging, Nutrition, and Stress*. Jackson: Sierra Publishing, 1984.

Andreopoulos, Spyros, and John R. Hogness, eds. *Health Care for an Aging Society*. New York: Churchill Livingstone, 1988.

Ansello, Edward F., and Thomas Rose. *Aging and Lifelong Disabilities: Partnership for the Twenty First Century*. Palm Springs: Elvirita Lewis Foundation, 1989.

Armbrecht, H. J., et al., eds. *Nutritional Intervention in the Aging Process*. New York: Springer, 1984.

Baker, Scott R., and Marvin Rogul, eds. *Environmental Toxicity and the Aging Process*. New York: Alan R. Liss, 1987.

Balazs, Andras, et al. *Reproduction and Aging*. New York: Irvington, 1974.

Balin, Arthur K., and Albert M. Kligman, eds. *Aging and the Skin*. New York: Raven P, 1988.

Beall, C. M., ed. *Cross-Cultural Studies of Biological Aging*. Elmsford: Pergamon, 1982.

Bearn, Alexander G., ed. *The Aging Process: Therapeutic Implications*. New York: Raven P, 1985.

Bittles, A. H., and K. J. Collins, eds. *The Biology of Human Aging*. New York: Cambridge UP, 1986.

Blest, A. D. "Longevity, Palatability and Natural Selection in Five Species of New World Saturniid Moths." *Nature* 197 (1963): 1183-86.

Breckenridge, Mary B. *Age, Time and Fertility: Applications of Exploratory Data Analysis*. San Diego: Academic P, 1983.

Briggs, Roger. "Biological Aging." *Aging in Society*. Ed. John Bond, and Peter Coleman. London: Sage, 1990. 48-61.

Brody, Jacob, and George L. Maddox, eds. *Epidemiology and Aging: An International Perspective*. New York: Springer, 1988.

Burdman, Geri M. *Healthful Aging*. Englewood Cliffs: Prentice-Hall, 1986.

Carlisle, Anthony. *An Essay on the Disorders of Old Age, and on the Means for Prolonging Human Life*. Ed. Robert Kastenbaum. 1818; reprint, Salem: Ayer, 1979.

Carp, Frances M., and Eunice Kataoka. "Health Care Problems of the

Elderly of San Francisco's Chinatown." *The Gerontologist* 16.1 Part 1 (February 1976): 30-38.

Cherkin, Arthur, et al., eds. *Physiology and Cell Biology of Aging*. New York: Raven P, 1979.

Cohen, Elias, and Dharam P. Singal. *Non-HLA Antigens in Health, Aging, and Malignancy*. New York: Alan R. Liss, 1983.

Coleman, Paul D., ed. *Neurobiology of Aging: Research on Age-Related Phenomena, Neurodegeneration and Neuropathology*. Elmsford: Pergamon, 1989.

Comfort, A. *The Biology of Senescence*. 3rd ed. New York: Elsevier, 1979.

Comfort, A., et al. *Aging in Cold Blooded Animals*. Vol. 2. New York: Irvington, 1974.

Corso, John F. *Aging Sensory Systems and Perception*. New York: Praeger, 1981.

Cowdry, E. V., and Robert Kastenbaum, eds. *Problems of Aging: Biological and Medical Aspects*. 1939; reprint, Salem: Ayer, 1979.

Cristofalo, Vincent J., ed. *CRC Handbook of Cell Biology of Aging*. Vol. 2. Chicago: CRC P, 1985.

Cristofalo, Vincent J., et al. *Altered Endocrine Status During Aging*. New York: Alan R. Liss, 1984.

Cumming, M. E. "New Thoughts on the Theory of Disengagement." *New Thoughts on Old Age*. Ed. Robert Kastenbaum. New York: Springer, 1964. 3-18.

Curtis, Howard J. *Biological Mechanisms of Aging*. Springfield: Charles C. Thomas, 1966.

Davies, Peter, and Caleb E. Finch, eds. *Molecular Neuropathology of Aging*. Cold Spring Harbor: Cold Spring Harbor Laboratory, 1988.

Davis, Bernard B., and W. Gibson Wood, eds. *Homeostatic Function and Aging*. New York: Raven P, 1985.

Dean, Ward. *Biological Aging Measurement: Clinical Applications*. Ed. Hans U. Weber, et al. 2nd revised ed. Thousand Oaks: The Center for Bio-Gerontology, 1988.

Dietz, A. A., ed. *Aging: Its Chemistry*. Washington: American Association for Clinical Chemistry, 1980.

Fabris, N. *Immunology and Aging*. Norwell: Kluwer Academic, 1982.

Fieschi, C., et al., eds. *Effects of Aging on Regulation of Cerebral Blood Flow and Metabolism*. Farmington: S. Karger, 1984.

Florini, James R., ed. *Handbook of Biochemistry in Aging*. Chicago: CRC P, 1981.

Florini, James R., et al., eds. *Rates of Protein Synthesis During Aging*. New York: Irvington, 1974.

Frolkis, V. V. *Aging and Life-Prolonging Processes*. New York: Springer, 1983.

——, ed. *Physiology of Cell Aging*. Farmington: S. Karger, 1984.

Garland, Jeffrey. "Environment and Behaviour: A Clinical Perspective." *Aging in Society*. Ed. John Bond, and Peter Coleman. London: Sage, 1990. 123-43.

Gelfand, Donald E. "Ethnicity, Aging and Mental Health." *International J of Aging and Human Development* 10.3 (1979-80): 289-98.

Gelfant, S., and J. G. Smith, Jr. "Aging: Noncycling Cells, an Explanation." *Science* 178 (1972): 357-61.

Gershon, D., ed. *The Effect of Exercise on Several Systems of Aging Mice*. Farmington: S. Karger, 1987.

Glass, Leon, and Michael C. Mackey. *From Clocks to Chaos: The Rhythms of Life*. Princeton: Princeton UP, 1988.

Goldstein, A. L., ed. *Biomedical Advances in Aging*. New York: Plenum, 1989.

Gubrium, Jaber F. "Death Worlds in a Nursing Home." *Urban Life* 4.3 (1975): 317-38.

Hampton, John K., Jr. *The Biology of Human Aging*. Vol. 1. Templeton: KinetoPress, 1989.

Harker, Judith O., and Walter H. Riege. "Aging and Delay of Effects on Recognition of Words and Designs." *J of Gerontology* 40.5 (1985): 601-04.

Harris, Raymond, and Sara Harris, eds. *Physical Activity, Aging and Sports, Vol. I: Scientific and Medical Research*. Albany: Center for the Study of Aging, 1989.

Harris, Raymond, et al., eds. *Physical Activity, Aging and Sports, Vol. II: Practice, Program and Policy*. Albany: Center for the Study of Aging, 1990.

Holding, Dennis H., et al. "Date Attribution, Age, and the Distribution of Life-Time Memories." *J of Gerontology* 41.4 (1986): 481-85.

Haug, Marie. "Aging and the Right to Terminate Medical Treatment." *J of Gerontology* 33.4 (July 1978): 586-91.

Johnson, Horton A., ed. *Relations Between Normal Aging and Disease*. New York: Raven P, 1985.

Johnson, John E., Jr., ed. *Aging and Cell Function*. New York: Plenum, 1984.

——, ed. *Aging and Cell Structure*. Vol. 1. New York: Plenum, 1981.

Johnson, John E., Jr., et al. *Free Radicals, Aging and Degenerative Diseases*. New York: Alan R. Liss, 1986.

Kahn, Carol. *Beyond the Helix: DNA and the Quest for Longevity*. New York: Times Books, 1985.

Kaldor, George, and William J. DiBattista, eds. *Aging in Muscle*. New York: Raven P, 1978.

Kastenbaum, Robert. "Theories of Human Aging." *The J of Social Issues* 21.4 (1965): 13-36.

Katzman, Robert, and Robert D. Terry. *The Neurology of Aging*. Philadelphia: F. A. Davis, 1983.

Kolkner, A., and P. Ahmed, eds. *Aging*. New York: Elsevier, 1982.

Korenman, S. G., ed. *Endocrine Aspects of Aging*. New York: Elsevier, 1982.

Leopold, A. C. "Senescence in Plant Development." *Science* 134 (1961): 1727-32.

Leutz, Walter N., et al. *Changing Health Care for an Aging Society: Planning for the Social Health Maintenance Organization*. Lexington: Lexington Books, 1985.

Lints, F. A. *Genetics and Aging*. Farmington: S. Karger, 1978.

——, ed. *Non-Mammalian Models for Research on Aging*. Farmington: S. Karger, 1985.

Macieira-Coelho, A. *The Biology of Normal Proliferating Cells in Vitro: Relevance for in Vivo Aging*. Farmington: S. Karger, 1987.

Makinodan, T., et al. *Aging and the Immune Function*. New York: Irvington, 1974.

March, James, and James McGaugh, eds. *Aging: Biology and Behavior*. San Diego: Academic P, 1981.

Meites, Joseph, ed. *Neuroendocrinology of Aging*. New York: Plenum, 1983.

Milligan, W. L., et al. "A Comparison of Physical Health and Psychosocial Variables as Predictors of Reaction Time and Serial Learning Performance in Elderly Men." *J of Gerontology* 39.6 (1984): 704-10.

Milne, L. J., and M. Milne. *The Ages of Life*. New York: Harcourt, Brace, 1968.

Morganti, P., and W. Montagna, eds. *A New Look at Old Skin: A Challenge to Cosmetology*. Port Washington: Scholium International, 1986.

Morin, Robert J., ed. *Frontiers in Longevity Research: Applications of Nutritional and Other Discoveries in the Prevention of the Age-Related Disorders*. Springfield: Charles C. Thomas, 1985.

Mortimer, James, and Francis J. Pirozzolo, eds. *The Aging Motor System*. New York: Praeger, 1982.

Muller, Werner E., and Johannes W. Rohen, eds. *Biochemical and Morphological Aspects of Aging*. Philadelphia: Coronet Books, 1981.

Nanda, B. S., et al. *Aging Pigment: Current Research*. Vol. 1. New York: Irvington, 1974.

Neumann, James W. *Listening to Your Own Body: A Guide to the Neurological Problems That Afflict Us As We Grow Older*. Bethesda: Adler and Adler, 1987.

Noüy, L. du. *Le temps et la vie*. Paris: Gallimard, 1936.

Ordy, J. Mark, and Kenneth Brizzee, eds. *Sensory Systems and Communication in the Elderly*. New York: Raven P, 1979.

Ostfeld, A., et al. *NIH Workshop on Blood Pressure Regulation and Aging*. Ed. M. Horan et al. New York: Biomedical Information, 1985.

Pearl, Raymond, and Ruth D. Pearl. *The Ancestry of the Long-Lived*. Ed. Robert Kastenbaum. 1934; reprint, Salem: Ayer, 1979.

Phillips, Harry T., and Susan Gaylord. *Aging and Public Health*. New York: Springer, 1985.

Pierpaoli, Walter, and Novera H. Spector, eds. *Neuroimmunomodulation: Interventions in Aging and Cancer*. New York: New York Academy of Sciences, 1988.

Robertson, O. H., et al. *Endocrines and Aging*. New York: Irvington, 1972.

Rockstein, Morris, and Marvin Sussman. *Biology of Aging*. Belmont: Wadsworth, 1979.

Roff, Lucinda L., and David L. Klemmock. "Sexual Activity Among Older Persons: A Comparative Analysis of Appropriateness." *Research on Aging* 1 (1979): 389-99.

Rosenbloom, Alfred A., and Meredith Morgan. *Vision and Aging: General and Clinical Perspectives*. New York: Professional P Books, 1986.

Rothstein, Morton. *Biochemical Approaches to Aging*. San Diego: Academic P, 1982.

——, ed. *Review of Biological Research in Aging*. Vol. 1. New York: Alan R. Liss, 1983.

——, ed. *Review of Biological Research in Aging*. Vol. 2. New York: Alan R. Liss, 1985.

Rothstein, Morton, et al., eds. *Review of Biological Research in Aging*. Vol. 3. New York: Alan R. Liss, 1987.

Roy, A. K., and B. Chatterjee, eds. *Molecular Basis of Aging*. San Diego: Academic P, 1984.

Sacher, G. A. "Entropic Contributions to Mortality and Aging." *Symposium on Information Theory in Biology*. New York: Pergamon, 1958.

——. "Relation of Lifespan to Brain Weight and Body Weight in Mammals." *The Lifespan of Animals*. Ed. G. E. W. Wolstenholme, and M. O'Connor. London: Churchill, 1959.

Sauer, H. W., ed. *Cellular Aging*. Farmington: S. Karger, 1984.

Scheff, Steven W., ed. *Aging and Recovery of Function in the Central Nervous System*. New York: Plenum, 1984.

Schneider, Edward L., ed. *The Aging Reproductive System*. New York: Raven P, 1978.

Schneider, Edward L., et al, eds. *Handbook of the Biology of Aging*. 3rd ed. San Diego: Academic P, 1990.

Sekular, Robert, et al. *Aging and Human Visual Function*. New York: Alan R. Liss, 1982.

Sohal, R. S., et al, eds. *Molecular Biology of Aging: Gene Stability and Gene Expression*. New York: Raven P, 1985.

Sowers, James R., and James V. Felicetta, eds. *Endocrinology of Aging*. New York: Raven P, 1988.

Spence, Alexander. *Biology of Human Aging*. Englewood Cliffs: Prentice-Hall, 1989.

Stein, D. G. *Psychobiology of Aging*. New York: Elsevier, 1980.

Strehler, Bernard Louis. *Time, Cells and Aging*. New York: Academic P, 1962.

Strehler, Bernard Louis, and A. S. Midvan. "General Theory of Mortality and Aging." *Science* 132 (1960): 14-21.

Strehler, Bernard Louis, et al., eds. *The Biology of Aging*. Washington: American Institute of Biological Sciences, 1960.

Tejmar, J. "Age Differences in Cyclic Motor Reaction." *Nature* 195 (1962): 813-14.

Thompson, D'Arcy. *On Growth and Form*. Cambridge: Cambridge UP, 1968.

Valenti, G., ed. *Psychoneuroendocrinology of Aging*. New York: Springer, 1989.

Van Toller, C., et al. *Aging and the Sense of Smell*. Springfield: Charles C. Thomas, 1985.

Vernadakis, A. *Hormones in Development and Aging*. Bridgeport: Robert B. Luce, 1981.

Walford, R. L., et al. *The Relation of Aging to Immunity*. New York: Irvington, 1973.

Warner, Huber R., et al., eds. *Modern Biological Theories of Aging*. New York: Raven P, 1987.

Whitbourne, S. Krauss. *The Aging Body*. New York: Springer, 1985.

Zappia, V., et al., eds. *Advances in Post-Translational Modifications of Proteins and Aging*. New York: Plenum, 1988.

C. IN LITERATURE

Ansello, Edward F. "Age and Ageism in Children's First Literature." *Educational Gerontology* 2.3 (1977): 255-74.

Blythe, R. *The View in Winter*. Harmondsworth: Penguin, 1979.

Gibbs, J. M. "Dear Diary: An Exploration of Time-Use Among Two Community Samples of Widows." *Death Education* 8.1 (1984): 29-46.

Laurence, Margaret. *The Stone Angel*. Toronto: McClelland and Stewart, 1964.

Lyell, Ruth G., ed. *Middle Age, Old Age: Short Stories, Poems, Plays and Essays on Aging*. San Diego: Harcourt, Brace, 1980.

Peterson, David A., and Donna Z. Eden. "Teenagers and Aging: Adolescent Literature as an Attitude Source." *Educational Gerontology* 2 (1977): 311-25.

Porter, Laurel, and Lawrence M. Porter, eds. *Aging in Literature*. Troy: International Book Center, 1984.

Robin, Ellen P. "Old Age in Elementary School Readers." *Educational Gerontology* 2 (1977): 275-92.

Rooke, Constance, ed. *Stories of Aging*. Toronto: Oxford UP, 1986.

Sackville-West, V. *All Passion Spent*. London: Hogarth P, 1931.

Schuerman, Laurell E., Donna Z. Eden, and David A. Peterson. "Older People in Women's Periodical Fiction." *Educational Gerontology* 2 (1977): 327-51.

Sokoloff, Janice. *The Margin That Remains: A Study of Aging in Literature*. New York: Peter Lang, 1987.

Tolstoy, Leo. *Last Diaries*. Trans. Lydia W. Kesich. Ed. Robert Kastenbaum. 1960; reprint, Salem: Ayer, 1979.

Vickery, Florence E. *Old and Growing: Conversations, Letters, Observations, and Reflections on Growing Old*. Springfield: Charles C. Thomas, 1978.

Wright, Jeff, ed. *Three Zero, Turning Thirty: An Anthology of Writers As They Turn Thirty*. New York: Hard P, 1985.

D. PSYCHOLOGY

Aisenberg, R. "What Happens to Old Psychologists." *New Thoughts on Old Age*. Ed. Robert Kastenbaum. New York: Springer, 1964.

Arenberg, D. "Cognition and Aging: Verbal Learning, Memory, Problem Solving, and Aging." *The Psychology of Adult Development and Aging*. Ed. C. Eisdorfer, and M. P. Lawton. Washington:

American Psychological Association, 1972. 74-92.

Ariès, Philippe. *The Hour of Our Death*. Harmondsworth: Penguin, 1983.

——. *Western Attitudes Towards Death*. Baltimore: Johns Hopkins UP, 1974.

Baum, S. K. "Older People's Anxiety About Afterlife." *Psychological Reports* 52 (June 1983): 895-98. Includes bibliography.

Bayles, Kathryn A., et al. *Communication and Cognition in Normal Aging and Dementia*. Boston: College-Hill P, 1987.

Bell, B. "Significance of Functional Age for Interdisciplinary and Longitudinal Research in Aging." *Aging and Human Development* 3 (1972): 145-47.

Birren, James E., and R. Bruce Sloane, eds. *Handbook of Mental Health and Aging*. Englewood Cliffs: Prentice-Hall, 1980.

Butler, R. N. "The Life Review: An Interpretation of Reminiscence in the Aged." *Psychiatry* 26 (1963): 65-76.

Cavan, R. *Personal Adjustment in Old Age*. Chicago: Science Research Associates, 1949.

Coleman, Peter. *Aging and Reminiscence Processes: Social and Clinical Implications*. Chichester: Wiley, 1986.

——. "Psychological Aging." *Aging in Society*. Ed. John Bond, and Peter Coleman. London: Sage, 1990. 62-88.

Costa, P. T., and Robert Kastenbaum. "Some Aspects of Memories and Ambitions in Centenarians." *J of Genetic Psychology* 110 (1967): 3-16.

Craik, Fergus, and Sandra Trehub, eds. *Aging and Cognitive Processes*. New York: Plenum, 1982.

Cumming, E., and W. Henry. *Growing Old*. New York: Basic Books, 1961.

Curtin, S. R. *Nobody Ever Died of Old Age*. Boston: Little, Brown, 1972.

Eisdorfer, C., and M. P. Lawton, eds. *The Psychology of Adult Development and Aging*. Washington: American Psychological Association, 1972.

Featherman, David L., and Trond Petersen. "Markers of Aging: Modeling the Clocks that Time Us." *Research on Aging* 8.3 (1986): 339-65.

Filsinger, Eric, and William J. Sauer. "An Empirical Typology of Adjustment to Aging." *J of Gerontology* 33.3 (1978): 437-45.

Fink, H. F. "The Relationship of Time Perspective to Age, Institutionalization, and Activity." *J of Gerontology* 12.4 Section B (October 1957): 414-17.

George, Linda K. "The Impact of Personality and Social Status Factors upon Levels of Activity and Psychological Well-Being." *J of Gerontology* 33.6 (November 1978): 840-47.

George, Linda K., and Lucille B. Bearon. *Quality of Life in Older Persons: Meaning and Measurement*. New York: Human Sciences P, 1980.

Green, H. B. "Temporal Stages in the Development of the Self." *The Study of Time II*. Ed. J. T. Fraser, and N. Lawrence. New York: Springer, 1975. 1-19.

Gurland, Barry J., et al. *The Mind and Mood of Aging: Mental Health Problems of the Community Elderly in New York and London*. Binghamton: Haworth P, 1983.

Holme, R. *Developmental Psychology Today*. Palo Alto: Communications Research Machines, 1971.

Hulbert, R. J., and W. Lens. "Time and Self-Identity in Later Life." *International J of Aging and Human Development* 27.4 (1988): 293-303.

Inman, W. S. "Emotion, Cancer and Time: Coincidence or Determinism." *British J of Medical Psychology* 40 (1967): 225-31.

Joubert, C. E. "Subjective Acceleration of Time: Death Anxiety and Sex Differences." *Perceptual and Motor Skills* 57 (August 1983): 49-50.

Kagan, J., and H. A. Moss. *Birth to Maturity*. New York: Wiley, 1962.

Kastenbaum, Robert. "As the Clock Runs Out." *Mental Hygiene* 50 (1966): 332-36.

——. "Cognitive and Personal Futurity in Later Life." *J of Individual Psychology* 19 (1963): 216-19.

——. "Engrossment and Perspective in Later Life." *Contributions to the Psychobiology of Aging*. Ed. Robert Kastenbaum. New York: Springer, 1965. 3-18.

——, ed. *New Thoughts on Old Age*. New York: Springer, 1964.

——. "Perspectives on the Development and Modification of Behavior in the Aged: A Developmental Perspective." *The Gerontologist* 8 (1968): 280-84.

——. "Time, Death and Ritual in Old Age." *The Study of Time II*. Ed. J. T. Fraser, and N. Lawrence. New York: Springer, 1975. 20-38.

——. "What Happens to the Man Who Is Inside the Aging Body? An Inquiry into the Developmental Psychology of Later Life." *Duke University Council on Aging and Human Development, Proceedings of Seminars, 1965-1969*. Ed. F. C. Jeffers. Durham: Duke UP, 1969. 99-112.

Kastenbaum, Robert, and R. B. Aisenberg. *The Psychology of Death*.

New York: Springer, 1972.

Kastenbaum, Robert, and N. Durkee. "Elderly People View Old Age." *New Thoughts on Old Age*. Ed. Robert Kastenbaum. New York: Springer, 1964. 250-64.

Kastenbaum, Robert, and B. K. Kastenbaum. "Hope, Survival, and the Caring Environment." *Prediction of Life Span*. Ed. E. Palmore, and F. C. Jeffers. Lexington: Heath, 1971. 249-72.

Kastenbaum, Robert, et al. "'The Ages of Me': Toward Personal and Interpersonal Definitions of Functional Aging." *Aging and Human Development* 3 (1972): 197-212.

Keith, P. M. "Perceptions of Time Remaining and Distance from Death." *Omega* 12.4 (1981-82): 307-18.

Kiernat, Jean M. "Retrospection as a Life Span Concept." *Physical and Occupational Therapy in Geriatrics* 3.2 (1983): 35-48.

Kübler-Ross, Elisabeth. "What Is it Like to Be Dying?" *Understanding Death and Dying*. Ed. Sandra G. Wilcox, and Marilyn Sutton. Port Washington: Alfred Publishing, 1977. 99-111.

——. *On Death and Dying*. London: Tavistock, 1986.

Lohmann, Nancy. "Correlations of Life Satisfaction, Morale and Adjustment Measures." *J of Gerontology* 32.1 (1977): 73-75.

——. "A Factor Analysis of Life Satisfaction, Adjustment and Morale Measures with Elderly Adults." *International J of Aging and Human Development* 11.1 (1980): 35-43.

McCarthy, S. Viterbo. "Perception of Life after Seventy." *Activities, Adaptation and Aging* 8.2 (1986): 39-47.

Multran, Elizabeth, and Peter J. Burke. "Feeling 'Useless': A Common Component of Young and Old Identities." *Research on Aging* 1 (1979): 187-212.

Neugarten, Bernice. "Acting One's Age: New Rules for Old." *Psychology Today* 13.11 (1980): 66-80.

——, ed. *Middle Age and Aging: A Reader in Social Psychology*. Chicago: U of Chicago P, 1968.

Newman, Margaret A., and J. K. Gaudiano. "Depression as an Explanation for Decreased Subjective Time in the Elderly." *Nursing Research* 33.3 (1984): 137-39.

Pollock, K., and Robert Kastenbaum. "Delay of Gratification in Later Life: An Experimental Analog." *New Thoughts on Old Age*. Ed. Robert Kastenbaum. New York: Springer, 1964. 281-90.

Quinn, P. K., and M. Reznikoff. "The Relationship Between Death Anxiety and the Subjective Experience of Time in the Elderly." *International J of Aging and Human Development* 21.3 (1985): 197-210.

Rakowski, William. "Temporal Context and the Perceptions of Geriatric Patients." *Social Science and Medicine* 16.3 (1982): 241-44.

Raynor, Joel O. *Motivation, Career Striving, and Aging*. Washington: Hemisphere Publications, 1982.

Rodin, Judith, and Ellen Langer. "Aging Labels: The Decline of Control and the Fall of Self-Esteem." *J of Social Issues* 36 (1980): 12-29.

Schaie, K. Warner, and Carmi Schooler. *Social Structures and Aging: Psychological Processes*. Hillsdale: Lawrence Erlbaum, 1988.

Shifflett, P. A. "Future Time Perspective, Past Experiences, and Negotiation of Food Use Patterns Among the Aged." *The Gerontologist* 27 (October 1987): 611-15. Includes bibliography.

Shifflett, Peggy A., and William A. McIntosh. "Food Habits and Future Time: An Exploratory Study of Age Appropriate Food Habits Among the Elderly." *International J of Aging and Human Development* 24.1 (1986): 1-17.

Surwillo, W. W. "Age and the Perception of Short Intervals of Time." *J of Gerontology* 19 (1964): 322-24.

Tobin, S. "Earliest Memory as Data for Research in Aging." *Research, Planning, and Action for the Elderly*. Ed. D. P. Kent, R. Kastenbaum, and S. Sherwood. New York: Behavioral Publications, 1972. 252-78.

Ward, Russell A. "The Impact of Subjective Age and Stigma on Older Persons." *J of Gerontology* 32.2 (March 1977): 227-32.

Zarit, Steven H. *Aging and Mental Disorders: Psychological Approaches to Assessment and Treatment*. New York: Free P, 1983.

E. SOCIAL GERONTOLOGY

Amoss, Pamela T., and Stevan Harrell, eds. *Other Ways of Growing Old: Anthropological Perspectives*. Stanford: Stanford UP, 1981.

Antonucci, Toni, Nancy Gillett, and Frances W. Hoyer. "Values and Self-Esteem in Three Generations of Men and Women." *J of Gerontology* 34.3 (1979): 415-22.

Atchley, Robert C. "Issues in Retirement Research." *The Gerontologist* 19.1 (1979): 44-54.

——. *Social Forces and Aging: An Introduction to Social Gerontology*. 5th ed. Belmont: Wadsworth, 1988.

——. *The Sociology of Retirement*. Cambridge: Schenkman Publishing, 1976.

Atchley, Robert C., and Mildred Seltzer, eds. *The Sociology of Aging:*

Selected Readings. Belmont: Wadsworth, 1976.

Back, Kurt W. "The Ambiguity of Retirement." *Behavior and Adaptation in Late Life*. Ed. Ewald W. Busse, and Eric Pfeiffer. 2nd ed. Boston: Little, Brown, 1977. 78-98.

Barrow, Georgia M., and Patricia A. Smith. *Aging, Ageism and Society*. St. Paul: West Publishing, 1979.

Baum, Martha, and Rainer C. Baum. *Growing Old: A Societal Perspective*. Englewood Cliffs: Prentice-Hall, 1980.

Bell, Bill D., ed. *Contemporary Social Gerontology: Significant Developments in the Field of Aging*. Springfield: Charles C. Thomas, 1976.

Bengtson, V. "Time, Aging and the Continuity of Social Structure." *J of Social Issues* 39.4 (1983): 45-72.

Bengtson, V., and J. Robertson, eds. *Grandparenthood*. Beverly Hills: Sage, 1985.

Blaikie, A., and J. MacNicol. "Aging and Social Policy: A Twentieth Century Dilemma." *Human Aging in Later Life: Multidisciplinary Perspectives*. Ed. A. Warnes. London: Arnold, 1989. 69-82.

Bond, John. "Living Arrangements of Elderly People." *Aging in Society*. Ed. John Bond, and Peter Coleman. London: Sage, 1990. 161-80.

Bynum, Jack E., B. L. Cooper, and F. Gene Acuff. "Retirement Reorientation: Senior Adult Education." *J of Gerontology* 33.2 (1978): 253-61.

Carver, Vida, and Penny Liddiard, eds. *An Aging Population: A Reader and Sourcebook*. New York: Holmes and Meier, 1979.

Christenson, James A. "Generational Value Differences." *The Gerontologist* 17.4 (1977): 367-73.

Cohen, G. *Social Change and the Life Course*. London: Tavistock, 1987.

Cohen, Stephen Z., and Bruce M. Gans. *The Other Generation Gap: The Middle-Aged and Their Aging Parents*. Expanded and updated ed. New York: Dodd, Mead, 1988.

Coleman, Peter. "Adjustment in Later Life." *Aging in Society*. Ed. John Bond, and Peter Coleman. London: Sage, 1990. 89-122.

Collette-Pratt, Clara. "Attitudinal Predictors of Devaluation of Old Age in a Multigenerational Sample." *J of Gerontology* 31.2 (1976): 193-97.

Dooghe, Gilbert, and Jan Helander, eds. *Family Life in Old Age*. Norwell: Kluwer Academic, 1979.

Dowd, James S., and Vern L. Bengtson. "Aging in Minority Populations: An Examination of the Double Jeopardy Hypothesis." *J of Gerontology* 33.3 (1978): 427-36.

Fogel, Robert, et al., eds. *Aging: Stability and Change in the Family*. San Diego: Academic P, 1981.

Fontana, Andrea. *The Last Frontier: The Social Meaning of Growing Old.* Beverly Hills: Sage, 1977.

Gelfand, Donald E., and Alfred A. Kutzik, eds. *Ethnicity and Aging.* New York: Springer, 1979.

George, Linda K. "The Happiness Syndrome: Methodological and Substantive Issues in the Study of Social Psychological Well-Being in Adulthood." *The Gerontologist* 19.2 (April 1979): 210-16.

——. *Role Transition in Later Life.* Monterey: Brooks-Cole, 1980.

Goldstein, Sidney. "Negro-White Differentials in Consumer Patterns of the Aged, 1960-1961." *The Gerontologist* 2.3 Part 1 (1971): 242-49.

Gubrium, Jaber F., ed. *Time, Roles, and Self in Old Age.* New York: Human Sciences P, 1976.

Hadjihristev, Argir. *Life-Styles for Long Life: Longevity in Bulgaria.* Trans. Gari Lesnoff-Caravaglia. Springfield: Charles C. Thomas, 1988.

Harper, D. Wood, Jr., and Joseph M. Garza. "Ethnicity, Family Generational Structure, and Intergenerational Solidarity." *Sociological Symposium* 2 (Spring 1969): 75-82.

Harris, Diana K., and William E. Cole. *The Sociology of Aging.* Boston: Houghton Mifflin, 1980.

Hendricks, Jon, and C. Davis Hendricks. *Aging in Mass Society: Myths and Realities.* 3rd ed. Glenview: Scott, Foresman, 1986.

Henretta, John C., Richard T. Campbell, and Gloria Gardocki. "Survey Research in Aging: An Evaluation of the Harris Survey." *The Gerontologist* 17.2 (April 1977): 160-67.

Hess, Beth B., and Elizabeth W. Markson. *Aging and Old Age: An Introduction to Social Gerontology.* New York: Macmillan, 1980.

Hickey, Tom, et al. "Attitudes toward Aging as a Function of In-Service Training and Practitioner Age." *J of Gerontology* 31.6 (1976): 681-86.

Hobman, David, ed. *The Social Challenge of Aging.* New York: St. Martin's, 1978.

Ikels, Charlotte. *Aging and Adaptation: Chinese in Hong Kong and the United States.* Hamden: Archon, 1983.

Invester, Connie, and Carl King. "Attitudes of Adolescents toward the Aged." *The Gerontologist* 17.1 (February 1977): 85-89.

Jackson, Jacquelyne J. "Marital Life among Aging Blacks." *The Family Coordinator* 21.1 (1972): 21-27.

——. *Minorities and Aging.* Belmont: Wadsworth, 1980.

——. "Negro Aged: Toward Needed Research in Social Gerontology." *The Gerontologist* 2.1 Part 2 (1971): 52-57.

——. "Social Gerontology and the Negro: A Review." *The Gerontologist*

7.3 (1967): 168-78.

Jaslow, Philip. "Employment, Retirement, and Morale among Older Women." *J of Gerontology* 31.2 (1976): 212-18.

Jefferys, Margot, ed. *Growing Old in the Twentieth Century*. London: Routledge, 1989.

Jerrome, Dorothy, ed. *Aging in Modern Society*. New York: St Martin's, 1983.

——. "Intimate Relationships." *Aging in Society*. Ed. John Bond, and Peter Coleman. London: Sage, 1990. 181-208.

Johnson, Elizabeth S., and John B. Williamson. *Growing Old: The Social Problems of Aging*. New York: Holt, Rinehart, 1980.

Johnson, Julie. *When Parents Grow Old*. Boise: Pacific P Publishing, 1986. 31 pages.

Johnson, Malcolm. "Dependency and Interdependency." *Aging in Society*. Ed. John Bond, and Peter Coleman. London: Sage, 1990. 209-28.

Kagan, Nathan. "Beliefs, Attitudes and Stereotypes about Old People: A New Look at Old Issues." *Research on Aging* 1 (1979): 11-36.

Kalish, Richard A. "The Aged and the Dying Process: The Inevitable Decisions." *J of Social Issues* 21.4 (1965): 87-96.

——. "Death and Dying in a Social Context." *Handbook of Aging and the Social Sciences*. Ed. Robert H. Binstock, and Ethel Shanas. New York: Van Nostrand Reinhold, 1976. 483-507.

Kalish, Richard A., and Sam Yuen. "Americans of East Asian Ancestry: Aging and the Aged." *The Gerontologist* 2.1 Part 2 (1971): 36-47.

Kasschau, Patricia L. "Age and Race Discrimination Reported by Middle-Aged and Older Persons." *Social Forces* 55.3 (March 1977): 728-42.

Kidwell, Jane, and Alan Booth. "Social Distance and Intergenerational Relations." *The Gerontologist* 17.5 (October 1977): 412-20.

Lacy, William B., and Jon Hendricks. "Developmental Models of Adult Life: Myth or Reality." *International J of Aging and Human Development* 11.2 (1980): 89-110.

Lawton, M. P. *Environment and Aging*. Monterey: Brooks-Cole Publishing, 1980.

Lawton, M. Powell, Morton H. Kleban, and Maurice Singer. "The Aged Jewish Person and the Slum Environment." *J of Gerontology* 26.2 (1971): 231-39.

Lee, Gary R., and John M. Finney. "Sampling in Social Gerontology: A Method of Locating Specialized Populations." *J of Gerontology* 32.6 (1977): 689-93.

Levin, Jack, and William C. Levin. *Ageism: Prejudice and Discrimination against the Elderly*. Belmont: Wadsworth, 1980.

Longino, Charles F., Jr., Kent A. McClelland, and Warren A. Peterson. "The Aged Subculture Hypothesis: Social Integration, Gerontophilia, and Self-Conception." *J of Gerontology* 35.5 (September 1980): 758-67.

McKinney, John C., and Frank T. DeVyver, eds. *Aging and Social Policy*. New York: Irvington, 1980.

McPherson, Barry D. *Aging As a Social Process*. New York: Routledge, Chapman and Hall, 1984.

Maddox, George L. "Sociology of Later Life." *Annual Review of Sociology, 5*. Ed. Alex Inkeles, James Coleman, and Ralph H. Turner. Palo Alto: Annual Reviews, Inc., 1979. 113-35.

Maddox, George L., and James Wiley. "Scope, Concepts and Methods in the Study of Aging." *Handbook of Aging and the Social Sciences*. Ed. Robert H. Binstock, and Ethel Shanas. New York: Van Nostrand Reinhold, 1976.

Magan, Geralyn G., ed. *Aging, Race and Culture: Issues in Long Term Care*. Washington: American Association of Homes for the Aging, 1982.

Makarushka, Julia L., and Robert D. McDonald. "Informed Consent, Research, and Geriatric Patients: The Responsibility of Institutional Review Committees." *The Gerontologist* 19.1 (February 1979): 61-66.

Markides, Kyriakos S., and Charles S. Mindel. *Aging and Ethnicity*. Newbury Park: Sage, 1987.

Marshall, Victor W. *Last Chapters: A Sociology of Aging and Dying*. Monterey: Brooks-Cole, 1980.

Matthews, Sarah. "Old Women and Identity Maintenance: Outwitting the Grim Reaper." *Urban Life* 4.3 (1975): 339-48.

Mizruchi, Ephraim H., et al. *Time and Aging: Conceptualization and Application in Sociological and Gerontological Research*. Dix Hills: General Hall, 1982.

Mueller, H. Gustav, and Virginia C. Geoffrey, eds. *Communication Disorders in Aging: Assessment and Management*. Washington: Gallaudet UP, 1987.

Palmore, Erdman B. "The Effects of Aging on Activities and Attitudes." *The Gerontologist* 8.4 (Winter 1968): 259-63.

——. *The Honorable Elders: A Cross-Cultural Analysis of Aging in Japan*. Durham: Duke UP, 1975.

——. *Social Patterns in Normal Aging: Findings from the Duke Longitudinal Studies*. Durham: Duke UP, 1981.

Palmore, Erdman B., and George L. Maddox. "Sociological Aspects of Aging." *Behavior and Adaptation in Late Life*. Ed. Ewald W. Busse, and Eric Pfeiffer. 2nd ed. Boston: Little, Brown, 1977. 31-58.

Payne, Barbara P. "The Older Volunteer: Social Role Continuity and Development." *The Gerontologist* 17.4 (August 1977): 355-61.

Phillipson, Chris. *Capitalism and the Construction of Old Age*. London: Macmillan, 1982.

——. "The Sociology of Retirement." *Aging in Society*. Ed. John Bond, and Peter Coleman. London: Sage, 1990. 144-60.

Quadagno, Jill S., ed. *Aging, the Individual and Society: Readings in Social Gerontology*. New York: St. Martin's, 1980.

Rifai, Marlene, A. Y., ed. *Justice and Older Americans*. Lexington: Lexington Books, 1977.

Riley, Matilda W., and Ronald P. Abeles, eds. *Aging from Birth to Death: Sociotemporal Perspectives*. Vol 2. Boulder: Westview P, 1982.

Rosenberg, George S. *The Worker Grows Old*. San Francisco: Jossey-Bass, 1970.

Rosenfeld, Jeffrey P. *The Legacy of Aging: Inheritance and Disinheritance in Social Perspective*. Norwood: Ablex Publishing, 1979.

Rosenthal, Saul H. *Sex Over Forty*. Los Angeles: Jeremy P. Tarcher, 1987.

Ross, Lorraine N., and George T. Beall. *Intergenerational Programming: Opportunities for National Organizations*. Washington: The National Council on the Aging, 1985. 24 pages.

Rowles, Graham D., ed. *Aging and Milieu: Environmental Perspectives on Growing Old*. San Diego: Academic P, 1982.

Rutzen, Robert S. "The Social Distribution of Primary Social Isolation among the Aged: A Subcultural Approach." *International J of Aging and Human Development* 1 (1980): 77-87.

Siegel, Barry, and Judith Lasher. "Deinstitutionalizing Elderly Patients: A Program of Resocialization." *The Gerontologist* 18.3 (June 1978): 293-300.

Sokolovsky, Jay. *Cultural Context of Aging: Worldwide Perspectives*. Westport: Bergin and Garvey, 1989.

Strange, Heather, et al. *Aging and Cultural Diversity: New Directions and Annotated Bibliography*. Westport: Bergin and Garvey, 1987.

Taira, Francis. *Aging: A Guide for the Family*. Lancaster: Technomic Publishing, 1983.

Townsend, Peter. "The Structured Dependency of the Elderly: Creation of Social Policy in the Twentieth Century." *Aging and Society* 1

(1981): 5-28.

Uhlenberg, Peter. "Older Women: The Growing Challenge to Design Constructive Roles." *The Gerontologist* 19.3 (June 1979): 236-41.

Walker, Alan. "Poverty and Inequality in Old Age." *Aging in Society*. Ed. John Bond, and Peter Coleman. London: Sage, 1990. 229-49.

Ward, Russell A. *The Aging Experience: An Introduction to Social Gerontology*. New York: Harper and Row, 1979.

——. "Aging-Group Consciousness: Implications in an Older Sample." *Sociology and Social Research* 61 (1977): 496-519.

——. "Aging, the Use of Time, and Social Change." *International J of Aging and Human Development* 14.13 (1981- 82): 177-87.

Watson, Wilbur H., and Robert J. Maxwell, eds. *Human Aging and Dying: A Study of Sociocultural Gerontology*. New York: St. Martin's, 1977.

Wenger, G. C. *The Supportive Network*. London: Allen and Unwin, 1984.

Williamson, John B., et al. *Aging and Society: An Introduction to Social Gerontology*. New York: Holt, Rinehart, 1980.

3. ARCHAEOLOGY

Angel, J. Lawrence. *The People of Lerna: Analysis of a Prehistoric Aegean Population*. Washington: Smithsonian, 1971.

Arnold, C. J. *The Archaeology of the Early Anglo-Saxon Kingdoms*. New York: Routledge, Chapman and Hall, 1988.

Ashmore, Wendy, and Robert J. Sharer. *Discovering Our Past: A Brief Introduction to Archaeology*. Mountain View: Mayfield Publishing, 1988.

Baker, Emerson W. *The Clarke and Lake Company: The Historical Archaeology of a Seventeenth Century Maine*. Augusta: Maine Historic Preservation Commission, 1985.

Baker, J. R., and D. R. Brothwell. *Animal Diseases in Archaeology*. San Diego: Academic P, 1980.

Barker, Graeme. *Landscape and Society: Prehistoric Central Italy*. San Diego: Academic P, 1981.

Beaudry, Mary C., ed. *Documentary Archaeology in the New World*. New York: Cambridge UP, 1989.

Bertman, Stephen. *Doorways Through Time: The Romance of Archaeology*. Los Angeles: Jeremy P. Tarcher, 1987.

Binford, Lewis Roberts. *Debating Archaeology*. San Diego: Academic P, 1989.

——. *In Pursuit of the Past: Decoding the Archaeological Record*. New York: Thames and Hudson, 1983.

Blinderman, Charles. *The Piltdown Inquest*. Buffalo: Prometheus Books, 1986.

Bonnischen, Robson. *Pleistocene Bone Technology in the Beringian Refugium*. Ottawa: National Museums of Canada, 1979.

Bradford, John. *Ancient Landscapes: Studies in Field Archaeology*. 1957; reprint, Westport: Greenwood, 1980.

Branigan, K., consultant ed. *The Atlas of Archaeology*. London: Macdonald, 1982.

Breschini, Gary S., et al. *California Radiocarbon Dates*. 5th ed. Salinas: Coyote P, 1988.

Broderick, M., and A. A. Morton. *Concise Dictionary of Egyptian Archaeology*. Chicago: Ares Publications, 1979.

Carlson, Gayle F., and Terry L. Steinacher. *Nebraska Highway Archaeological and Historical Investigations, 1969-1975*. Lincoln: Nebraska State Historical Society, 1984.

Caso, Antonio. *Thirteen Masterpieces of Mexican Archaeology*. New York: Gordon P, 1976.

Ceram, C. W. [Kurt W. Marek] *Gods, Graves, and Scholars: The Story of Archaeology*. Trans. E. B. Garside. London: Victor Gollancz, 1952.

Chapman, Carl H. *Preliminary Salvage Archaeology in the Pomme de Terre Reservoir Area, Missouri*. Columbia: Missouri Archaeological Society, 1954.

Daniel, G. *150 Years of Archaeology*. London: Duckworth, 1975.

——. *The Three Ages*. Cambridge: Cambridge UP, 1943.

Eddy, Frank W. *Archaeology, a Cultural Evolutionary Approach*. Englewood Cliffs: Prentice-Hall, 1984.

Flegenheimer, Arnold J. *Archaeological Adventure in Israel: A Practical Guide*. Great Neck: Roth Publishing, 1987.

Fleure, H. J., and Harold Peake. *Hunters and Artists*. Northford: Elliot's Books, 1927.

——. *Peasants and Potters*. Northford: Elliot's Books, 1927.

Foster, Michael S., and Phil C. Weigand, eds. *Archaeology of West and Northwest Mesoamerica*. Boulder: Westview P, 1985.

Gardin, J. C., et al. *Archaeological Constructs*. New York: Cambridge UP, 1980.

Gibbon, Guy. *Anthropological Archaeology*. New York: Columbia UP, 1984.

——, ed. *Prairie Archaeology: Papers in Honor of David A. Baerreis*. Minneapolis: Department of Anthropology, U of Minnesota, 1983.

Goetze, Albrecht. *Old Babylonian Omen Texts*. Northford: Elliot's Books, 1947.

Gowlett, John. *Ascent to Civilization: The Archaeology of Early Man*. New York: McGraw-Hill, 1984.

Grayson, Donald K. *Quantitative Zooarchaeology: Topics in the Analysis of Archaeological Faunas*. Ed. G. W. Dimbleby. San Diego: Academic P, 1984.

Grebinger, P. *Discovering Past Behavior: Experiments in the Archaeology of the American Southwest*. New York: Gordon and Breach, 1978.

Green, Stanton W., and Stephen M. Perlman, eds. *The Archaeology of Frontiers and Boundaries*. San Diego: Academic P, 1985.

Hackman, George G., ed. *Temple Documents of the Third Dynasty of Ur From Umma*. Northford: Elliot's Books, 1937.

Hammond, Norman, ed. *South Asian Archaeology*. Park Ridge: Noyes, 1974.

Harris, Edward, ed. *Principles of Archaeological Stratigraphy*. 2nd ed. San Diego: Academic P, 1989.

Harrison, R. M., et al. *Excavations at Sarachane in Istanbul*. Vol. 1.

Princeton: Princeton UP, 1986.

Hassan, Fekri A. *Demographic Archaeology*. San Diego: Academic P, 1981.

Hedges, John W. *Tomb of the Eagles: Death and Life in a Stone Age Tribe*. New York: New Amsterdam Books, 1987.

Heizer, Robert Fleming, Thomas R. Hester, and Carol Graves. *Archaeology, a Bibliographical Guide to the Basic Literature*. New York: Garland, 1980.

Hodder, Ian, ed. *Archaeology as Long-Term History*. Cambridge: Cambridge UP, 1987. Includes bibliography.

——. *Reading the Past: Current Approaches to Interpretation in Archaeology*. New York: Cambridge UP, 1986.

Hopkirk, Peter. *Foreign Devils on the Silk Road: The Search for the Lost Cities and Treasures of Chinese Central Asia*. Amherst: U of Massachusetts P, 1984.

Hudson, Kenneth. *Guide to the Industrial Archaeology of Europe*. Cranbury: Fairleigh Dickinson, 1971.

Kalinowski, Hans-Otto, et al. *Carbon-13 NMR Spectroscopy*. New York: Wiley, 1987.

Karol, I. L., et al. *Radioactive Isotopes in the Atmosphere and Their Use in Meteorology*. Philadelphia: Coronet Books, 1967.

Klindt-Jensen, O. A. *A History of Scandinavian Archaeology*. London: Thames and Hudson, 1975.

Knudson, S. J. *Culture in Retrospect: An Introduction to Archaeology*. Prospect Heights: Waveland P, 1985.

Krupp, Edwin C., ed. *Archaeoastronomy and the Roots of Science*. Boulder: Westview P, 1984.

——. *Echoes of the Ancient Skies: The Astronomy of Lost Civilizations*. New York: Harper and Row, 1983. Includes bibliography.

Kuang-Chi, Chang. *The Archaeology of Ancient China*. New Haven: Yale UP, 1963.

Kuo-Wei, Wang. "Archaeology in the Sung Dynasty." *Chinese J* (Shanghai) 6.5 (1927): 222-31.

Leonard, Robert D., and George T. Jones, eds. *Quantifying Diversity in Archaeology*. New York: Cambridge UP, 1989.

Macalister, R. A. S. *Textbook of European Archaeology*. London, 1921.

Mason, Ronald J. *Great Lakes Archaeology*. San Diego: Academic P, 1981.

Matheson, Susan B. *Dura-Europos: The Ancient City and the Yale Collection*. New Haven: Yale Art Gallery, 1982. 48 pages.

Meyers, Thomas, and Mark Denies. *Longterm and Peakscan: Neutron Activation Analysis Computer Programs*. Ann Arbor: U of Michigan,

Museum of Anthropology, 1972.
Michael, H. N., and E. K. Ralph, eds. *Dating Techniques for Archaeologists*. Cambridge: MIT P, 1971.
Miller, Daniel, and Christopher Tilley. *Ideology, Power and Prehistory*. New York: Cambridge UP, 1984.
Polach, Dilette, ed. *Radiocarbon Dating Literature: The First 21 Years, 1947-68—An Annotated Bibliography*. San Diego: Academic P, 1988.
Rix, Michael. *Industrial Archaeology*. London: Historical Association, 1967. Includes bibliography.
Schmidt, Peter R. *Historical Archaeology: A Structural Approach in an African Culture*. Westport: Greenwood, 1978.
Sharer, Robert J. *Archaeology: Discovering Our Past*. Mountain View: Mayfield Publishing, 1987.
Sheets, Payson D. *Archaeological Studies of Disaster: Their Range and Value*. Toronto: Natural Hazard Research, 1980.
Shennan, Stephen. *Quantifying Archaeology*. New York: Columbia UP, 1987.
Sherratt, Andrew, ed. *The Cambridge Encyclopaedia of Archaeology*. New York: Cambridge UP, 1980. Includes bibliography.
Silverberg, Robert. *Clocks for the Ages: How Scientists Date the Past*. New York: Macmillan, 1971.
——. *The Moundbuilders of Ancient America: The Archaeology of a Myth*. Athens: Ohio UP, 1986.
Smith, Philip E. *Paleolothic Archaeology in Iran*. Philadelphia: U of Pennsylvania, University Museum, 1986.
Spriggs, Matthew J., ed. *Marxist Perspectives in Archaeology*. New York: Cambridge UP, 1984.
Tainter, Joseph A. *The Collapse of Complex Societies*. New York: Cambridge UP, 1988.
Taylor, R. E., ed. *Advances in Obsidian Glass Studies: Archaeological and Geochemical Perspectives*. Park Ridge: Noyes, 1977.
——. *Radiocarbon Dating*. San Diego: Academic P, 1987.
Taylor, R. E., and Clement W. Meighan, eds. *Chronologies in New World Archaeology*. San Diego: Academic P, 1978.
Thompson, Henry O. *Archaeology in Jordan*. New York: Peter Lang, 1989.
——. *Biblical Archaeology: The World, the Mediterranean, the Bible*. New York: Paragon House, 1987.
Treistman, J. M. *The Prehistory of China: An Archaeological Exploration*. Newton Abbot: David and Charles, 1972.
Trigger, Bruce G. *A History of Archaeological Interpretation*. New York: Cambridge UP, 1989.

——. *Time and Traditions*. New York: Columbia UP, 1978.

Ubelaker, Douglas H. *Human Skeletal Remains: Excavation, Analysis, Interpretation*. 2nd ed. Washington: Taraxacum, 1989.

Vinsrygg, Synnove. "Time in Archaeological Thought." *Time, Science, and Society in China and the West: The Study of Time V*. Ed. J. T. Fraser, N. Lawrence, and F. C. Haber. Amherst: U of Massachusetts P, 1986. 225-40.

Wagstaff, J. M., ed. *Landscape and Culture: Geographical and Archaeological Perspectives*. New York: Blackwell, 1987.

Wang, Chih H., and David L. Willis. *Radiotracer Methodology in the Biological, Environmental and Physical Sciences*. Englewood Cliffs: Prentice-Hall, 1975.

Watts, Gordon P., Jr., ed. *Underwater Archaeology: The Challenge Before Us*. San Marino: Fathom Eight, 1981.

Weaver, Muriel P. *Aztecs, Maya and Their Predecessors: Archaeology of Mesoamerica*. San Diego: Academic P, 1981.

Wenke, Robert J. *Patterns in Prehistory: Humankind's First Three Million Years*. 2nd ed. New York: Oxford UP, 1984.

Whitehouse, Ruth, and John Wilkins. *The Making of Civilization: History Discovered Through Archaeology*. New York: Knopf, 1986.

The World Atlas of Archaeology. London: M. Beazley, 1985. Includes bibliography.

4. ART AND ARCHITECTURE

A. GENERAL

Arnheim, Rudolf. *Entropy and Art*. Berkeley: U of California P, 1971.

Art and Time. London: Barbican Art Gallery, 1986.

Aznar, José Camón. *El tiempo en el arte*. Madrid: Sociedad de Estudios y Publicaciones, 1958.

Baker, Will. *Backward: An Essay on Indians, Time, and Photography*. Berkeley: North Atlantic Books, 1983.

Baudson, Michel, ed. *Zeit, die vierte Dimension in der Kunst*. Weinheim: Acta Humaniora der VCH, 1985.

Blatt, S. J., and E. S. Blatt. *Continuity and Change in Art: The Development of Modes of Representation*. Hillsdale: Erlbaum, 1984.

Bluestone, George. "Of Time and Space." In his *Novels into Film*. Berkeley: U of California P, 1961. Ch. 1. Includes bibliographies.

Daum, J. "Der Jahreslauf in alten Kalendern." *Kunst* 98 (December 1986): 878-85.

Embler, Weller B. "Flight: A Study of Time and Philosophy and the Arts in the 20th Century." *Arts in Society* (U of Wisconsin) 8 (1971): 306-23.

Engelmann, H. U. "Rhythmus und bildnerisches Denken." *Melos* 33 (1966): 261-67.

Fleming, William. "Newer Concepts of Time and Their Relation to the Temporal Arts." *J of Aesthetics and Art Criticism* 4 (1945): 101-106.

Furlan, G., et al., eds. *New Paths in the Use of Nuclear Techniques for Art and Archaeology*. Teaneck: World Scientific Publishing, 1986.

Gamzu, H. "Le temps et l'espace dans l'oeuvre de Yaacov Agam." *XXième Siècle* 56 (September 1980): 113-15.

Giedion, Sigfried. *Architecture and the Phenomena of Transition: The Three Space Conceptions in Architecture*. Cambridge: Harvard UP, 1971.

——. *The Eternal Present: A Contribution on Constancy and Change*. 2 vols. New York: Pantheon, 1962-64.

——. *Space, Time and Architecture: The Growth of a New Tradition*. 5th revised ed. Cambridge: Harvard UP, 1967. Bibliographical footnotes.

Gombrich, E. H. "Moment and Movement in Art." *J of Warburg and Courtauld Institutes* 27 (1964): 292-306.

Kirby, Michael. *The Art of Time: Essays on the Avant-Garde*. New York: Dutton, 1969.

Lessing, Gotthold Ephraim. *Laocoon: An Essay on the Limits of Painting and Poetry* (1766). Trans. Edward A. McCormick. Baltimore: Johns Hopkins UP, 1984. Locus classicus for the argument concerning the temporal nature of poetry and the spatial nature of the plastic arts.

Lowry, Bates. "Time and Motion." *The Visual Experience*. Englewood Cliffs: Prentice-Hall, 1966. Ch. 14.

Luchting, Wolfgang A. "*Hiroshima, mon amour*, Time, and Proust." *J of Aesthetics and Art Criticism* 21.3 (Spring 1963): 299-313.

McClain, J. "Time in the Visual Arts: Lessing and Modern Criticism." *J of Aesthetics and Art Criticism* 44.1 (Fall 1985): 41-58.

Macey, Samuel L. "Hogarth and the Iconography of Time." *Studies in Eighteenth-Century Culture*. Vol. 5. Ed. Ronald C. Rosbottom. Madison: U of Wisconsin P, 1976. 41-53, and Plates 1-4.

Meklakh, B. S., et al., eds. *Rhythm, Space and Time and Art* (in Russian). Leningrad: Nauka, 1974.

Moholy-Nagy, Lázló. *Vision in Motion*. 2nd ed. Chicago: P. Theobald, 1956.

Morgan, R. C. "The Spectacle in Time: A New Look at Art and Technology." *Arts Magazine* 62 (October 1987): 76-79.

Neumann, Erich. "Art and Time." *Man and Time: Papers from the Eranos Yearbooks*. Trans. Ralph Manheim. Ed. Joseph Campbell. New York: Pantheon, 1957. 3-37.

Paoletti, J. T. "Rita Myers and Deep Time." *Arts Magazine* 62 (February 1988): 84-85.

Praz, Mario. "Spatial and Temporal Interpenetration." *Mnemosyne: The Parallel between Literature and the Visual Arts*. Princeton: Princeton UP, 1970. Ch. 7.

Souriau, Etienne. "Time in the Plastic Arts." Trans. Marjorie Kupersmith. *J of Aesthetics and Art Criticism* 7.4 (June 1949): 294-307.

Stechow, W. "Raum und Zeit in der graphischen und musikalischen Illustration." *Zeitschrift für Asthetik und allgemeine Kunstwissenschaft* 25 (1931): 118-26.

B. CINEMA AND TELEVISION

Anderson, D. R., et al. "Estimates of Young Children's Time with Television: A Methodological Comparison of Parent Reports with Time-Lapse Video Home Observation." *Child Development* 56 (October 1985): 1345-57.

Arnheim, R. "Film et réalité." *Revue d'Esthétique* 26.4 (1973): 27-45.

Bluestone, George. "Time in Film and Fiction." *J of Aesthetics and Art Criticism* 19 (1961): 311-15.

Breslow, L. "TV and Our Sense of Time." *ETC* 38 (Spring 1981): 84-90.

Brett, Guy. *Kinetic Art: The Language of Movement*. London: Studio-Vista, 1968. Includes bibliography.

Buchman, L. M. "Orson Welles's *Othello*: A Study of Time in Shakespeare's Tragedy." *Shakespeare Survey* 39 (1987): 53-65.

Coates, P. "Chris Marker and the Cinema as Time Machine." *Science-Fiction Studies* 14 (November 1987): 307-15.

Deleuze, Gilles. *Cinema II: The Time Image*. Minneapolis: U of Minneapolis P, 1989.

Eisenstein, Sergei. "The Filmic Fourth Dimension" (1922). *Film Form*. Trans. Jay Leyda. New York: Harcourt, Brace, 1949. 64-71.

Gessner, Robert. "The Faces of Time." *The Moving Image*. New York: Dutton, 1968. Ch. 10.

——. "Seven Faces of Time: An Aesthetic for Cinema." *The Nature and Art of Motion*. Ed. Gyorgy Kepes. New York: Braziller, 1965. 158-67.

Hauser, Arnold. "Space and Time in the Film." *The Social History of Art*. 2 vols. London: Routledge and Kegan Paul, 1951. 2:940-48.

Hein, Birgit, et al., eds. *Film als Film: 1910 bis heute; vom Animationsfilm der 20er zum Filmenvironment der 70er Jahre*. Stuttgart: Hatje, 1978.

Jacobs, Lewis, ed. *The Movies as Medium*. New York: Farrar, Straus, and Giroux, 1970. Includes bibliography.

Kawin, Bruce F. "The Continuous Present." *Telling It Again and Again: Repetition in Literature and Film*. Ithaca: Cornell UP, 1972. Ch. 4.

Lawson, John H. "Time and Space." *Film: The Creative Process*. New York: Hill and Wang, 1964. Ch. 26.

Panofsky, Erwin. "Style and Medium in the Motion Pictures." *Critique* 1.3 (January-February 1947): 5-28.

Robbe-Grillet, Alain. *Last Year at Marienbad*. Trans. Richard Howard. New York: Grove P, 1962. Text for the film by Alain Resnais.

Slade, Mark. "The Screen as a Clock without Hands." *Language of Change*. Toronto: Holt, Rinehart, 1970. 13-27.

Stephenson, Ralph, and J. R. Debrix. *The Cinema as Art*. 2nd ed. Harmondsworth: Penguin, 1976.

Wachhorst, W. "Time-Travel Romance on Film: Archetypes and Structures." *Extrapolation* 25 (Winter 1984): 340-59.

C. TIME AND DEATH PERSONIFIED

Brandon, S. G. F. "Time as God and Devil." *Bulletin of the John Rylands Library, Manchester* 47 (1964): 12-31.

Burgess, Frederick. *English Churchyard Memorials*. London: Lutterworth, 1963.

Caldwell, Marilyn. "Amor Vincit Tempus." *North Carolina Museum of Art Bulletin* 8 (1968): 26-35.

Castelli, Enrico, ed. *Il simbolismo del tempo, studi di filosofia dell'arte*. Padua: CEDAM, 1973.

Cohen, Kathleen. *Metamorphosis of a Death Symbol: The Transi Tomb in the Late Middle Ages and the Renaissance*. Berkeley: U of California P, 1973.

Esdaile, Katherine A. *English Monumental Sculpture Since the Renaissance*. London: Society for Promoting Christian Knowledge, 1927.

Gordon, Donald. "Veritas Filia Temporis." *J of the Warburg and Courtauld Institutes* 3 (1939-40): 228-40.

Herberger, Charles F. *The Riddle of the Sphinx: Calendric Symbolism in Myth and Icon*. 2nd ed. Athens: Efstathiades Group, 1988.

Hollstein, F. W. H., ed. "Claesz's *Death Dance* (7 plates, 1562)." In his *Dutch and Flemish Etchings, Engravings, and Woodcuts ca. 1450-1700*. Amsterdam: Menno Hertzberger, n.d.

Klibansky, Raymond, Erwin Panofsky, and Fritz Saxl. *Saturn and Melancholy*. London: Nelson, 1964.

Lucian of Samosata. *Saturnalia* and *Cronosolon* in *The Works*. Trans. H. W. Fowler, and F. G. Fowler. Oxford: Clarendon P, 1905. Vol. 4.

Lydgate, John. *The Daunce of Machabree*. London, 1554.

Macey, Samuel L. "The Changing Iconography of Father Time." *The Study of Time III*. Ed. J. T. Fraser, N. Lawrence, and D. Park. New York: Springer, 1978. 540-75.

Macrobius. *The Saturnalia*. Trans. Percival Vaughan Davies. New York: Columbia UP, 1969.

Panofsky, Erwin. "Father Time." *Studies in Iconology: Humanistic Themes in the Art of the Renaissance*. Oxford: Oxford UP, 1939. Ch. 3.

——. *Tomb Sculpture*. New York: Abrams, n.d.

Petrarch, Francesco. *The Triumphs* (with Mr. Colvin's Notes upon the Engravings). London: John Murray, 1906. Includes the *Triumph of Time*.

Roscher, W. H. "Chronos." In his *Ausführliches Lexikon der griechischen und römischen Mythologie*. Leipzig: Teubner, 1884-90. 1:899.

Rosenfeld, Hellmut. *Der mittelalterliche Totentanz*. 2nd ed. Cologne: Böhlau, 1968.

Saxl, Fritz. "Veritas Filia Temporis." *Philosophy and History: Essays Presented to Ernst Cassirer*. Ed. R. Klibansky and H. J. Paton. New York: Harper and Row, 1963. 197-222.

Whittick, Arnold. *Symbols, Signs, and Their Meaning*. London: Leonard Hill, 1960.

Wittkower, Rudolf. "Chance, Time, and Virtue." *J of the Warburg Institute* 1 (1937-38): 313-21.

Zaehner, R. C. *Zurvan: A Zoroastrian Dilemma*. Oxford: Oxford UP, 1955.

Zimmer, Heinrich. *Myths and Symbols in Indian Art and Civilization*. Ed. Joseph Campbell. New York: Pantheon, 1946.

5. BIOLOGY

A. GENERAL

Beer, Gavin de. *Embryos and Ancestors*. New York: Oxford UP, 1962.

Beresford, C. H. "Time: A Biological Dimension." *J of the Royal College of Physicians, London* 22.2 (April 1988): 94-96.

Bodman, K. von. "Die Wirkung von Pharmaka auf Kurzzeitschätzung und circadiane Periodik von Säugetieren." *Zeitschrift für Vergleichende Physiologie* 68 (1970): 276-92.

Chance, B. "Enzyme Mechanisms in Living Cells." *The Mechanism of Enzyme Action*. Ed. W. D. McElroy, and B. Glass. Baltimore: Johns Hopkins P, 1954.

Cloudsley-Thompson, J. L. *Animal Conflict and Adaptation*. London: Foulis, 1965.

Dawkins, Richard. *The Blind Watchmaker*. Harlow: Longman Scientific, 1986.

——. *The Selfish Gene*. Oxford: Oxford UP, 1976.

Driesch, H. *Science and Philosophy of the Organism*. London: A. and C. Black, 1908.

Dubos, R. *Man Adapting*. New Haven: Yale UP, 1965.

Finerty, James P. *The Population Ecology of Cycles in Small Mammals: Mathematical Theory and Biological Fact*. New Haven: Yale UP, 1981.

Foerster, Heinz von. "Quantum Theory of Memory." In *6th Conference on Cybernetics of the Josiah Macy, Jr. Foundation*. New York, 1949.

Gladwin, Harold Sterling. *Tree-Ring Analysis: Problems of Dating*. Globe: Gila Pueblo, 1944.

Gmitro, J. I., and L. E. Scriven. "A Physiochemical Basis for Pattern and Rhythm." *Towards a Theoretical Biology*. Edinburgh: Edinburgh UP, 1969. 184-203.

Goodwin, Brian C. "Biological Control Processes in Time." In *Interdisciplinary Perspectives of Time*. Ed. Roland Fischer. *Annals of the New York Academy of Sciences* 138, Part 2 (1967): 752-55.

——. "Pattern Formation and Its Regeneration in the Protozoa." *Society for General Microbiology* 30 (1980): 377-404.

——. "Temporal Order as the Origin of Spatial Order in Embryos." *The Study of Time I*. Ed. J. T. Fraser, F. C. Haber, and G. H. Müller. New York: Springer, 1972. 190-99.

——. *Temporal Organization in Cells*. London: Academic P, 1963.

Godwin, Brian C., and M. H. Cohen. "A Phase-Shift Model for the

Spatial and Temporal Organization of Developing Systems." *J of Theoretical Biology* 25 (1969): 49-107.

Hardy, A. *The Open Sea—Its Natural History. The World of Plankton*. London: Collins, 1956.

Hinshelwood, C. M. *The Kinetics of Chemical Change*. Oxford: Clarendon P, 1940.

Hoagland, Hudson. "Some Biochemical Considerations of Time." *The Voices of Time*. Ed. J. T. Fraser. 2nd ed. Amherst: U of Massachusetts P, 1981. 312-29.

Kalmus, Hans. "Axioms and Theorems in Biology." *Nature* 198 (1963): 240-43.

Krebs, J. R., et al. "Time Horizons of Foraging Animals." *Annals of the New York Academy of Sciences* 423 (1984): 278-91.

Lecomte du Noüy, Pierre. *Biological Time*. New York: Macmillan, 1937.

Levinton, Jeffrey S. *Genetics, Paleontology, and Macroevolution*. Cambridge: Cambridge UP, 1988. Includes extensive bibliography.

Luyet, B. J., and P. M. Gehenio. "Life and Death at Low Temperature." *Biodynamica* (1940): 341.

Magnan, A. *Cinematographie jusqu'à 12000 vues par seconde (avec application à l'étude du vol des insectes)*. Paris: Herman, 1932.

Menaker, M., ed. *Biochronometry*. Washington: National Academy of Science, 1971.

———. "Synchronization with the Photic Environment Via Extra-Retinal Receptors in the Avian Brain." *Symposium on Biochronometry*. Washington: National Academy of Sciences, 1971. 315-32.

Muses, C. E. "Aspects of Some Problems in Biological and Medical Cybernetics." *Progress in Biocybernetics*. Ed. N. Wiener, and J. B. S. Haldane. New York: Elsevier, 1965. 243-48.

Needham, A. E. "How Living Organisms Repair Themselves." *New Scientist* 259 (1961): 284-87.

Needham, Joseph. *Order and Life*. Cambridge: MIT P, 1968.

Pagel, W. "J. B. Van Helmont, De Tempore, and Biological Time." *Osiris* 8 (1949): 346-417.

Pittendrigh, C. S. "On Temporal Organization in Living Systems." *The Harvey Lectures* (Series 56). New York: Academic P, 1961. 93-125.

Polanyi, M. "Life's Irreducible Structure." *Science* 160 (1968): 1308-12.

Pöppel, E. "Oscillations as Possible Basis for Time Perception." *The Study of Time I*. Ed. J. T. Fraser, F. C. Haber, and G. H. Müller. New York: Springer, 1972. 219-41.

Portmann, Adolf. "Time in the Life of the Organism." *Man and Time:*

Papers from the Eranos Yearbooks. Trans. Ralph Manheim. Ed. Joseph Campbell. New York: Pantheon, 1957. 308-23.
Post, R. H. "Population Differences in Red and Green Colour Vision Deficiency: A Review, and a Query on Selection Relaxation." *Eugenics Quarterly* 9 (1962): 131-35.
Smith, A. U. *Biological Effects of Freezing and Supercooling*. Baltimore: Williams and Wilkins, 1961.
Sonneborn, T. M. "Gene Action in Development." *Proceedings of the Royal Society of London. Series B. Biological Sciences* 176 (1970): 347-66.
Thom, R. *Stabilité structurelle et morphogénèse*. Reading: W. A. Benjamin, 1972.
Thorpe, W. H. *Birdsong: The Biology of Vocal Communication and Expression in Birds*. Cambridge: Cambridge UP, 1961.
——. *Learning and Instinct in Animals*. London: Methuen, 1956.
Uexküll, J. von, and G. Kriszat. *Streifzüge durch die Umwelten von Tieren und Menschen*. Frankfurt: S. Fischer, 1970.
Waddington, C. H. *Principles of Embryology*. London: Allen and Unwin, 1956.
——. *The Strategy of the Genes*. London: Allen and Unwin, 1957.
Weismann, A. *Essays on Heredity*. Trans. A. E. Shiply. Oxford: Oxford UP, 1891.
Wells, G. P. *The Sources of Animal Behaviour*. London: H. K. Lewis, 1955.
Wever, R. "Spontanperiodik des Menschen bei Ausschluss aller Zeitgeber." *Naturwissenschaften* 49 (1962): 337-42.
White, L. "The Historical Roots of Our Ecological Crisis." *Science* 155 (1967): 1203-07.
Wilson, E. O. *Sociobiology: The New Synthesis*. Cambridge: The Bellknap P, Harvard UP, 1975.
Wilson, J. A. "Increasing Entropy of Biological Systems." *Nature* 219 (1968): 534-35.

B. CHRONOBIOLOGY (INCLUDING BIOLOGICAL CLOCKS AND DYSRHYTHMIA)

Aanonsen, A. "Medical Problems of Shift Work." *Industrial Medicine and Surgery* 28 (1959): 422-27.
Accary, J. P., et al. "Small Bowel and Plasma Gastrin Rhythms in Cats." *Chronobiology International* (Oxford) 4.1 (1987): 53-58.
Albertsson-Wikland, K., et al. "Analyses of 24-Hour Growth Hormone

Profiles in Children: Relation to Growth." *J of Clinical Endocrinology and Metabolism* 67.3 (September 1988): 493-500.

Alluisi, E. A., and W. D. Chiles. "Sustained Performance, Work-Rest Scheduling, and Diurnal Rhythms in Man." *Acta Psychologica* 27 (1967): 436-42.

Apfelbaum, M., et al. "Rhythmes circadiens de l'alternance veille-sommeil pendant l'isolement souterrain de sept jeunes femmes." *Presse Médicale* 77 (1969): 879-82.

Apostol, S. "The Diurnal and Seasonal Biorhythms of the Immunological Responses." *Revista Medico-Chirurgicala A Societatii de Medici Si Naturalisti din Iasi* 92.4 (October-December 1988): 729-31.

Arendt, J., et al. "Light and Melatonin as Zeitgebers in Man." *Chronobiology International* (Oxford) 4.2 (1987): 273-82.

Aschoff, Jürgen, ed. *Biological Rhythms*. New York: Plenum, 1981.

——. *Circadian Clocks*. Amsterdam: North-Holland, 1964.

——. "Circadian Rhythms in Man: A Self-Sustained Oscillator with an Inherent Frequency Underlies Human 24-Hour Periodicity." *Science* 148 (1965): 1427-32.

——. "Circadian System Properties." *Advances in Physiological Sciences*. Ed. F. Obal, and G. Benedek. Budapest: Academiai Kiado, 1981. 18:1-17.

——. "Circadian Timing." *Timing and Time Perception*. Ed. John Gibbon, and Lorraine Allan. New York: New York Academy of Sciences, 1984. 442-68.

——. "Comparative Physiology: Diurnal Rhythms." *Annual Review of Physiology* 25 (1963): 581-600.

——. "Die innere Uhr des Menschen." *Die Zeit*. Ed. A. Peisl, and A. Mohler. Munich: Oldenburg, 1983. 133-44.

——. "Exogenous and Endogenous Components in Circadian Rhythms." *Cold Spring Harbor Symposia on Quantitative Biology* 25 (1960): 11-28.

——. "Timegivers of 24-Hour Physiological Cycles." *Man's Dependence on the Earthly Atmosphere*. Ed. K. E. Schaefer. New York: Macmillan, 1962.

——. "Zeitgeber der tierischen Tagesperiodik." *Naturwissenschaft* 41 (1954): 49-56.

Aschoff, Jürgen, and J. Meyer-Lohmann. "Angeborene 24-Stunden-Periodik beim Kuecken." *Pflügers Archiv* 260 (1954): 170-76.

Aschoff, Jürgen, and R. Wever. "Resynchronisation der Tagesperiodik von Vögeln nach Phasensprung des Zeitgebers." *Zeitschrift für Vergleichende Physiologie* 46 (1963): 321-35.

Aschoff, J., et al., eds. *Vertebrate Circadian Systems: Structure and*

Physiology. New York: Springer, 1982.

Bailleul, F., et al. "Interindividual Differences in the Circadian Hematologic Time Structure of Cancer Patients." *Chronobiology International* (Oxford) 3.1 (1986): 47-54.

Barnwell, F. H. "Daily and Tidal Patterns of Activity in Individual Fiddler Crabs (Genus *Uca*) from the Woods Hole Region." *Biological Bulletin* 130 (1966): 1-17.

——. "The Role of Rhythmic Systems in the Adaptation of Fiddler Crabs to the Intertidal Zone." *American Zoologist* 8 (1968): 569-83.

——. "Variation in the Form of the Tide and Some Problems It Poses for Biological Timing Systems." *Biological Rhythms in the Marine Environment*. Ed. P. J. DeCoursey. Columbia: South Carolina UP, 1976. 161-87.

Bassi, C. J., et al. "Circadian Rhythm in Goldfish Visual Sensitivity." *Investigative Ophthalmology and Visual Science* 28.11 (November 1987): 1811-15.

Beck, S. D. *Insect Photoperiodism*. New York: Academic P, 1968.

Bell, G. M., et al. "Abnormal Diurnal Urinary Sodium and Water Excretion in Diabetic Autonomic Neuropathy." *Clinical Science* (London) 73.3 (September 1987): 259-65.

Bennett, M. F. *Living Clocks in the Animal World*. Springfield: Charles C. Thomas, 1974.

Bennett, M. F., J. Shriner, and R. A. Brown. "Persistent Tidal Cycles of Spontaneous Motor Activity in the Fiddler Crab, *Uca Pugnax*." *Biological Bulletin* 112 (1957): 267-75.

Beresford, C. H. "Time: A Biological Dimension." *J of the Royal College of Physicians of London* 22.2 (April 1988): 94-96.

Bernard, C. *Leçons sur phénomènes de la vie communs aux animaux et végétaux*. 2 vols. Paris: Baillère, 1878-79.

Bicakova-Rocher, A., et al. "Seasonal Variations in Socially and Legally Unacceptable Sexual Behaviour." *Chronobiology International* (Oxford) 2.3 (1985): 203-08.

Biological Rhythms in Psychiatry and Medicine. Chevy Chase: U. S. Department of Health, Education and Welfare, National Institute of Mental Health, Public Health Publication 2088, 1970.

Bjerner, B. "Diurnal Variation in Mental Performance." *British J of Industrial Medicine* (1955): 103-10.

Bobbert, A. C., et al. "Effects of Illumination on the Rabbit's EEG During the Two Phases of Its Circadian Rhythm." *Chronobiology International* (Oxford) 4.2 (1987): 225-34.

Bogaty, P., et al. "Circadian Patterns in Coronary Disease: The Mournfulness of Morning." *Canadian Journal of Cardiology* 4.1

(January-February 1988): 5-11.

Boulos, Z., A. M. Rosenwasser, and M. Terman. "Feeding Schedules and the Circadian Organization of Behavior in the Rat." *Behavior and Brain Research* 1 (1980): 39-65.

Boulos, Z., and M. Terman. "Food Availability and Daily Biological Rhythms." *Neuroscience and Biobehavior Review* 4 (1980): 119-31.

Brady, J., ed. *Biological Timekeeping.* New York: Cambridge UP, 1982.

——. "How Are Insect Circadian Rhythms Controlled?" *Nature* 223 (1969): 781-84.

Brahmachary, R. L. "Physiological Clocks." *International Review of Cytology* 21 (1967): 65-89.

Broten, T. P., et al. "Statistical Criteria for Using Short-Term Measurements as an Index of 24-Hour Mean Arterial Pressure in Unanesthetized Unrestrained Dogs." *Life Sciences* 42.17 (1988): 1625-33.

Brown, F. A., Jr. "Biological Clocks: Endogenous Cycles Synchronized by Subtle Geophysical Rhythms." *BioSystems* 8 (1976): 67-81.

——. "Response to Pervasive Geophysical Factors and the Biological Clock Problems." *Cold Spring Harbor Symposia on Quantitative Biology* 25 (1960): 57-71.

Brown, F. A., Jr., J. W. Hastings, and J. D. Palmer. *The Biological Clock: Two Views.* New York: Academic P, 1970.

Bruce, V. G. "Cell Division Rhythms and the Circadian Clock." *Circadian Clocks.* Ed. J. Aschoff. Amsterdam: North-Holland, 1965. 125-38.

——. "Environmental Entrainment of Circadian Rhythms." *Cold Spring Harbor Symposia on Quantitative Biology* 25 (1960): 29-48.

Bünning, Erwin. "Die Endogene Tagesrhythmik als Grundlage der photoperiodischen Reaktion." *Berichte. Deutsche Botanischen Gesellschaft* 54 (1936): 590-607.

——. *The Physiological Clock: Circadian Clocks and Biological Chronometry.* 3rd ed. New York: Springer, 1973.

Burgard, P., and M. Henry. "Quelques aspects de la fatigue dans l'aviation de transport." *Presse Médicale* 89 (1961): 1903-06.

Cambras, T., et al. "Generational Variability in the Patterns of Motor Activity Circadian Rhythm in the Rats." *Revista Espanola de Fisiologia* 44.3 (September 1988): 243-46.

Campbell, Jeremy. *Winston Churchill's Afternoon Nap: A Wide-Awake Inquiry into the Human Nature of Time.* New York: Simon and Schuster, 1986.

Caruana, M. P., et al. "Effects of Chronic Congestive Heart Failure Secondary to Coronary Artery Disease on the Circadian Rhythm of Blood Pressure and Heart Rate." *American J of Cardiology* 62.10

Part 1 (October 1, 1988): 755-59.

Chachati, A., et al. "Variability of Albumin Excretion in Insulin-Dependent Diabetics." *Diabetic Medicine* 4.5 (September-October, 1987): 441-45.

Christensen, A. P., et al. "The Relationship Between Premenstrual Dysphoria and Daily Ratings Dimensions." *J of Affective Disorders* 16.2-3 (March-June 1989): 127-32.

Church, Russell M. "Properties of the Internal Clock." *Timing and Time Perception*. Ed. John Gibbon and Lorraine Allan. New York: New York Academy of Sciences, 1984. 566-82.

Cipolla-Neto, J., et al. "Remarkable Similarities Between the Temporal Organization of Neocortical Electrographic Sleep Patterns of Rats and Humans." *Brazilian J of Medical and Biological Research* (São Paulo) 21.3 (1988): 599-601.

Clodoré, M., et al. "Diurnal Variation in Subjective and Objective Measures of Sleepiness: The Effects of Sleep Reduction and Circadian Type." *Chronobiology International* (Oxford) 3.4 (1986): 255-63.

Cloudsley-Thompson, John Leonard. "Adaptive Functions of Circadian Rhythms." *Cold Spring Harbor Symposia on Quantitative Biology* 25 (1960): 345-55.

——. "Biological Clocks and Their Synchronizers." *The Study of Time III*. Ed. J. T. Fraser, N. Lawrence, and D. Park. New York: Springer, 1978. 188-213.

——. *Biological Clocks: Their Functions in Nature*. London: Weidenfeld and Nicolson, 1980.

——. "Diurnal Rhythms of Locomotory Activity in Isolated Desert Locusts, *Schistocerca Gregaria* (Forsk)." *J of Interdisciplinary Cycle Research* 8 (1977): 27-36.

——. *Rhythmic Activity in Animal Physiology and Behavior*. New York: Academic P, 1961.

——. "Studies in Diurnal Rhythms—VIII. The Endogenous Chronometer in *Gryllus Compestris L. (Orthoptera: Gryllidae)*." *J of Insect Physiology* 2 (1958): 275-80.

——. "Time Sense of Animals." *The Voices of Time*. Ed. J. T. Fraser. 2nd ed. Amherst: U of Massachusetts P, 1981. 296-311.

Cold Spring Harbor Symposia on Quantitative Biology. *Biological Clocks*. Vol. 25. Cold Spring Harbor: The Biological Laboratory, 1960.

Colquhoun, W. P., ed. *Biological Rhythm and Human Performance*. New York: Academic P, 1971.

Condon, William S. "Cultural Microrhythms." *Interaction Rhythms: Periodicity in Communicative Behavior*. Ed. M. Davis. New York: Human Sciences P, 1982. 53-77.

Conroy, R. T. W. L., and J. N. Mills. *Human Circadian Rhythms*. London: Churchill, 1970.

Cosnard, M., et al., eds. *Rhythms in Biology and Other Fields of Application: Deterministic and Stochastic Approaches*. New York: Springer, 1983.

Costa, G., et al. "Usual Meal Times in Relation to Age, Sex, Work Activity and Morningness-Eveningness." *Chronobiologia* (Milan) 14.4 (October-December 1987): 383-91.

Crane, J. E. "The Time Zone Fatigue Syndrome." *Flying Physician* 7 (1963): 19-22.

Crook, Frederick. *Natural Rhythms, Organic and Inorganic*. Guernsey: F. Crook, 1987.

Csanyi, Vilmos. *Evolutionary Systems and Society: A General Theory*. Durham: Duke UP, 1988.

Daan, Serge, and Jürgen Aschoff. "Circadian Contributions to Survival." *Vertebrate Circadian Systems: Structure and Physiology*. Ed. Jürgen Aschoff, Serge Daan, and Gerard A. Groos. New York: Springer, 1982. 305-21.

——. "Short-Term Rhythms in Activity." *Biological Rhythms*. Ed. Jürgen Aschoff. New York: Plenum, 1981. 491-98.

Daan, Serge, and Eberhard Gwinner, eds. *Biological Clocks and Environmental Time*. New York: Guilford P, 1989.

Davis, F. C. "Daily Variations in Maternal and Fetal Weight Gain in Mice and Hamsters." *J of Experimental Zoology* 250.3 (June 1989): 273-82.

——. "Ontogeny of Circadian Rhythms." *Handbook of Behavioral Neurobiology*. Ed. Jürgen Aschoff. New York: Plenum, 1981. 4:187-90.

Davis, F. C., et al. "Development of Hamster Circadian Rhythms: Role of the Maternal Suprachiasmatic Nucleus." *J of Comparative Physiology. A, Sensory, Neural, and Behavioral Physiology* 162.5 (April 1988): 601-10.

Diez-Noguera, A., et al. "Detection of Changes in the Mouse Circadian Rhythm Induced by Stress." *Revista Espanola de Fisiologia* 44.3 (September 1988): 247-55.

Dowse, H. B., and J. D. Palmer. "The Chronomutagenic Effect of Deuterium Oxide on the Period and Entrainment of a Biological Rhythm." *Biological Bulletin* 143 (1972): 513-24.

——. "Entrainment of Circadian Activity Rhythms in Mice by Electrostatic Fields." *Nature* 222 (1969): 564-66.

Edmunds, L. N. *Cellular and Molecular Bases of Biological Clocks*. New York: Springer, 1987.

———. "Persistent Circadian Rhythms of Cell Division in Euglena." *Symposium on Biochronometry*. Ed. M. Menaker. Washington: National Academy of Sciences, 1971. 594-611.

Ehret, C. F., and E. Trucco. "Molecular Models for the Circadian Clock, I: The Chronon Concept." *J of Theoretical Biology* 15 (1967): 240-62.

Elliott, J. A., and B. D. Goldman. "Seasonal Reproduction: Photoperiodism and Biological Clocks." *Neuroendocrinology of Reproduction*. Ed. N. T. Adler. New York: Plenum, 1981. 377-423.

Endoh, D., et al. "Circadian Variation in Lung Tumor Induction with X-Rays in Mice." *J of Radiation Research* (Tokyo) 28.2 (June 1987): 186-89.

Enright, J. T. "Synchronization and Ranges of Entrainment." *Circadian Clocks*. Ed. J. Aschoff. Amsterdam: North-Holland, 1965. 112-24.

Evans, James R. *Rhythm in Psychological, Linguistic and Musical Processes*. Springfield: C. C. Thomas, 1986.

Farr, L. A., et al. "Circadian Disruption and Surgical Recovery." *Nursing Research* 37.3 (May-June 1988): 170-75.

Fingermen, M. "Tidal Rhythmicity in Marine Organisms." *Cold Spring Harbor Symposia on Quantitative Biology* 25 (1960): 481-89.

Fischman, A. J., et al. "Constant Light and Dark Affect the Circadian Rhythm of the Hypothalamic-Pituitary-Adrenal Axis." *Neuroendocrinology* 47.4 (April 1988): 309-16.

Fisher, K. C., et al. *Mammalian Hibernation III*. Edinburgh: Oliver and Boyd, 1967.

Fleisner, Gunther, and Gerda Fleisner. *Efferent Control of Visual Sensitivity in Arthropod Eyes: With Emphasis on Circadian Rhythms*. New York: VCH Publications, 1988.

Flink, E. B. "Effect of Sudden Time Displacement by Air Travel on Synchronization of Adrenal Function." *Proceedings. Society for Experimental Biology and Medicine* 100 (1959): 498-501.

Folkard, S., et al. "Night Shift Paralysis in Air Traffic Control Officers." *Ergonomics* 30.9 (September 1987): 1353-63.

Foret, J., et al. "Sleep and Body Temperature in 'Morning' and 'Evening' People." *Sleep* 8.4 (December 1985): 311-18.

Fraser, G., et al. "Effect of Sleep and Circadian Cycle on Sleep Period Energy Expenditure." *J of Applied Physiology* 66.2 (February 1989): 830-36.

Fraser, W. D., et al. "Diurnal Variation in Admission to Hospital of Women in Labour." *Canadian J of Surgery* 32.1 (January 1989): 33-35.

Frazier, T. W., J. A. Rummel, and H. S. Lipscomb. "Circadian Variability in Vigilance Performance." *J of Aerospace Medicine* 39 (1968): 383-95.

Freeman, G. L., and C. T. Hovland. "Diurnal Variations in Performance and Related Physiological Processes." *Psychological Bulletin* 31 (1934): 777-99.

Fukagawa, K., et al. "Disruption of Light-Dark Cycle of Feeding and Drinking Behavior, and Ambulatory Activity Induced by Development of Obesity in the Zucker Rat." *International J of Obesity* 12.5 (1988): 481-90.

Garner, W. W., and H. A. Allard. "Effect of the Relative Length of Day and Night and Other Factors of the Environment on Growth and Reproduction in Plants." *J of Agricultural Research* 18 (1920): 553.

Gerbes, A. L., et al. "The Influence of Timeshift on Circadian Rhythm of Sensitivity to X-Irradiation in Mice." *Chronobiology International* (Oxford) 1.3 (1984): 177-84.

Gerritzen, F. "The Diurnal Rhythm in Water, Chloride, Sodium and Potassium Excretion During a Rapid Displacement from East to West and Vice Versa." *J of Aerospace Medicine* 33 (1962): 697-701.

Gerritzen, F., T. Strengers, and S. Esser. "Studies on the Influence of Fast Transportation on the Circadian Excretion Pattern of the Kidney in Humans." *J of Aerospace Medicine* 40 (1969): 264-71.

Ghata, J., et al. "Rythmes circadiens désynchronisés (17-hydroxy-corticostéroïdes température rectale, veille-sommeil) chez deux sujets adultes sains." *Annales d'Endocrinologie* 30 (1969): 245-60.

Gibson, C. J. "Diurnal Alterations in Retinal Tyrosine Level and Dopamine Turnover in Diabetic Rats." *Brain Research* 454.1-2 (June 28, 1988): 60-66.

Gibson, R. B. "The Effects of Transposition of the Daily Routine on the Rhythm of Temperature Variation." *American J of Medical Science* 129 (1905): 1048-59.

Gittelson, Bernard. *Biorhythm: A Personal Science*. Revised ed. New York: Warner Books, 1988.

Glasgow, M. S., et al. "The Use of Successive Blood-Pressure Measurements to Estimate Blood-Pressure Variability." *J of Behavioral Medicine* 11.5 (October 1988): 435-46.

Glass, Leon. *From Clocks to Chaos: The Rhythm of Life*. Princeton: Princeton UP, 1988.

Gooddy, W. "Outside Time and Inside Time." *Perspectives in Biology and Medicine* 12 (1969): 239-53.

——. "Some Factors in the Production of Dysrhythmia and Disorientation Associated with Rapid Latitudinal Transfer." *The Study of*

Time I. Ed. J. T. Fraser, F. C. Haber, and G. H. Müller. New York: Springer, 1972. 533-46.

Gosse, P., et al. "High Night Blood Pressure in Treated Hypertensive Patients Not Harmless." *American J of Hypertension* 1.3 Part 3 (July 1988): 195S-98S.

Grebe, J. J. "Time: Its Breadth and Depth in Biological Rhythms." *Annals of the New York Academy of Sciences* 112 (1962): 1206-10.

Griffiths, R. A. "Natural Environmental Cues and Circadian Rhythms of Behaviour—A Perspective." *Chronobiology International* (Oxford) 3.4 (1986): 247-53.

Groh, K. R., et al. "Circadian Manifestations of Barbiturate Habituation, Addiction and Withdrawal in the Rat." *Chronobiology International* (Oxford) 5.2 (1988): 153-66.

Groos, Gerard A., and Serge Daan. "The Use of Biological Clocks in Time Perception." *Time, Mind, and Behavior*. Ed. John A. Michon, and Janet L. Jackson. New York: Springer, 1985. 65-74.

Gullett, C. C. "Jet Planes and the Circadian Cycle." *J of the American Medical Association* 197 (1966): 935.

Gundel, A., et al. "Transition Between Advance and Delay Responses to Eastbound Transmeridian Flights." *Chronobiology International* (Oxford) 6.2 (1989): 147-56.

Gwinner, E. *Circannual Rhythms*. New York: Springer, 1986.

——. "Circannual Rhythms: Their Dependence on the Circadian System." *Biological Clocks in Seasonal Reproductive Cycles*. Ed. B. K. Follett, and D. E. Follett. Bristol: John Wright, 1981. 153-70.

——. "A Comparative Study of Circannual Rhythms in Warblers." *Symposium on Biochronometry*. Ed. M. Menaker. Washington: National Academy of Sciences, 1971.

Halberg, Franz. "Beobachtungen über 24-Stunden-Periodik in standardisierter Versuchsanordnung vor und nach Epinephrektomie und bilateraler optischer Enukleation." *Berichte über die Gesamte Physiologie* 162 (1954): 354-55.

——. "Chronobiology." *Annual Review of Physiology* 31 (1969): 675-725.

——. *Circadian Temporal Organization and Experimental Pathology*. Turin: Minerva Medica, 1962.

——. "Physiologic Rhythms." *Physiological Problems in Space Travel*. Ed. J. D. Hardy. Springfield: Charles C. Thomas, 1964. 298-322.

——. "Physiologic Rhythms and Bioastronautics." *Bioastronautics*. Ed. K. E. Schaeffer. New York: Macmillan, 1964. 181-95.

——. "Physiologic 24-Hour Rhythms: A Determinant of Response to Environmental Agents." *Man's Dependence on the Earthly Atmosphere*. Ed. K. E. Schaeffer. New York: Macmillan, 1962. 48-89.

———. "Pouvoir de résolution des calculatrices électroniques en pathologie temporelle: Analogie avec la microscopie." *Scientia* 101, supplement (1966): 172-79.

———. "Temporal Coordination of Physiologic Function." *Cold Spring Harbor Symposia on Quantitative Biology* 25 (1960): 289-310.

———. "The 24-Hour Scale: A Time Dimension of Adaptive Functional Organization." *Perspectives in Biology and Medicine* 3 (1960): 491-527.

Halberg, F., P. G. Albrecht, and C. P. Barnum. "Phase Shifting of Liver Glycogen Rhythm in Intact Mice." *American J of Physiology* 199 (1960): 400-02.

Halberg, F., M. Engeli, and C. Hamburger. "The 17-Ketosteroid Excretion of a Healthy Man on Weekdays and Weekends." *Experimental Medicine and Surgery* 23 (1965): 61-69.

Halberg, F., and A. Reinberg. "Rythmes circadiens et rythmes des basses fréquences en physiologie humaine." *J de Physiologie* 59 (1967): 117-200.

Halberg, F., Y. L. Tong, and E. A. Johnson. "Circadian System Phase: An Aspect of Temporal Morphology; Procedures and Illustrative Examples." *The Cellular Aspects of Biorhythms*. Ed. H. von Mayersbach. Berlin: Springer, 1967.

Halberg, F., M. B. Visscher, and J. J. Bittner. "Relation of Visual Factors to Eosinophil Rhythms in Mice." *American J of Physiology* 179 (1954): 229-35.

Halberg, F., et al. "Agreement in Endpoints from Circadian Rhythmometry on Healthy Human Beings Living on Different Continents." *Experientia* 25 (1969): 107-12.

———. "Computer Techniques in the Study of Biologic Rhythms." *Annals of the New York Academy of Sciences* 115 (1964): 695-720.

———. "Étude en libre-cours des rythmes circadiens du pouls de l'alternance veille-sommeil et de l'estimation du temps pendant les deux mois de séjour souterrain d'un homme adulte jeune." *Comptes Rendus Académie de Sciences* 260 (1965): 1259-62.

———, eds. *Glossary of Chronobiology*. Milan: Il Ponte, 1977.

———. "Physiologic 24-Hour Periodicity in Human Beings and Mice, the Lighting Regimen and Daily Routine." *Photoperiodism and Related Phenomena in Plants and Animals*. Ed. A. P. Wittrow. Washington: American Association for the Advancement of Science, 1959. 803-78.

———. "Spectral Resolution of Low-Frequency, Small-Amplitude Rhythms in Excreted Ketosteroid: Probable Androgen Induced Circaseptan Desynchronization." *Acta d'Endocrinologie* 103 (1965): 1-54.

Halberg, J., et al. "Further Steps Toward a Neonatal Chronocardiology." *Chronobiologia* (Milan) 14.3 (July-September 1987): 297-99.

Hamner, Karl C. "Experimental Evidence for the Biological Clock." *The Voices of Time*. Ed. J. T. Fraser. 2nd ed. Amherst: U of Massachusetts P, 1981. 281-95.

Handyside, Joann. *Bio Rhythms in Man*. Lorain: Dayton Laboratories, 1978.

Harker, J. E. "Diurnal Rhythms in the Animal Kingdom." *Biological Reviews* 33 (1958): 1-52.

——. *The Physiology of Diurnal Rhythms*. Cambridge: Cambridge UP, 1964.

Hartman, B. O., and G. K. Cantrell. "Sustained Pilot Performance Requires More Than Skill." *J of Aerospace Medicine* 38 (1967): 801-03.

Hasler, A. D., and H. O. Schwassmann. "Environmental Cues in the Orientation Rhythm of Fish." *Cold Spring Harbor Symposia on Quantitative Biology* 25 (1960): 429-31.

Hastings, J. Woodward. "Biochemical Aspects of Rhythms: Phase Shifting by Chemicals." *Cold Spring Harbor Symposia on Quantitative Biology* 25 (1960): 131-43.

Hastings, J. Woodward, et al., eds. *The Molecular Basis of Circadian Rhythms*. Berlin: Dahlem Konferenzen, 1976.

Hauty, G. T., and T. Adams. "Phase Shifting of the Human Circadian System." *Circadian Clocks*. Ed. J. Aschoff. Amsterdam: North-Holland, 1965. 413-25.

——. "Phase Shifts of the Human Circadian System and Performance Deficit During the Periods of Transition: 1—East-West Flight." *J of Aerospace Medicine* 37 (1966): 664-68.

Heckert, Hilmar. *Lunationsrhythmen des menschlichen Organismus, Methodisches und Ergebnisse*. Leipzig: Akademische Verlagsgesellschaft Geest und Portig, 1961.

Helfrich, C., et al. "Circadian Activity Rhythm in the House Fly Continues After Optic Tract Severance and Lobectomy." *Chronobiology International* (Oxford) 2.1 (1985): 19-32.

Henderson, A. J., et al. "Circadian Rhythm of Peak Expiratory Flow in Asthmatic and Normal Children." *Thorax* 44.5 (May 1989): 410-14.

Hoffmann, K. "Experimental Manipulation of the Orientational Clock in Birds." *Cold Spring Harbor Symposia on Quantitative Biology* 25 (1960): 379-87.

Holubar, Josef. *The Sense of Time: An Electrophysiological Study of Its Mechanisms in Man*. Trans. John S. Barlow. Cambridge: MIT P,

1969.

Hyde, Margaret Oldroyd. *Animal Clocks and Compasses: From Animal Migration to Space Travel.* New York: Whittlesey House, 1960.

Iranmanesh, A., et al. "24-Hour Pulsatile and Circadian Patterns of Cortisol Secretion in Alcoholic Men." *J of Andrology* 10.1 (January-February 1989): 54-63.

Jackson, D. A., et al. "Circadian Variation in Fat Concentration of Breast-Milk in a Rural Northern Thai Population." *British J of Nutrition* 59.3 (May 1988): 349-63.

Jilge, B., et al. "The Internal Synchronization of Five Circadian Functions of the Rabbit." *Chronobiology International* (Oxford) 1.3 (1984): 195-204.

Kafka, M. S., et al. "Relationships Between Behavioral Rhythms, Plasma Corticosterone and Hypothalamic Circadian Rhythms." *Chronobiology International* (Oxford) 3.2 (1986): 117-22.

Kalmus, Hans. "Über das Problem der sogenannten exogenen und endogenen, sowie der orblichen Rhythmik und über organische Periodezität überhaupt." *Rivista di Biologia* 24 (1938): 191-225.

——. "Über die Natur des Zeitgedächtnisses der Bienen." *Zeitschrift für vergleichende Physiologie* 20 (1934): 405-19.

Kanabrocki, E. L., et al. "Ten-Year-Replicated Circadian Profiles for 36 Physiological, Serological and Urinary Variables in Healthy Men." *Chronobiology International* (Oxford) 5.3 (1988): 237-84.

Karakashian, M. W., and H. G. Schweiger. "Circadian Properties of the Rhythmic System in Individual Nucleated and Enucleated Cells of *Acetabularia Mediterranea.*" *Experimental Cell Research* 97 (1976): 366-77.

Kato, K., et al. "Programmed Induction of Labor for Primiparous Women to Ensure Daytime Delivery." *Asia-Oceania J of Obstetrics and Gynaecology* (Tokyo) 13.4 (December 1987): 405-15.

Kimberly, R. P. "Rhythmic Patterns in Human Interaction." *Nature* 228 (1970): 88-90.

Kirlin, P. C., et al. "Spontaneous Neurohormonal Fluctuation in Chronic Congestive Heart Failure." *American J of Cardiology* 62.1 (July 1, 1988): 150-51.

Klein, E., et al. "24-Hour Motor Activity in Genetically Nervous and Normal Pointer Dogs." *Biological Psychiatry* 23.5 (March 1, 1988): 523-27.

Klein, K. E., H. Brüner, and S. Ruff. "Untersuchungen zur Belastung des Bordpersonals auf Fernflügen mit Düsenmaschinen." *Zeitschrift für Flugwissenschaften* 14 (1966): 109-21.

Klein, K. E., H. M. Wegmann, and H. Brüner. "Circadian Rhythm in

Indices of Human Performance, Physical Fitness, and Stress Resistance." *J of Aerospace Medicine* 39 (1968): 512-18.

Kleitman, N. *Sleep and Wakefulness*. Revised ed. Chicago: U of Chicago P, 1963.

Klevecz, R. R., et al. "Circadian and Ultradian Rhythms of Proliferation in Human Ovarian Cancer." *Chronobiology International* (Oxford) 4.4 (1987): 513-23.

———. "Circadian Gating of S Phase in Human Ovarian Cancer." *Cancer Research* 47.23 (December 1, 1987): 6267-71.

Knutsson, A., et al. "Prevalence of Risk Factors for Coronary Artery Disease Among Day and Shift Workers." *Scandinavian J of Work, Environment and Health* 14.5 (October 1988): 317-21.

Kupfer, David J., et al., eds. *Biological Rhythms and Mental Disorders*. New York: Guilford P, 1989.

Laerum, O. D., et al. "Circadian and Circannual Variations of Cell Cycle Distribution in the Mouse Bone Marrow." *Chronobiology International* (Oxford) 5.1 (1988): 19-35.

Lafontaine, E., et al. "Influence of Air Travel East-West and Vice Versa on Circadian Rhythms of Urinary Elimination of Potassium and 17 Hydroxycorticosteriods." *J of Aerospace Medicine* 38 (1967): 944-47.

Langlois, P. H., et al. "Temporal Patterns of Reported Single-Vehicle Car and Truck Accidents in Texas, U.S.A. During 1980-83." *Chronobiology International* (Oxford) 2.2 (1985): 131-40.

Lauber, J. K., et al. "Sleepiness, Circadian Dysrhythmia, and Fatigue in Transportation System Accidents." *Sleep* 11.6 (December 1988): 503-12.

Lavernhe, J., and C. Blanc. "Sommeil, insomnie et dette de sommeil." *Presse Médicale* 75 (1967): 391-95.

Lesser, Carolyn. *The Goodnight Circle*. San Diego: Harcourt, Brace, 1984.

Lewis, P. R., and M. C. Lobban. "Dissociation of Diurnal Rhythms in Human Subjects Living in Abnormal Time Routines." *Quarterly J of Experimental Physiology* 42 (1957): 371-86.

Lindauer, M. "Time Compensated Sun Orientation in Bees." *Cold Spring Harbor Symposia on Quantitative Biology* 25 (1960): 371-77.

Livermore, A. H., Jr., et al. "Light Transducer for the Biological Clock: A Function for Rapid Eye Movements." *J of Neural Transmission* 72.1 (1988): 37-42.

Lodeesen, M., and J. E. Crane. "Tired Jet Pilots." *Flying* 72 (1963): 52.

Lofts, B., and K. R. Murton. "Photoperiodic and Physiological Adaptations Regulating Avian Breeding Cycles and Their Ecological Significance." *J of Zoology, London* 155 (1968): 327-94.

Luce, Gay Gaer. *Biological Rhythms in Human and Animal Physiology*. New York: Dover, 1971.

——. *Biological Rhythms in Psychiatry and Medicine*. Chevy Chase: National Institute of Mental Health, 1970.

——. *Body Time*. New York: Pantheon, 1971.

McClintock, M. K. "Menstrual Synchrony and Suppression." *Nature* 229 (1971): 244-45.

McIntyre, I. M., et al. "Human Melatonin Response to Light at Different Times of the Night." *Psychoneuroendocrinology* 14.3 (1989): 187-93.

McMurry, L., and J. W. Hastings. "No Desynchronization among Four Circadian Rhythms in the Unicellular Alga, *Gonyaulax Polyedra*." *Science* 175 (1972): 1137-38.

Marshall, A. J. "Annual Periodicity in Migration and Reproduction of Birds." *Cold Spring Harbor Symposia on Quantitative Biology* 25 (1960): 499-505.

Meis, P. J., et al. "Pregnancy Alters Diurnal Variation of Plasma Glucose Concentration." *Chronobiology International* (Oxford) 1.2 (1984): 145-49.

Mejean, L., et al. "Circadian and Ultradian Rhythms in Blood Glucose and Plasma Insulin of Healthy Adults." *Chronobiology International* (Oxford) 5.3 (1988): 227-36.

Mellerup, B., et al. "Seasonal Variation in Urinary Excretion of Calcium, Magnesium and Phosphate in Manic-Melancholic Patients." *Chronobiology International* (Oxford) 1.1 (1984): 81-86.

Menaker, M., ed. *Symposium on Biochronometry*. Washington: National Academy of Sciences, 1971.

Menaker, M., and A. Eskin. "Entrainment of Circadian Rhythms by Sound in Passer Domesticus." *Science* 154 (1966): 1579-81.

Mendlewicz, J., and H. M. Van Pragg. *Biological Rhythms and Behavior*. Farmington: S. Karger, 1983.

Mergenhagen, D., and H. G. Schweiger. "Circadian Rhythm of Oxygen Evolution in Cell Fragments of *Acetabularia Mediterranea*." *Experimental Cell Research* 92 (1975): 127-30.

Minors, D. S., A. R. Scott, and J. M. Waterhouse. "Circadian Arrhythmia: Shiftwork, Travel and Health." *J. Society of Occupational Medicine* 36 (1986): 39-44.

Minors, D. S., and J. M. Waterhouse. *Circadian Rhythms and the Human*. Stoneham: Butterworth Publications, 1981.

——. "Circadian Rhythms and Their Mechanisms." *Experiention* 42

(1986): 1-3.

———. "Does 'Anchor Sleep' Entrain Circadian Rhythms?" *J of Physiology* 345 (1983): 451-67.

Minors, D. S., et al. "Mathematical and Statistical Analysis of Circadian Rhythms." *Psychoneuroendocrinology* 13.6 (1988): 443-64.

Minten, J., et al. "Half-Life of Thyroxine in Dogs: Explanation for the Diurnal Plasma Level Variations?" *Chronobiology International* (Oxford) 4.2 (1987): 127-28.

Moeller, H., et al. "Seasonal and Diurnal Changes of Prostatic Androgen Receptor and Circulating Testosterone in Young Mature Rats." *Research in Experimental Medicine* 188.6 (1988): 451-62.

Monk, T. H., et al. "Task Variables Determine Which Biological Clock Controls Circadian Rhythms in Human Performance." *Nature* 304 (1983): 543-45.

———. "The Timing of Bedtime and Waketime Decisions in Free-Running Subjects." *Psychophysiology* 26.3 (May 1989): 304-10.

Moore-Ede, Martin C., and C. A. Czeisler, eds. *Mathematical Models of the Circadian Sleep-Awake Cycle*. New York: Raven P, 1984.

Moore-Ede, Martin C., et al. *The Clocks That Time Us: Physiology of the Circadian Timing System*. Cambridge: Harvard UP, 1984.

Mrosovsky, N. "Phase Response Curves for Social Entrainment." *J of Comparative Physiology. A, Sensory, Neural, and Behavioral Physiology* 162.1 (January 1988): 35-46.

Mrosovsky, N., et al. "A Behavioral Method for Accelerating Re-entrainment of Rhythms to New Light-Dark Cycles." *Nature* 330 (November 26-December 2 1987): 372-73.

Muller, J. E. "Morning Increase of Onset of Myocardial Infarction. Implications Concerning Triggering Events." *Cardiology* (Basel) 76.2 (1989): 96-104.

Naylor, E. "Temperature Relationships of the Locomotor Rhythm of *Carcinus*." *J of Experimental Biology* 40 (1963): 669-79.

———. "Tidal and Diurnal Rhythm of Locomotory Activity in *Carcinus Maenas*." *J of Experimental Biology* 35 (1958): 602-10.

Naylor, E., and R. J. Atkinson. "Pressure and the Rhythms of Inshore Animals." *The Effects of Pressure on Organisms*. New York: Academic P, 1972. 395-415.

Nelson, W. "Food Restriction, Circadian Disorder and Longevity of Rats and Mice." *J of Nutrition* 118.3 (March 1988): 286-89.

Nelson, W., and F. Halberg. "Phase Relations of Circadian Rhythms: Animals." *Handbook of Environmental Biology*. Ed. P. L. Altman, and D. S. Dittmer. Bethesda: American Society for Experimental Biology, 1966. 586-96.

Nelson, W., et al. "Rhythm Scrambling: Good, Bad or Indifferent?" *Chronobiologia* (Milan) 14.3 (July-September 1987): 291-95.

Neumann, D. "Tidal and Lunar Rhythms." *Handbook of Behavioral Neurobiology*. Ed. Jürgen Aschoff. New York: Plenum, 1981. 4:351-80.

Nihei, M., et al. "Similarity in Amplitude of the Nocturnal Fall in Blood Pressure in Old and Young Patients with Essential Hypertension." *Tohoku J of Experimental Medicine* 158.2 (June 1989): 163-69.

Okada, A., et al. "Circadian Variations of Sweating in Man Exposed to Thermal Stimuli." *Annals of Physiological Anthropology* 6.3 (July 1987): 175-77.

Palmer, J. D. *Biological Clocks in Marine Organisms: The Control of Physiological and Behavioral Tidal Rhythms*. New York: Wiley, 1974.

——. "Clock-Controlled Vertical Migration Rhythms in Intertidal Organisms." *Biological Rhythms in the Marine Environment*. Ed. P. J. DeCoursey. Columbia: South Carolina UP, 1976. 239-55.

——. "Daily and Tidal Components in the Persistent Rhythmic Activity of the Crab, *Sesarma*." *Nature* 215 (1967): 64-66.

——. "How a Bird Tells the Time of Day." *Natural History* 75.3 (1966): 48-53.

——. *Human Biological Rhythms*. Ed. J. J. Head. Burlington: Carolina Biological, 1983.

——. "Human Rhythms." *BioScience* 27 (1977): 93-99.

——. *An Introduction to Biological Rhythms*. New York: Academic P, 1976.

——. "The Living Clocks of Marine Organisms." *The Study of Time III*. Ed. J. T. Fraser, N. Lawrence, and D. Park. New York: Springer, 1978. 216-35.

Palmer, J. D., and F. E. Round. "Persistent, Vertical-Migration Rhythms in Benthic Microflora. I. The Effect of Light and Temperature on the Rhythmic Behaviour of *Euglena Obtusa*." *J of the Marine Biological Association of the United Kingdom* 45 (1965): 567-82.

——. "Persistent, Vertical-Migration Rhythms in Benthic Microflora. VI. The Tidal and Diurnal Nature of the Rhythm in the Diatom *Hantzschia Virgata*." *Biological Bulletin* 132 (1967): 44-55.

Pancheri, Paolo, and Lucio Zichella, eds. *Biorhythms and Stress in the Physiopathology of Reproduction*. New York: Hemisphere Publishing, 1988.

Park, O. "Nocturnalism—The Development of a Problem." *Ecological Monographs* 10 (1940): 485-536.

Pavlidis, Theodosios. *Biological Oscillators: Their Mathematical Analysis*. San Diego: Academic P, 1973.

Pengelley, E. T., ed. *Circannual Clocks*. New York: Academic P, 1974.

Pengelley, E. T., and S. J. Asmundson. "Annual Biological Clocks." *Scientific American* (April 1971): 72-80.

Pittendrigh, C. S. "Circadian Organization and the Photoperiodic Phenomena." *Biological Clocks in Seasonal Reproductive Cycles*. Ed. B. K. Follett, and D. E. Follet. Bristol: John Wright, 1981. 1-35.

———. "Circadian Rhythms and the Circadian Organisation of Living Systems." *Cold Spring Harbor Symposia on Quantitative Biology* 25 (1960): 159-84.

———. "Perspectives in the Study of Biological Clocks." *Symposium on Perspectives in Marine Biology*. Berkeley: U of California P, 1958. 239-68.

Pittendrigh, C. S., and V. G. Bruce. "Daily Rhythms as Coupled Oscillator Systems and Their Relation to Thermoperiodism and Photoperiodism." *Photoperiodism and Related Phenomena in Plants and Animals*. Ed. A. P. Wittrow. Washington: American Association for Advanced Science, 1959. 475-505.

Pittendrigh, C. S., and P. Kaus. "On the Significance of Transients in Daily Rhythms." *Proceedings of the National Academy of Science* 44 (1958): 965-73.

Pittendrigh, C. S., and D. H. Minis. "The Entrainment of Circadian Oscillations by Light and Their Role as Photoperiodic Clocks." *American Naturalist* 98 (1964): 261-93.

Pocock, S. J., et al. "Diurnal Variations in Serum Biochemical and Haematological Measurements." *J of Clinical Pathology* 42.2 (February 1989): 172-79.

Pohl, R. "Tagesrhythmus im phototaktischen Verhalten der *Euglena Gracilis*." *Zeitschrift für Naturforschung* 3b (1948): 367-78.

Poirel, C. "Rhythmometric Analyses of Circadian Variations of Grooming Behavior in the Mouse: Research Perspectives in Behavioral Chronobiology." *J of General Psychology* 115.2 (April 1988): 187-201.

Pöppel, E. "Desynchronisationen circadianer Rhythmen innerhalb einer isolierten Gruppe." *Pflügers Archiv* 299 (1968): 364-70.

Pöppel, E., and R. Wever. "Circadiane Periodik des Menschen unter dem Einfluss von Licht-Dunkel-Wechseln unterschiedlicher Periode." *Pflügers Archiv* 306 (1969): 58-70.

Portalppi, F., et al. "Does Placebo Lower Blood Pressure in Hypertensive Patients? A Noninvasive Chronobiological Study." *Japanese Heart J* 29.2 (March 1988): 189-97.

Possidente, B., et al. "Circadian Period in Mice: Analysis of Genetic and Maternal Contributions to Inbred Strain Differences." *Behavior Genetics* 18.1 (January 1988): 109-17.

Poulsen, E. U., et al. "Home-Monitoring of Uroflow in Normal Male Adolescents. Relation Between Flow-Curve, Voided Volume and Time of Day." *Scandinavian J of Urology and Nephrology. Supplementum* 114 (1988): 58-62.

Pownall, R., et al. "Circadian Rhythmicity in Back Pain: Its Relationship to Short and Long Interval Ibuprofen Therapy." *British J of Clinical Practice* 40.10 (October 1986): 429-33.

Preston, A. P. "Location in the Cage and Diurnal Distribution of Feather Pecking by Caged Layers." *British Poultry Science* 28.4 (December 1987): 653-58.

Ramsey, P. W. "Hyperglycemia at Dawn." *American J of Nursing* 87.11 (November 1987): 1424-26.

Rashotte, M. E., et al. "Coping With Rising Food Costs in a Closed Economy: Feeding Behavior and Nocturnal Hypothermia in Pigeons." *J of Experimental Analysis of Behavior* 50.3 (November 1988): 441-56.

Recabarren, S. E., et al. "Circadian Variation of Rectal Temperature in Newborn Sheep." *J of Developmental Physiology* 9.5 (October 1987): 399-408.

Reilly, T., et al. "Circadian Variation in Human Stature." *Chronobiology International* (Oxford) 1.2 (1984): 121-26.

Reimann, H. A. *Periodic Diseases*. Philadelphia: Davis, 1963.

Reinberg, A. "Biorythmes et chronobiologie." *Presse Médicale* 77 (1969): 877-78.

——. "Evaluation of Circadian Dyschronism during Transmeridian Flights." *The Study of Time I.* Ed. J. T. Fraser, F. C. Haber, and G. H. Müller. New York: Springer, 1972. 523-32.

——. "L'homme et les rythmes circadiens." *Cahiers Sandoz* 9 (1966): 1-50.

——. "Rhythmes biologiques (rappel de quelques données récentes)." *Revue Médicale Aéronautique Spatiale* 7 (1968): 127-30.

Reinberg, A., and J. Ghata. *Biological Rhythms*. Trans. C. J. Cameron. New York: Walker, 1964.

Reinberg, A., et al. "Circadian Rhythms in Competitive Sabre Fencers: Internal Desynchronization and Performance." *Chronobiology International* (Oxford) 2.3 (1985): 195-201.

——. "Spectre thermique (rythmes de la température rectale) d'une femme adulte saine, avant, pendant et aprés son isolement souterrain de trois mois." *Comptes Rendus, Académie de Sciences*

262 (1966): 782-85.

Richter, Curt P. "Astronomical References in Biological Rhythms." *The Study of Time II*. Ed. J. T. Fraser, and N. Lawrence. New York: Springer, 1975. 39-53.

——. "Clock Mechanism Esotropia in Children's Alternate Day Squint." *The Johns Hopkins Medical J* 122 (1968): 218-23.

——. "Inborn Nature of the Rat's 24-Hour Clock." *J of Comparative and Physiological Psychology* 75.1 (1971): 1-4.

——. "Inherent Twenty-Four Hour and Lunar Clocks in Primates." *Communications in Behavioral Biology* 5 Part A (1968): 305-31.

——. "Periodic Phenomena in Man and Animals: Their Relation to Neuroendocrine Mechanisms (A Monthly or Nearly Monthly Cycle)." *Endocrinology and Human Behaviour*. London: Oxford UP, 1968. 284-309.

——. "Two Day Cycles of Alternating Good and Bad Behavior in Psychotic Patients." *Archives of Neurology and Psychiatry* 39 (1938): 587-698.

Rivest, R. W., et al. "Differences Between Circadian and Ultradian Organization of Cortisol and Melatonin Rhythms During Activity and Rest." *J of Clinical Endocrinology and Metabolism* 68.4 (April 1989): 721-29.

Roenneberg, T., et al. "Two Photoreceptors Control the Circadian Clock of a Unicellular Alga." *Naturwissenschaften* 75.4 (April 1988): 206-07.

Rose, Kenneth J. *The Body in Time*. New York: Wiley, 1988.

Rosenthal, M. J., et al. "Nocturnal Regulation of Free Fatty Acids in Healthy Young and Elderly Men." *Metabolism: Clinical and Experimental* 37.7 (July 1988): 645-48.

Round, F. E., and J. D. Palmer. "Persistent, Vertical-Migration Rhythms in Benthic Microflora. II. Field and Laboratory Studies of Diatoms from the Banks of the River Avon." *J of the Marine Biological Association of the United Kingdom* 46 (1966): 191-214.

Rusak, B., et al. "Daily Hoarding Opportunity Entrains the Pacemaker for Hamster Activity Rhythms." *J of Comparative Physiology. A, Sensory, Neural, and Behavioral Physiology* 164.2 (December 1988): 165-71.

Ryan, T., et al. "Oxygen Consumption During Sleep: Influence of Sleep Stage and Time of Night." *Sleep* 12.3 (June 1989): 201-10.

Sack, D. A., et al. "The Timing and Duration of Sleep in Partial Sleep Deprivation Therapy of Depression." *Acta Psychiatrica Scandinavica* (Copenhagen) 77.2 (February 1988): 219-24.

Samel, A., et al. "Sleep and Circadian Rhythms of an Airline Pilot

Operating on the Polar Route: A Case Study." *Aviation Space and Environmental Medicine* 59.5 (May 1988): 443-47.

Saunders, D. S. *Insect Clocks*. 2nd ed. Elmsford: Pergamon, 1982.

Schaltenbrand, G. "Cyclic States as Biological Space-Time Fields." *The Study of Time II*. Ed. J. T. Fraser, and N. Lawrence. New York: Springer, 1975. 54-68.

Scharf, J. H., and H. von Mayersbach, eds. *Chronobiologie*. Halle an der Saale: Deutsche Academie der Naturforscher, 1977.

Schelling, H. von. "Experienced Space and Time." *Bioastronautics*. Ed. K. E. Schaeffer. New York: Macmillan, 1964.

Scheving, Lawrence E., and Franz Halberg. *Chronobiology: Principles and Applications to Shifts in Schedules*. Norwell: Kluwer Academic, 1981.

Schmidt-König, K. "The Sun Azimuth Compass." *Science* 131 (1960): 826-28.

Schulz, R., et al. "On the Role of Energy Metabolism in Neurospora Circadian Clock Function." *Chronobiology International* (Oxford) 2.4 (1985): 223-33.

Seïte, R., et al. "Chronobiological Studies on the Nucleolus." *Chronobiology International* (Oxford) 2.2 (1985): 69-91.

Shibata, H., et al. "Regulatory Alterations of Daily Energy Expenditure Induced by Fasting or Overfeeding in Unrestrained Rats." *J of Applied Physiology* 63.2 (August 1987): 465-70.

Shibata, S., et al. "Development of a Fetal Circadian Rhythm After Disruption of the Maternal Circadian System." *Brain Research* 469.1-2 (June 1, 1988): 313-17.

Siegel, P. V., S. J. Gerathewohl, and S. R. Mohler. "Time Zone Effects." *Science* 164 (1969): 1249-55.

Simon, C., et al. "Absence of the Dawn Phenomenon in Normal Subjects." *J of Clinical Endocrinology and Metabolism* 67.1 (July 1988): 203-05.

Simpson, H. W., and M. C. Lobban. "Effects of 21-Hour Day on the Human Circadian Excretory Rhythms of 17 OH Corticosteroids and Electrolytes." *J of Aerospace Medicine* 38 (1967): 1205-13.

Smith, Anthony. *The Seasons: Life and Its Rhythms*. New York: Harcourt, Brace, 1970.

Sollberger, Arne. *Biological Rhythm Research*. Amsterdam: Elsevier, 1965. Includes bibliography.

Souêtre, E., et al. "Circadian Rhythms in Depression and Recovery: Evidence for Blunted Amplitude as the Main Chronobiological Abnormality." *Psychiatry Research* 28.3 (June 1989): 263-78.

Strughold, H. "Day-Night in Atmospheric Flight, Space Flight and

Other Celestial Bodies." *Annals of the New York Academy of Sciences* 98 (1962): 1109-15.

———. "The Physiological Clock in Aeronautics and Astronautics." *Annals of the New York Academy of Sciences* 134 (1965): 413-22.

———. "Physiological Day-Night Cycle After Global Flight." *J of Aviation Medicine* 63 (1952): 464-73.

———. *Your Body Clock*. New York: Charles Scribner's, 1971.

Sudo, A. "Adrenaline in Various Organs of the Rat: Its Origin, Location and Diurnal Fluctuation." *Life Sciences* 41.22 (November 30, 1987): 2477-84.

Sundberg, S., et al. "Rapid Reversal of Circadian Blood Pressure Rhythm in Shift Workers." *J of Hypertension* 6.5 (May 1988): 393-96.

Sweeney, Beatrice M. "Circadian Rhythms in Unicellular Organisms." *Circadian Rhythmicity*. Wageningen, Netherlands: Centre for Agricultural Publishing and Documents, 1972. 137-56.

———. *Rhythmic Phenomena in Plants*. 2nd ed. New York: Academic P, 1987.

Sweeney, B. M., and J. W. Hastings. "Effects of Temperature upon Diurnal Rhythms." *Cold Spring Harbor Symposia on Quantitative Biology* 25 (1960): 87-104.

Sweeney, B. M., and F. T. Haxo. "Persistence of a Photosynthetic Rhythm in Enucleated *Acetabularia*." *Science* 134 (1961): 1361-63.

Tarquini, B., et al. "Temporal Aspects of Pathophysiology of Human Ulcer Disease." *Chronobiology International* (Oxford) 4.1 (1987): 75-89.

Thase, M. E., et al. "Diagnostic Performance of Nocturnal Penile Tumescence Studies in Healthy, Dysfunctional (Impotent), and Depressed Men." *Psychiatry Research* 26.1 (October 1988): 79-87.

Thor, D. H., and D. L. Hoats. "A Circadian Variable in Self-Exposure to Light by the Rat." *Psychonomic Science* 12.1 (1968): 1-2.

Tochikubo, O., et al. "Mathematical Evaluation of 24-Hour Blood-Pressure Variability in Young, Middle-Aged and Elderly Hypertensive Patients." *Japanese Circulation J* 51.10 (October 1987): 1123-30.

Torsvall, L., et al. "Sleep on the Night Shift: 24-Hour EEG Monitoring of Spontaneous Sleep/Wake Behavior." *Psychophysiology* 26.3 (May 1989): 352-58.

Tripathi, P. M., et al. "Circadian Periodicity in Yellow-Throated Sparrow as Demonstrated by Intermittent Light Experiments." *Acta Physiologica Hungarica* (Budapest) 72.1 (1988): 79-84.

Tsai, T. H., et al. "Synchronization of the Circadian Activity Rhythm in Hamsters Following Intermittent Schedule Shifts." *Chronobiology International* (Oxford) 3.4 (1986): 237-46.

Udry, J. R., and N. M. Morris. "Distribution of Coitus in the Menstrual Cycle." *Nature* 220 (1968): 593-96.

Van Cauter, E., et al. "Nocturnal Decrease in Glucose Tolerance During Constant Glucose Infusion." *J of Clinical Endocrinology and Metabolism* 69.3 (September 1989): 604-11.

Vaughan, M. K., et al. "Pineal and Retinal Lysosomal Enzyme Rhythms." *Brain Research* 417.2 (August 11, 1987): 321-26.

Vichyanond, P., et al. "Circadian Variation of Skin Reactivity and Allergy Skin Tests." *J of Allergy and Clinical Immunology* 83.6 (June 1989): 1101-06.

Vidrio, Enrique A., et al., eds. *Eleventh International Congress of Anatomy, Part C: Biological Rhythms in Structure and Function.* New York: A. R. Liss, 1981.

Vokac, Z., et al. "Phase-Shifts of Apparent Circadian Rhythms Due to West and East Transmeridian Flights or to Corresponding Night-Shift Sleep Displacements." *Chronobiology International* (Oxford) 1.2 (1984): 139-44.

Ward, Ritchie R. *The Living Clocks*. London: Collins, 1972.

Weaver, D. R., et al. "Maternal-Fetal Communication of Circadian Phase in a Precocious Rodent, the Spiny Mouse." *American J of Physiology* 253.4 Part 1 (October 1987): E401-09.

Webb, H. M. "Interactions of Daily and Tidal Rhythms." *Biological Rhythms in the Marine Environment*. Ed. P. J. DeCoursey. Columbia: U of South Carolina P, 1976. 129-35.

Webb, H. M., and F. A. Brown, Jr. "Interaction of Diurnal and Tidal Rhythms in the Fiddler Crab, *Uca Pugnax*." *Biological Bulletin* 129 (1965): 582-91.

Wilkins, M. B. "The Influence of Temperature and Temperature Changes in Biological Clocks." *Circadian Clocks*. Ed. J. Aschoff. Amsterdam: North-Holland, 1965. 146-63.

Williams, B. G., and E. Naylor. "Spontaneously Induced Rhythm of Tidal Periodicity in Laboratory-Reared *Carcinus*." *J of Experimental Biology* 47 (1967): 229-34.

———. "Synchronization of the Locomotor Tidal Rhythm of *Carcinus*." *J of Experimental Biology* 51 (1969): 715-25.

Willich, S. N., et al. "Circadian Variation in the Incidence of Sudden Cardiac Death in the Framingham Heart Study Population." *American J of Cardiology* 60.10 (October 1, 1987): 801-06.

Wilson, D. W., et al. "Circadian Breast Skin Temperature Rhythms: Overt and Occult Benign and Occult Primary Malignant Breast Disease." *Chronobiology International* (Oxford) 1.2 (1984): 167-72.

Windt, C. van der, et al. "Cerebral Infarction Does Not Occur

Typically at Night." *J of Neurology, Neurosurgery and Psychiatry* 51.1 (January 1988): 109-11.

Winfree, Arthur T. *The Geometry of Biological Time*. New York: Springer, 1980.

——. *The Timing of Biological Clocks*. New York: W. H. Freeman, 1987.

——. "Unclocklike Behavior of Biological Clocks." *Nature* 253 (1975): 315-19.

——. *When Time Breaks Down: The Three-Dimensional Dynamics of Electrochemical Waves and Cardiac Arrythmias*. Princeton: Princeton UP, 1987.

Young, M., and J. Ziman. "Cycles in Social Behavior." *Nature* 229 (1971): 91-95.

C. EVOLUTION

Aarts, H. J., et al. "Different Evolution Rates Within the Lens-Specific Beta-Crystallin Gene Family." *J of Molecular Evolution* 28.4 (April 1989): 313-21.

Abelson, P. H. "Chemical Events on the Primitive Earth." *Proceedings of the U.S. National Academy of Science* 55 (1966): 1365-72.

Abugov, R. "A Sex-Specific Quantitative Genetic Theory for Life History and Development." *J of Theoretical Biology* 132.4 (June 22, 1988): 437-47.

Altmüller, A., et al. "Biological and Genetic Evolution of the Nucleoprotein Gene of Human Influenza A Virus." *J of General Virology* 70 Part 8 (August 1989): 2111-19.

Aoki, K., et al. "Pleiotropy and Preadaptation in the Evolution of Human Language Capacity." *Theoretical Population Biology* 35.2 (April 1989): 181-94.

Avers, Charlotte J. *Process and Pattern in Evolution*. New York: Oxford UP, 1989.

Baker, J. R. "The Evolution of Breeding Seasons." *Evolution: Essays on Aspects of Evolutionary Biology*. Ed. G. R. de Beer. London: Oxford UP, 1938. 161-77.

Beer, Gavin de. *Atlas of Evolution*. London: Nelson, 1964.

——. *Charles Darwin*. New York: Doubleday, 1967.

Bendall, D. S. *Evolution from Molecules to Men*. New York: Cambridge UP, 1986.

Benman, B. E., and L. Persson, eds. *Size-Structured Populations: Ecology and Evolution*. New York: Springer, 1989.

Bernal, J. D. *The Physical Basis of Life*. London: Routledge and Kegan Paul, 1951.

Bonner, John T. *The Evolution of Complexity by Means of Natural Selection*. Princeton: Princeton UP, 1988.

Boucher, Douglas H., ed. *The Biology of Mutualism: Ecology and Evolution*. New York: Oxford UP, 1988.

Bounous, G., et al. "Evolutionary Traits in Human Milk Proteins." *Medical Hypotheses* 27.2 (October 1988): 133-40.

Boyd, R., et al. "The Evolution of Reciprocity in Sizable Groups." *J of Theoretical Biology* 132.3 (June 7, 1988): 337-56.

Bowler, Peter J. *Evolution: The History of an Idea*. Revised ed. Berkeley: U of California P, 1989.

Breuer, H. P. "Darwinism in Victorian Letters." *Literature and Medicine* 6 (1987): 128-38.

Brooks, Daniel R., and E. O. Wiley. *Evolution As Entropy: Toward a Unified Theory of Biology*. 2nd ed. Chicago: U of Chicago P, 1988.

Bull, A. T. "Microbial Growth." *Companion to Biochemistry*. Ed. A. T. Bull, et al. London: Longman, 1974. 415-42.

Bull, J. J. "Sex Determining Mechanisms: An Evolutionary Perspective." *Experientia* (Supplement) 55 (1987): 93-115.

Burnet, F. M. *The Clonal Selection Theory of Acquired Immunity*. Cambridge: Cambridge UP, 1959.

Cairns-Smith, A. G. "Beginnings of Organic Evolution." *The Study of Time IV*. Ed. J. T. Fraser, N. Lawrence, and D. Park. New York: Springer, 1981. 15-33.

——. "A Case for an Alien Ancestry." *Proceedings of the Royal Society of London. Series B. Biological Sciences* 189 (1975): 249-74.

——. *Genetic Takeover and the Mineral Origins of Life*. Cambridge: Cambridge UP, 1982.

——. "The Origin of Life and the Nature of the Primitive Gene." *J of Theoretical Biology* 10 (1966): 53-88.

——. "Takeover Mechanisms and Early Biochemical Evolution." *BioSystems* 9 (1977): 105-09.

Cairns-Smith, A. G., and G. L. Walker. "Primitive Metabolism." *BioSystems* 5 (1974): 173-86.

Campbell, K. S., and M. F. Day, eds. *Rates of Evolution*. Winchester: Unwin Hyman, 1987.

Cantatore, P., et al. "Organization, Structure, and Evolution of Mammalian Mitochondrial Genes." *International Review of Cytology* 108 (1987): 149-208.

Cedergren, R., et al. "The Evolutionary Relationships Among Known Life Forms." *J of Molecular Evolution* 28.1-2 (December 1988-February 1989): 98-112.

Ciochon, Russell L., and A. B. Chiarelli, eds. *Evolutionary Biology of the New World Monkeys and Continental Drift*. New York: Plenum, 1980.

Corballis, M. C. "Laterality and Human Evolution." *Psychological Review* 96.3 (July 1989): 492-505.

Crerar, M. M., et al. "Studies on the Expression and Evolution of the Glycogen Phosphorylase Gene Family in the Rat." *Genome* 30.4 (August 1988): 582-90.

Curnow, R. N. "The Use of Markov Chain Models in Studying the Evolution of the Proteins." *J of Theoretical Biology* 134.1 (September 9, 1988): 51-57.

Darwin, Charles. *Origin of the Species*. New York: Collier Books, 1967.

Darwin, Erasmus. *Zoonomia*. London: J. Johnson, 1794. Chapter 39 contains early ideas on evolution, but lacks the element of the survival of the fittest that derives from Malthus.

Dickerson, R. E. "Chemical Evolution and the Origin of Life." *Scientific American* 239.3 (March 1978): 62-78.

Di Giulio, M. "Some Aspects of the Organization and Evolution of the Genetic Code." *J of Molecular Evolution* 29.3 (September 1989): 191-201.

Dobzhansky, T. *The Biology of Ultimate Concern*. New York: American Library, 1967.

——. *Genetics and the Evolutionary Process*. New York: Columbia UP, 1970.

——. *Mankind Evolving*. New Haven: Yale UP, 1962.

Doolittle, R. F., et al. "Origins and Evolutionary Relationships of Retroviruses." *Quarterly Review of Biology* 64.1 (March 1989): 1-30.

Dover, G. A. "DNA Turnover and the Molecular Clock." *J of Molecular Evolution* 26.1-2 (1987): 47-58.

Dye, S. F. "An Evolutionary Perspective of the Knee." *J of Bone and Joint Surgery* 69.7 (September 1987): 976-83.

Easteal, S. "Generation Time and the Rate of Molecular Evolution." *Molecular Biology and Evolution* 2.5 (September 1985): 450-53.

Eigen, Manfred. "Evolution und Zeitlichkeit." *Die Zeit*. Ed. A. Peisl, and A. Mohler. Munich: R. Oldenburg, 1983.

Eigen, Manfred, et al. "The Origin of Genetic Information." *Scientific American* 244.4 (April 1981): 88-118.

Eiseley, Loren. "Darwin and Malthus." *Darwin's Century*. New York: Doubleday, 1958. 178-82.

Eldredge, N., and S. J. Gould. "Punctuated Equilibria: An Alternative to Phyletic Gradualism." *Models in Paleobiology*. Ed. T. J. M.

Schopf. San Francisco: Freeman, Cooper, 1972. 82-115.

Erickson, Jon S. *The Living Earth: The Coevolution of the Planet and Life*. Blue Ridge Summit: TAB Books, 1989.

Fischer, K. "Malignancy May Contribute to Rapid Evolution." *Medical Hypotheses* 28.1 (January 1989): 35-38.

——. "Multiple Sclerosis and the Evolution of Growth Hormone Mechanisms in Man." *Medical Hypotheses* 27.2 (October 1988): 99-106.

Fischer, Roland. "Biological Time." *The Voices of Time*. Ed. J. T. Fraser. 2nd ed. Amherst: U of Massachusetts P, 1981. 357-82.

Foley, P. "Molecular Clock Rates at Loci Under Stabilizing Selection." *Proceedings of the National Academy of Sciences of the USA* 84.22 (November 1987): 7996-8000.

Fothergill, P. G. *Historical Aspects of Organic Evolution*. London: Hollis and Carter, 1952.

Fox, S. W. *The Origins of Prebiological Systems and of Their Molecular Matrices*. New York: Academic P, 1965.

Galili, U., et al. "Man, Apes, and Old World Monkeys Differ from Other Mammals in the Expression of Alpha-Galactosyl Epitopes on Nucleated Cells." *J of Biological Chemistry* 263.33 (November 25, 1988): 17755-62.

Gentry, G. A., et al. "Evolution of Herpes and Pox Viruses and Their Hosts: A Problem with the Molecular Clock." *Intervirology* 29.5 (1988): 277-80.

Gillespie, C. G. *Genesis and Geology*. Cambridge: Harvard UP, 1951.

Gingerich, P.D. "Temporal Scaling of Molecular Evolution in Primates and Other Mammals." *Molecular Biology and Evolution* 3.3 (May 1986): 205-21.

Goodwin, B. C. "A Structuralist View of Biological Origins." *The Study of Time IV*. Ed. J. T. Fraser, N. Lawrence, and D. Park. New York: Springer, 1981. 73-89.

Gould, S. J. "Evolution and the Triumph of Homology, or Why History Matters." *American Scientist* 74 (1986): 60-69.

Gruber, Howard E., and Paul H. Barrett. *Darwin on Man*. New York: Dutton, 1974.

Haldane, J. B. S. "Biological Possibilities in the Next Ten Thousand Years." *Man and His Future*. Ed. G. Wolstenholme. London: Churchill, 1963.

——. "The Cost of Natural Selection." *J of Genetics* 55 (1957): 511-24.

——. "Suggestions as to Quantitative Measurements of Rates of Evolution." *Evolution* 3 (1947): 51-56.

Harborne, J. B., ed. *Biochemical Aspects of Plant and Animal Coevolution*. San Diego: Academic P, 1978.

Harris, H., and Hans Kalmus. "The Measurement of Taste Sensitivity to Phenylthiourea (P.T.C.)." *Annals of Eugenics* 15 (1949): 24-31.

Hartwell, L. H. "Genetic Control of the Cell Division Cycle in Yeast." *J of Molecular Biology* 59 (1971): 183-94.

Hasegawa, M., et al. "Man's Place in Hominoidea as Inferred from Molecular Clocks of DNA." *J of Molecular Evolution* 26.1-2 (1987): 132-47.

Hayashida, H., et al. "Evolution of Influenza Virus Genes." *Molecular Biology and Evolution* 2.4 (July 1985): 289-303.

Hershey, A. D. "Genes and Hereditary Characteristics." *Nature* 226 (1970): 697-700.

Hoekstra, R. F. "The Evolution of Sexes." *Experientia* (Supplement) 55 (1987): 59-91.

Hofbauer, J., and K. Sigmund. *The Theory of Evolution and Dynamical Systems: Mathematical Aspects of Selection*. New York: Cambridge UP, 1988.

Hoffman, Antoni. *Arguments on Evolution: A Paleontologist's Perspective*. New York: Oxford UP, 1988.

Hofman, M. A. "On the Evolution and Geometry of the Brain in Mammals." *Progress in Neurobiology* 32.2 (1989): 137-58.

Holmquist, G. P. "Evolution of Chromosome Bands: Molecular Ecology of Noncoding DNA." *J of Molecular Evolution* 28.6 (June 1989): 469-86.

Huxley, J. S. *Evolution, the Synthesis*. London: Murray, 1949.

Iwasa, Y. "Free Fitness that Always Increases in Evolution." *J of Theoretical Biology* 135.3 (December 7, 1988): 265-81.

Kalmus, Hans. "The Measurement of Biological and Social Changes." *The Study of Time III*. Ed. J. T. Fraser, N. Lawrence, and D. Park. New York: Springer, 1978. 237-64.

——. "Organic Evolution and Time." *The Voices of Time*. Ed. J. T. Fraser. 2nd ed. Amherst: U of Massachusetts P, 1981. 330-52.

Kawano, Shoichi, et al., eds. *Evolution and Coadaptation in Biotic Communities*. New York: Columbia UP, 1988.

Kemble, J. V. "Man's Hand in Evolution." *J of Hand Surgery* 12.3 (October 1987): 396-99.

Kerkut, G. A. "Possible Evolutionary Futures for Mankind." *Comparative Biochemistry and Physiology. A: Comparative Physiology* 90.1 (1988): 5-10.

Kettlewell, H. B. D. "Further Selection Experiments on Industrial Melanism in the Lepidoptera." *Heredity* 10 (1956): 287-303.

——. "A Résumé of Investigations on the Evolution of Melanism in Lepidoptera." *Proceedings of the Royal Society of London. Series B.*

Biological Sciences 145 (1956): 297-303.

Kozawa, Y., et al. "Evolution of Tooth Structure in the Equoidea." *J of Nihon University School of Dentistry* 30.4 (December 1988): 287-96.

Krüger, J., et al. "The Problem of Our Common Mitochondrial Mother." *Human Genetics* 82.4 (July 1989): 308-12.

Kuczynski, R. R. *Population Movements*. Oxford: Clarendon P, 1936.

Kueppers, B. O. *Molecular Theory of Evolution: Outline of a Physico-Chemical Theory of the Origin of Life*. New York: Springer, 1983.

Kvenvolden, K. A. "Natural Evidence for Chemical and Early Biological Evolution." *Origins of Life* 5 (1974): 71-86.

Layzer, David. *Cosmogenesis: The Growth of Order in the Universe*. New York: Oxford UP, 1989.

Lazcano, A., et al. "The Evolutionary Transition from RNA to DNA in Early Cells." *J of Molecular Evolution* 27.4 (1988): 283-90.

Lemmon, R. M. "Chemical Evolution." *Chemical Reviews* 70 (1970): 95-109.

Lewin, R. "DNA Clock Conflict Continues." *Science* 241 (September 30, 1988): 1756-59.

Lewontin, R. C. "Inferring the Number of Evolutionary Events from DNA Coding Sequence Differences." *Molecular Biology and Evolution* 6.1 (January 1989): 15-32.

Li, W. H., et al. "Rapid Evolution of Goat and Sheep Globin Genes Following Gene Duplication." *Molecular Biology and Evolution* 1.1 (December 1983): 94-108.

——. "Reconstruction of Phylogenetic Trees and Estimation of Divergence Times Under Nonconstant Rates of Evolution." *Cold Spring Harbor Symposia on Quantitative Biology* 52 (1987): 847-56.

Lipmann, F. "Projecting Backwards from the Present State of Evolution of Biosynthesis." *The Origins of Prebiological Systems*. Ed. S. W. Fox. New York: Academic P, 1965. 259-80.

Lovejoy, C. O. "Evolution of Human Walking." *Scientific American* 259.5 (November 1988): 118-25.

McGeoch, D. J. "The Genome of Herpes Simplex Virus: Structure, Replication, and Evolution." *J of Cell Science* (Supplement) 7 (1987): 67-94.

Martin, S. L., et al. "Tempo and Mode of Concerted Evolution in the L1 Repeat Family of Mice." *Molecular Biology and Evolution* 2.2 (March 1985): 127-40.

Mason, I. L. *Evolution of Domesticated Animals*. New York: Wiley, 1984.

Mayr, E. *Animal Species and Evolution*. Cambridge: Harvard UP, 1963.

Merrell, D. J. *Evolution and Genetics.* New York: Holt, Rinehart, 1962.

Mettler, Lawrence E., et al. *Population Genetics and Evolution.* 2nd ed. Englewood Cliffs: Prentice-Hall, 1988.

Miller, S. L. "A Production of Amino Acids Under Possible Primitive Earth Conditions." *Science* 117 (1953): 528-29.

Miller, S. L., and L. W. Orgel. *The Origins of Life on the Earth.* Englewood Cliffs: Prentice-Hall, 1974.

Miyata, T., et al. "Male-Driven Molecular Evolution: A Model and Nucleotide Sequence Analysis." *Cold Spring Harbor Symposia on Quantitative Biology* 52 (1987): 863-67.

Mortlock, Robert P., ed. *Microorganisms as Model Systems for Studying Evolution.* New York: Plenum, 1984.

Muller, H. J. "Should We Weaken or Strengthen Our Genetic Heritage?" *Evolution and Man's Progress.* Ed. H. Hoagland, and R. W. Burhoe. New York: Columbia UP, 1962.

Murray, J. A., et al. "Unexpected Divergence and Molecular Coevolution in Yeast Plasmids." *J of Molecular Biology* 200.3 (April 5, 1988): 601-07.

Nakamura, D., et al. "On the Evolution of Sex Determination." *Proceedings of the Royal Society of London. Series B: Biological Sciences* 232 (November 23, 1987): 159-80.

Needham, Joseph. "Evolution and Thermodynamics." In his *Time, the Refreshing River.* London: Allen and Unwin, 1943.

Nierhaus, K. H., et al. "New Aspects of Structure, Assembly, Evolution, and Function of Ribosomes." *Cold Spring Harbor Symposia on Quantitative Biology* 52 (1987): 665-74.

Nogueira, Neto P. "The Spread of a Fierce African Bee in Brazil." *Bee World* 45.3 (1964): 119-21.

Oparin, A. I. *Life: Its Nature, Origins and Development.* Trans. Ann Synge. New York: Academic P, 1966.

——. *The Origin of Life on the Earth.* Edinburgh: Oliver and Boyd, 1957.

Orgel, L. E. "Evolution of the Genetic Apparatus: A Review." *Cold Spring Harbor Symposia on Quantitative Biology* 52 (1987): 9-16.

Pääbo, S., et al. "Ancient DNA and the Polymerase Chain Reaction. The Emerging Field of Molecular Archaeology." *J of Biological Chemistry* 264.17 (June 15, 1989): 9709-12.

Pamilo, P., et al. "Relationships Between Gene Trees and Species Trees." *Molecular Biology and Evolution* 5.5 (September 1988): 568-83.

Penny, D., et al. "Reliability of Evolutionary Trees." *Cold Spring*

Harbor Symposia on Quantitative Biology 52 (1987): 857-62.
Penrose, L. S. *The Biology of Mental Defect*. 4th ed. London: Sidwick and Jackson, 1967.
Quentin, Y. "Successive Waves of Fixation of B1 Variants in Rodent Lineage History." *J of Molecular Evolution* 28.4 (April 1989): 299-305.
Qumsiyeh, M. B., et al. "Differential Rates of Genic and Chromosomal Evolution in Bats of the Family Rhinolophidae." *Genome* 30.3 (June 1988): 326-35.
Rein, R., S. Nir, and M. N. Stamatiadou. "Photochemical Survival Principal in Molecular Evolution." *J of Theoretical Biology* 33 (1971): 309-18.
Rose, M. R., et al. "What Evolutionary Biology Can Do for Gerontology." *J of Gerontology* 44.2 (March 1989): B27-29.
Rushton, J. P., et al. "Population Differences in Susceptibility to AIDS: An Evolutionary Analysis." *Social Science and Medicine* 28.12 (1989): 1211-20.
Sagai, T., et al. "New Evidence for Trans-Species Evolution of the H-2 Class I Polymorphism." *Immunogenetics* 30.2 (1989): 89-98.
Saunders, P. T., and M. W. Ho. "On the Increase of Complexity in Evolution, Part 1." *J of Theoretical Biology* 63 (1976): 375-84.
——. "On the Increase of Complexity in Evolution, Part 2." *J of Theoretical Biology* 90 (1981): 515-30.
Schweber, S. S. "The Origin of the *Origin Revisited*." *J of the History of Biology* 10 (1977): 229-316.
Simpson, G. G. *The Major Features of Evolution*. New York: Columbia UP, 1953.
——. *Tempo and Mode in Evolution*. New York: Columbia UP, 1944.
Smith, J. Maynard. *The Evolution of Sex*. New York: Cambridge UP, 1978.
——. *The Theory of Evolution*. 3rd ed. Harmondsworth: Penguin, 1975.
——. "Time in the Evolutionary Process." *The Study of Time I*. Ed. J. T. Fraser, F. C. Haber, and G. H. Müller. New York: Springer, 1972. 200-06.
——. "What Use Is Sex?" *Nature* 230 (1971): 209-10.
Stearns, S. C. "Why Sex Evolved and the Differences It Makes." *Experientia* (Supplement) 55 (1987): 15-31.
Steiner, D. F., et al. "An Overview of Insulin Evolution." *Hormone and Metabolic Research* 20.7 (July 1988): 443-44.
Steward, R. C. "Preference of Melanic Moths to Dark Background." *J of Zoology* 181 (1977): 371-95.
Stoff, Jesse A., and Charles R. Pellegrino. *Darwin's Universe: Origins*

and Crises in the History of Life. 2nd ed. Blue Ridge Summit: TAB Books, 1986.

Taussig, H. B. "Evolutionary Origin of Cardiac Malformations." *J of the American College of Cardiology* 12.4 (October 1988): 1079-86.

Todes, D. P. "Darwin's Malthusian Metaphor and Russian Evolutionary Thought, 1859-1917." *Isis* 78 (December 1987): 537-51.

Tucker, P. K., et al. "Y Chromosome Evolution in the Subgenus Mus (Genus Mus)." *Genetics* 122.1 (May 1989): 169-79.

Veizer, J. "The Earth and Its Life: Systems Perspective." *Origins of Life and Evolution of the Biosphere* 18.1-2 (1988): 13-39.

Vincent, K. A., et al. "Evolution and Transcription of Old World Monkey Globin Genes." *J of Molecular Biology* 207.3 (June 5, 1989): 465-80.

Voipio, P. "What Did Mendel Say About Evolution?" *Hereditas* 107.1 (1987): 103-05.

Vorzimmer, Peter. "Darwin, Malthus, and the Theory of Natural Selection." *J of the History of Ideas* 30 (1969): 527-42.

Yokoyama, S. "Molecular Evolution of the Human and Simian Immunodeficiency Viruses." *Molecular Biology and Evolution* 5.6 (November 1988): 645-59.

Whyte, L. L. *Internal Factors in Evolution*. New York: Braziller, 1965.

6. CHEMISTRY

Ahlberg, P., and L. Sundelof. *Structure and Dynamics in Chemistry*. Philadelphia: Coronet Books, 1978.

Albright, Thomas A., et al. *Orbital Interactions in Chemistry*. New York: Wiley, 1985.

Allen, Ralph O., ed. *Archaeological Chemistry IV*. Washington: American Chemical Society, 1989.

Andrews, D. L. *Lasers in Chemistry*. New York: Springer, 1986.

Armstrong, Roger, et al. *Laboratory Chemistry: A Life Science Approach*. New York: Macmillan, 1980.

Beck, Curt W., ed. *Archaeological Chemistry*. Washington: American Chemical Society, 1974.

Berlin, Margulis L. *Origins of Life: Chemistry and Radiochemistry*. Berlin: Springer, 1973.

Bohr, Neils. "Chemistry and the Quantum Theory of Atomic Constitution." *J. Chemical Society* 1932.1 (1932): 349-84.

Calvin, M. *Chemical Evolution: Molecular Evolution towards the Origin of Living Systems on the Earth and Elsewhere*. Oxford: Clarendon P, 1969.

——, ed. *Organic Chemistry of Life: Readings from* Scientific American. New York: W. H. Freeman, 1973.

Carter, Giles F., ed. *Archaeological Chemistry, Vol. II*. Washington: American Chemical Society, 1978.

Chapman, O. L., ed. *Organic Photochemistry, Vol. 3*. New York: Dekker, 1973.

Coyle, J. D. *Introduction to Organic Photochemistry*. New York: Wiley, 1988.

Dauben, William G. *Organic Reactions Mechanism*. New York: Wiley, 1984.

De Mayo, Paul, ed. *Rearrangements in Ground and Excited States*. Vol. 1. San Diego: Academic P, 1980.

Delauney, J., F. Creusot, and G. Schapira. "Evolution of Ribosomal Proteins." *European J of Biochemistry* 39.1 (1973): 305-12.

Dorn, James M., and Barbara Hopkins. *Thanatochemistry: A Survey of General, Organic and Biochemistry for Funeral Service Professionals*. Englewood Cliffs: Prentice-Hall, 1985.

Eakin, R. E. "An Approach to the Evolution of Metabolism." *Proceedings of the National Academy of Sciences* 49.3 (1963): 360-66.

Epiotis, N. D. *Theory of Organic Reactions*. New York: Springer, 1978.

Fiechter, A., ed. *New Technological Concepts* (Advances in Biochemical

Engineering Series). New York: Springer, 1980.

Getoff, N. "Effect of Ferrous Ion on Formation of Oxalic Acid from Aqueous Carbonic Acid by Ultraviolet Irradiation." *Zeitschrift für Naturforschung* 18B (1963): 169-70.

Gilleland, Martha J. *Introduction to General, Organic and Biological Chemistry*. Saint Paul: West Publishing, 1982.

Glas, E. *Chemistry and Physiology in Their Historical and Philosophical Relations*. Philadelphia: Coronet Books, 1979.

Goldberg, E. D., ed. *Atmospheric Chemistry*. New York: Springer, 1983.

Goldstein, Irving S. *Organic Chemicals from Biomass*. Boca Raton: CRC P, 1981.

Goodman, Murray, and Frank Morehouse. *Organic Molecules in Action*. New York: Gordon and Breach, 1973.

Greger, R., et al., eds. *Renal Transport of Organic Substances*. New York: Springer, 1981.

Gupta, R. R. *Physical Methods in Heterocyclic Chemistry*. New York: Wiley, 1984.

Gutman, I., and O. E. Polansky. *Mathematical Concepts in Organic Chemistry*. New York: Springer, 1986.

Hassan, Saad S. *Organic Analysis Using Atomic Absorption Spectrometry*. New York: Halsted P, 1984.

Higgins, I. J., and R. G. Burns. *The Chemistry and Microbiology of Pollution*. San Diego: Academic P, 1975.

Hoagland, H. "The Chemistry of Time." *Scientific Monthly* 56 (1943): 56-61.

Holum, John R. *Elements of General and Biological Chemistry*. 7th ed. New York: Wiley, 1986.

Hori, H. "Evolution of 5S RNA." *J of Molecular Evolution* 7.1 (1975): 75-86.

Hutzinger, O., ed. *Reactions and Processes* (Handbook of Environmental Chemistry Series). New York: Springer, 1982.

Izatt, Reed M., and James J. Christensen. *Synthesis of Macrocycles: The Design of Selective Complexing Agent*. New York: Wiley, 1986.

Katritzky, Alan R., ed. *Advances in Heterocyclic Chemistry*. San Diego: Academic P, 1983.

Larson, Dewey B. *Nothing but Motion*. Portland: North Pacific, 1979.

Lefever, R., G. Nicolis, and I. Prigogine. "On the Occurrence of Oscillations Around the Steady State in Systems of Chemical Reactions Far from Equilibrium." *J of Chemical Physics* 47 (1967): 1045-51.

Leisinger, T., et al., eds. *Microbial Degradation of Xenobiotics and*

Recalcitrant Compounds. San Diego: Academic P, 1982.

McLafferty, F. W., ed. *Mass Spectrometry of Organic Ions*. San Diego: Academic P, 1963.

McMurry, John. *Essentials of General, Organic, and Biological Chemistry*. Englewood Cliffs: Prentice-Hall, 1989.

McTigue, P. T., ed. *Chemistry: Key to the Earth*. 2nd ed. Portland: International Specialized Book Services, 1983.

Malanowski, S., and R. Stephenson. *Handbooks of Thermodynamics of Organic Compounds for Chemical Engineers*. New York: Elsevier, 1987.

Marvel, Elliot N. *Thermal Electrocyclic Reactions*. San Diego: Academic P, 1980.

Matta, Michael S., and Antony C. Wilbraham. *Atoms, Molecules, Life*. Redwood City: Benjamin-Cummings, 1981.

Miller, B. *Organic Chemistry: The Basis of Life*. Reading: Addison-Wesley, 1980.

Miller, F. M. *Chemistry: Structure and Dynamics*. New York: McGraw-Hill, 1984.

Mills, John S., and Raymond White. *The Organic Chemistry of Museum Objects*. Stoneham: Butterworth, 1987.

Nakanishi, Koji, and Philippa H. Solomon. *Infrared Absorption Spectroscopy*. 2nd ed. Oakland: Holden-Day, 1977.

Ovchinnikov, Yu, ed. *Progress in Bio-Organic Chemistry and Molecular Biology*. Miami: ICSU P, 1984.

Owen, E. D. *Degradation and Stabilisation of PVC*. New York: Elsevier, 1984.

Papousek, D., and M. R. Aliev. *Molecular Vibrational-Rotational Spectra*. New York: Elsevier, 1982.

Parker, B. *Invisible Matter and the Fate of the Universe*. New York: Plenum, 1989.

Pullman, Bernard, and Nathan Goldblum, eds. *Excited States in Organic Chemistry and Biochemistry*. Norwell: Kluwer Academic, 1977.

Ranganathan, S. *The Appreciation of Molecular Transformations in Organic Chemistry*. Columbia: South Asia Books, 1975.

Ridley, Mark. *The Explanation of Organic Diversity: The Comparative Methods of Adaptations for Mating*. New York: Oxford UP, 1983.

Romero-Herrera, A. E., et al. "Molecular Evolution of Myoglobin and the Fossil Record: A Phylogenetic Synthesis." *Nature* 146 (1973): 389-95.

Silverstein, Robert M., et al. *Spectrometric Identification of Organic Compounds*. 4th ed. New York: Wiley, 1981.

Stephen-Sherwood, E., and J. Oro. "Chemical Evolution: Recent Syntheses of Bioorganic Molecules." *Space Life Sciences* 4 (1973): 5-31.

Stull, Daniel R., et al. *The Chemical Thermodynamics of Organic Compounds*. Melbourne: Krieger, 1987.

Tedder, J. M., and A. Nechvatal. *Basic Organic Chemistry: A Mechanistic Approach, Part 1*. 2nd ed. New York: Wiley, 1987.

Theng, B. K. G. *The Chemistry of Clay-Organic Reactions*. Bristol: Hilger, 1974.

Tinsley, B. M. "Nucleochronology and Chemical Evolution." *Astrophysical J* 198.1 Part 1 (1975): 145-50.

Traynham, James G., ed. *Essays on the History of Organic Chemistry*. Baton Rouge: Louisiana State UP, 1986.

Van Kampen, N. G. *Stochastic Processes in Physics and Chemistry*. New York: Elsevier, 1982.

Voegtle, F., and F. L. Boschke, eds. *Host Guest Complex Chemistry*. New York: Springer, 1982.

Vowles, P. D., and D. W. Connell. *Experiments in Environmental Chemistry: A Laboratory Manual*. Elmsford: Pergamon, 1980.

West, Michael A., ed. *Lasers in Chemistry*. New York: Elsevier, 1977.

West, M. W., and E. D. Gill. "Chemical Evolution and the Origin of Life: A Bibliography." *Origins of Life* 6.1-2 (1975): 285-300.

West, M. W., et al. "Chemical Evolution and the Origin of Life." *Origins of Life* 5.3-4 (1974): 507-27.

Whittingham, M. Stanley, and Allen J. Jacobson, eds. *Intercalation Chemistry*. San Diego: Academic P, 1982.

Wilkinson, Geoffrey, ed. *Comprehensive Coordination Chemistry: The Synthesis, Reactions, Properties and Applications of Coordination Compounds*. 7 vols. Elmsford: Pergamon, 1987.

Wilkinson, Geoffrey, et al., eds. *Comprehensive Coordination Chemistry, Vol. 1: Theory and Background*. Elmsford: Pergamon, 1989.

Wilson, S., ed. *Methods in Computational Chemistry, Vol. 2: Relativistic Effects in Atoms and Molecules*. New York: Plenum, 1988.

7. ECONOMICS

Abouchar, A. "The Time Factor and Soviet Investment Methodology." *Soviet Studies* 37 (July 1985): 417-27.

Alcock, J. E. "Cooperation, Competition, and the Effects of Time Pressure in Canada and India." *J of Conflict Resolution* 18.2 (1974): 171-97.

Anderson, R. G., et al. "New Look at the Relationship between Time-Series and Structural Econometric Models." *J of Econometrics* 23 (1983): 325-51.

Anderson, S. "Time Pattern of Hedging and the Volatility of Future Prices." *Review of Economic Studies* 50 (1983): 249-66.

Atkinson, A. B. "The Timescale of Economic Models: How Long Is the Long Run?" *Review of Economic Studies* 36.2 (1969): 137-52.

Babin, N. "Military Spending, Economic Growth, and the Time Factor." *Armed Forces and Society* 15 (Winter 1989): 249-62.

Barth, J. R., and J. T. Bennett. "Cyclical Behaviour, Seasonality, and Trend in Economic Time Series." *Nebraska J of Economics and Business* 13.1 (1974): 48-69.

Bausor, R. "Rational-Expectations Hypothesis and the Epistemics of Time." *Cambridge J of Economics* 7 (March 1983): 1-10.

Becker, G. S. "A Theory of the Allocation of Time." *Economic J* 75.9 (1965): 493-517.

Beckerman, W. *In Defence of Economic Growth*. London: Cape, 1974.

Bender, L. D., and L. C. Parcels. "Structural Differences and the Time Pattern of Basic Employment." *Land Economics* 59 (May 1983): 220-34.

Blakeney, R. N., and J. F. MacNaughton. "Effects of Temporal Placement of Unfavourable Information on Decision Making during the Selection Interview." *J of Applied Psychology* 55.2 (1971): 138-42.

Blyton, P. "Workplace Democracy, Unemployment and the Reduction of Working Time." *Economic and Industrial Democracy* 6.1 (1985): 113-20.

Bogachev, V., and L. Kantorovitch. "The Price of Time." *Problems of Economics* 12.10 (1970): 3-27.

Bradford, D., and H. Kelejian. "An Econometric Model of the Flight to the Suburbs." *J of Political Economy* 81 (1973): 560-89.

Brito, D. L. "Becker's Theory of the Allocation of Time and the St. Petersburg Paradox." *J of Economic Theory* 10.1 (1975): 123-26.

Calvo, G. A., and M. Obstfeld. "Optimal Time-Consistent Fiscal Policy with Finite Lifetimes." *Econometrica* 56 (March 1988): 411-32.

Includes bibliography.

Cipolla, Carlo M. *Before the Industrial Revolution: European Society and Economy, 1000-1700*. New York: Norton, 1976.

Clark, K. B., and R. B. Freeman. "How Elastic Is the Demand for Labor?" *Review of Economics and Statistics* 62 (November 1980): 509-20. Includes bibliography.

Clark, Norman, and Calestous Juma. *Long-Run Economics: An Evolutionary Approach to Economic Growth*. London: Pinter, 1987.

Contini, B. "The Value of Time in Bargaining Negotiations: A Dynamic Analysis." *Kyklos* 23.2 (1970): 255-71.

"Corporate Strategy: Time Is Money." *Economist* 309 (October 8, 1988): 68-69.

D'Ambrosio, C. A. "Asset Pricing, Time, and Causality: An Introspective View of Capital Theory." *Review of Social Economy* 2.1 (1969): 1-12.

Darrough, M. N., et al. "Dynamic and Stochastic Structure: An Analysis of Three Time Series of Household Budget Studies." *Review of Economics and Statistics* 65 (May 1983): 274-81.

De Serpa, A. C. "On the Comparative Statics of Time Allocation Theory." *Canadian J of Economics* 8.1 (1975): 101-11.

——. "A Theory of the Economics of Time." *Economic J* 81 (1971): 828-46.

Dorfman, R., P. A. Samuelson, and R. M. Solow. *Linear Programming and Economic Analysis*. New York: McGraw-Hill, 1953.

Epstein, L. G., and J. A. Haynes. "Rate of Time Preference and Dynamic Economic Analysis." *J of Political Economy* 91 (August 1983): 611-35.

Evans, A. W. "On the Theory of Valuation and Allocation of Time." *Scottish J of Political Economy* 19.1 (1972): 1-17.

Fama, E. "Behavior of Stock Market Prices." *J of Business*, U of Chicago 38 (1965): 34-105.

Gallant, A. R., and R. W. Koenker. "Costs and Benefits of Peak-Load Pricing of Electricity: A Continuous-Time Econometric Approach." *J of Econometrics* 26 (September-October 1984): 83-113. Includes bibliography.

Georgescu-Roegen, Nicholas. *The Entropy Law and the Economic Process*. Cambridge: Harvard UP, 1971.

——. *H. H. Gossen: The Laws of Human Relations*. Cambridge: MIT P, 1983.

Geweke, J., et al. "Comparing Alternative Tests of Causality in Temporal Systems: Analytic Results and Experimental Evidence." *J of Econometrics* 21 (February 1983): 161-94.

Goldman, S. M. "Consumption Behavior and Time Preference." *J of Economic Theory* 1.1 (1969): 39-47.

Gutmann, P. M. "Indifference Curves, Consumer Decisions and Time." *Rivista Internazionale di Scienze Economiche e Commerciali* 8 (1967): 816-24.

Hayden, F. G. "Project Evaluation in a Future Real Time System." *J of Economic Issues* 16 (June 1982): 401-11.

Heintel, Peter, and Thomas Macho. *Zeit und Arbeit: Hundert Jahre nach Marx.* Vienna: Verband der wissenschaftliches Gesellschaft Österreichs, 1985.

Hong, W. "Time Preference in Dynamic Trade Models: An Empirical Critique." *Economic Development and Cultural Change* 36 (July 1988): 741-51.

Howrey, E. P., and H. R. Varian. "Estimating the Distributional Impact of Time-of-Day Pricing of Electricity." *J of Econometrics* 26 (September-October 1984): 65-82.

Huberman, G. "Optimality of Periodicity." *Review of Economic Studies* 55 (January 1988): 127-38.

Kaufman, Herbert. *Time, Chance and Organizations: Natural Selection in a Perilous Environment.* Chatham: Chatham House, 1985.

Kiefer, N. M. "Economic Duration Data and Hazard Functions." *J of Economic Literature* 26 (June 1988): 646-79.

Koopmans, T. C., ed. *Statistical Inference in Dynamic Economic Models.* New York: Wiley, 1950.

Kotlikoff, L. J., et al. "Social Contracts as Assets: A Possible Solution to the Time-Consistency Problem." *American Economic Review* 78 (September 1988): 662-77.

Kuznets, Simon. "Retardation of Economic Growth." *J of Economic and Business History* 1 (1929): 534-60.

Lapan, H. E. "The Optimal Tariff, Production Lags, and Time Consistency." *American Economic Review* 78 (June 1988): 395-401.

Lebergott, Stanley. *Manpower in Economic Growth.* New York: McGraw-Hill, 1964.

Le Goff, Jacques. *Time, Work, and Culture in the Middle Ages.* Trans. Arthur Goldhammer. Chicago: U of Chicago P, 1980.

Mueller, D. C. "A Life Cycle Theory of the Firm." *J of Industrial Economics* 20.3 (1972): 199-219.

Nelson, R. R., and S. G. Winter. "Toward an Evolutionary Theory of Economic Capabilities." *American Economic Review* 63.2 (1973): 440-49.

Nichols, D., E. Smolensky, and T. N. Tideman. "Discrimination by Waiting Time in Merit Goods." *American Economic Review* 61.3

(1971): 312-23.

Park, R. E., and J. P. Acton. "Large Business Customer Response to Time-of-Day Electricity Rates." *J of Econometrics* 26 (September-October 1984): 229-52. Includes bibliography.

Parks, R. W., and D. Weitzel. "Measuring the Consumer Welfare Effects of Time-Differentiated Electricity Prices." *J of Econometrics* 26 (September-October 1984): 35-64. Includes bibliography.

Petty, Sir William. *The Petty Papers*. London: Constable, 1927.

Pierce, D. A., and L. D. Haugh. "Causality in Temporal Systems." *J of Econometrics* 5 (1977): 265-93.

Roth, A. E., et al. "The Deadline Effect in Bargaining: Some Experimental Evidence." *American Economic Review* 78 (September 1988): 806-23.

Salkever, D. S. "The Use of Dummy Variables to Compute Predictions, Prediction Errors and Confidence Intervals." *J of Econometrics* 4 (1976): 393-97.

Sargan, J. D. "Some Tests of Dynamic Specification for a Single Equation." *Econometrica* 48 (1980): 879-97.

Sargent, T. J., and C. A. Sims. "Business Cycle Modelling Without Pretending to Have Too Much A Priori Economic Theory." *New Methods of Business Cycle Research*. Ed. C. A. Sims. Minneapolis: Federal Reserve Bank of Minneapolis, 1977.

Schary, P. B. "Consumption and the Problem of Time." *J of Marketing* 35.2 (1971): 50-55.

Shackle, George L. S. *Time in Economics*. Westport: Greenwood, 1983.

Sharp, Clifford Henry. *The Economics of Time*. New York: Wiley, 1981.

Smith, Adam. *An Inquiry into the Nature and Causes of the Wealth of Nations*. 2 vols. 1776; reprint, New York: Kelley, 1966.

Solterer, J. "Asset Pricing, Time and Causality." *Review of Social Economy* 25.2 (1967): 55-66.

Tang, Anthony M., Fred M. Westfield, and James S. Worley, eds. *Evolution, Welfare, and Time in Economics*. Lexington: Lexington Books, 1976.

Thompson, C. B., and I. Vertinsky. "A Note on the Theory of Time Allocation." *Economia Internazionale* 25.2 (1972): 286-91.

Thompson, E. P. "Time, Work-Discipline and Industrial Capitalism." *Past and Present* 38 (December 1967): 61-84.

Ure, Andrew. *The Philosophy of Manufactures; or, An Exposition of the Scientific, Moral, and Commercial Economy of the Factory System of Great Britain*. 1835; reprint, London: Frank Cass, 1967.

Weber, Max. *The Protestant Ethic and the Spirit of Capitalism*. Trans.

Talcott Parsons. New York: Scribner's, 1904.

Winston, Gordon C. *The Timing of Economic Activities: Firms, Households, and Markets in Time-Specific Analysis*. Cambridge: Cambridge UP, 1982.

Wolfe, B., and R. Haverman. "Time Allocations, Market Work, and Changes in Female Health." *American Economic Review, Papers and Proceedings* 73 (May 1983): 134-39.

Zupko, Ronald Edward. *British Weights and Measures: A History from Antiquity to the Seventeenth Century*. Madison: U of Wisconsin P, 1977.

——. *A Dictionary of English Weights and Measures from Anglo-Saxon Times to the Nineteenth Century*. Madison: U of Wisconsin P, 1968.

8. GEOGRAPHY

A. GENERAL

Allan, W. "Space-Time Budgets and Activity Studies in Urban Geography and Planning." *Environment and Planning* 3 (1971): 353-68.

Carlstein, Tommy, Don Parkes, and Nigel Thrift, eds. *Timing Space and Spacing Time*. 3 vols. Vol. 1, *Making Sense of Time*. Vol. 2, *Human Activity and Time Geography*. Vol. 3, *Time and Regional Dynamics*. London: Arnold, 1978.

Gregson, Nicky. "On Duality and Dualism: The Case of Structuration and Time Geography." *Progress in Human Geography* 10.2 (1986): 184-205.

Grossman, L. S. "Collecting Time-Use Data in Third World Rural Communities." *Professional Geographer* 36 (November 1984): 444-54.

Lawton, R. "Space, Place and Time." *Geography* 68 (June 1983): 193-207.

Myrdal, G. "The Transfer of Technology to Underdeveloped Countries." *Scientific American* 231.3 (1974): 173-82.

Philbrick, A. K. "Hierarchical Nodality in Geographical Time-Space." *Economic Geography* 58 (January 1982): 1-19.

Thrift, Nigel. "Time and Theory in Human Geography, Part I." *Progress in Human Geography* 1 (1977): 65-101.

B. DEMOGRAPHY (HUMAN GEOGRAPHY)

Abu-Lughod, J. "Urban-Rural Differences as a Function of the Demographic Transition." *American J of Sociology* 69 (1964): 476-90.

Adegbola, O. "New Estimates of Fertility and Child Mortality in Africa South of the Sahara." *Population Studies* 31 (1977): 467-86.

Akhtar, R., and A. Learmonth. "The Resurgence of Malaria in India, 1965-76." *GeoJournal* 1.5 (1977): 69-79.

Albuquerque, K. De, P. Mader, and W. Stinner. "Modernization, Delayed Marriage and Fertility in Puerto Rico 1950 to 1970." *Social and Economic Studies* 25 (1976): 55-65.

Amin, S., ed. *Modern Migrations in Western Africa*. London: Oxford UP, 1974.

Anderson, J., M. Cheng, and W. Fook-Kee. "A Component Analysis of Recent Fertility Decline in Singapore." *Studies in Family Planning* 8 (1977): 282-87.

Antonovsky, A. "Social Class, Life Expectancy and Overall Mortality." *Milbank Memorial Fund Quarterly* 45 (1967): 31-73.

Benjamin, B. *Social and Economic Factors Affecting Mortality*. Paris: Mouton, 1965.

Bogue, D. *Principles of Demography*. New York: Wiley, 1969.

Böhning, W. R. "International Migration in Western Europe: Reflections on the Past Five Years." *International Labour Review* 118 (1979): 401-14.

Boserup, E. *Population and Technological Change*. Chicago: U of Chicago P, 1981.

Brass, W., et al. *The Demography of Tropical Africa*. Princeton: Princeton UP, 1968.

Brookfield, H. "Mauritius: Demographic Upsurge and Prospect." *Population Studies* 11 (1957-58): 102-22.

Brown, L. R. "World Population Growth, Soil Erosion, and Food Insecurity." *Science* 214 (1981): 995-1002.

Bryant, C., B. Stephens, and S. McIver. "Rural to Urban Migration: Some Data from Botswana." *African Studies Review* 21 (1978): 85-99.

Buchanan, K. "The White North and the Population Explosion." *Antipode* 5.3 (1973): 7-15.

Byrne, J. *Levels of Fertility in Commonwealth Caribbean 1921-65*. Jamaica: Institute for Social and Economic Research, U of West Indies, 1972.

Carlsson, G. "The Decline of Fertility: Innovation or Adjustment Process." *Population Studies* 20 (1966-67): 149-74.

Carrothers, G. "An Historical Review of the Gravity and Potential Concepts of Human Interaction." *J of American Institute of Planners* 22 (1956): 94-102.

Chambers, J. D. *Population, Economy and Society in Pre-Industrial Britain*. London: Oxford UP, 1972.

Champion, A. "Evolving Patterns of Population Distribution in England and Wales 1951-71." *Transactions of Institute of British Geographers* 1 (1976): 401-20.

Chandrasekhar, S., ed. *Asia's Population Problems*. London: Allen, 1967.

Chen, P. "Lessons from the Chinese Experience: China's Planned Birth Program and Its Transferability." *Studies in Family Planning* 6 (1975): 354-66.

Cherry, S. "The Hospitals and Population Growth." *Population Studies*

34 (1980): 59-75.

Chisholm, M. *Human Geography: Evolution or Revolution?* Harmondsworth: Penguin, 1975.

Clarke, J. I. *Population Geography*. Oxford: Pergamon, 1965.

——. "Population in Movement." *Studies in Human Geography*. Ed. M. Chisholm, and B. Rodgers. London: Heinemann, 1973. 85-124.

Coale, A. J. "The Decline of Fertility in Europe from the French Revolution to World War II." *Fertility and Family Planning: A World View*. Ed. S. Berman, et al. Ann Arbor: U of Michigan P, 1969. 3-24.

——. "The History of the Human Population." *Scientific American* 231.3 (1974): 41-51.

Coale, A. J., and E. Hoover. *Population Growth and Economic Development in Low-Income Countries*. Princeton: Princeton UP, 1958.

Coale, A. J., and M. Zelnick. *New Estimates of Fertility and Population in the United States*. Princeton: Princeton UP, 1963.

Cohen, M. *The Food Crisis in Prehistory: Overpopulation and the Origins of Agriculture*. New Haven: Yale UP, 1977.

Compton, P. "Religious Affiliation and Demographic Variability in Northern Ireland." *Transactions of Institute of British Geographers* 1 (1976): 433-52.

Condran, G., and E. Crimmins-Gardner. "Public Health Measures and Mortality in U.S. Cities in the Late Nineteenth Century." *Human Ecology* 6 (1978): 27-54.

Connell, K. *The Population of Ireland 1750-1845*. Oxford: Clarendon P, 1950.

Council of Europe. *Population Decline in Europe: Implications of a Declining or Stationary Population*. London: Arnold, 1978.

Cowan, H. *British Emigration to British North America*. Toronto: U of Toronto P, 1961.

Cox, K. *Man, Location and Behaviour: An Introduction to Human Geography*. New York: Wiley, 1972.

Cunliffe, B. *Rome and Her Empire*. London: Bodley Head, 1978.

Curtin, P. *The Atlantic Slave Trade: A Census*. Madison: U of Wisconsin P, 1969.

Dando, W. *The Geography of Famine*. London: Arnold, 1980.

Davis, K. "Asia's Cities: Problems and Options." *Population and Development Review* 1 (1975): 71-86.

Day, L. "Natality and Ethnocentrism: Some Relationships Suggested by an Analysis of Catholic-Protestant Differentials." *Population Studies* 22 (1968): 27-50.

Easterlin, R., ed. *Population and Economic Change in Developing*

Countries. Chicago: U of Chicago P, 1980. 209-65.

Ehrlich, P. R. *The Population Bomb*. London: Pan, 1971.

Federici, N. "The Influence of Women's Employment on Fertility." *World Views of Population Problems*. Ed. E. Szabady. Budapest: Akadémiai Kaidó, 1968. 77-82.

Flinn, M. W. *The European Demographic System, 1500-1820*. Baltimore: Johns Hopkins UP, 1981.

——, ed. *Scottish Population History from the 17th Century to the 1930s*. Cambridge: Cambridge UP, 1977.

Forrester, J. *World Dynamics*. Cambridge: Wright-Allen, 1971.

Forster, C., and G. Tucker. *Economic Opportunity and White American Fertility Ratios 1800-60*. New Haven: Yale UP, 1972.

Forster, F. "Use of a Demographic Base Map for the Presentation of Areal Data in Epidemiology." *British J of Preventive and Social Medicine* 20 (1966): 156-71.

Frankel, F. *India's Green Revolution: Economic Gains and Political Costs*. Princeton: Princeton UP, 1971.

Freedman, D. "The Relation of Economic Status to Fertility." *American Economic Review* 53 (1963): 414-26.

Freedman, R., P. Whelpton, and J. Smit. "Socio-Economic Factors in Religious Differentials in Fertility." *American Sociological Review* 26 (1961): 608-14.

Freeman, C., and M. Jahoda, eds. *World Futures: The Great Debate*. London: Martin Robertson, 1978.

Frey, W. "Status Selective White Flight and Central City Population Change." *J of Regional Science* 20 (1980): 71-89.

Gershuny, J., and J. P. Robinson. "Historical Changes in the Household Division of Labor." *Demography* 25 (November 1988): 537-52. Includes bibliography.

Gille, H. "The Demographic History of the Northern European Countries in the Eighteenth Century." *Population Studies* 3 (1949-50): 3-65.

Glass, D., and D. Eversley, eds. *Population History*. London: Arnold, 1965.

Griffith, G. T. *Population Problems of the Age of Malthus*. London: Cass, 1926.

Grigg, D. *Population Growth and Agrarian Change*. Cambridge: Cambridge UP, 1980.

Heer, D. "What Is the Annual Net Flow of Undocumented Mexican Immigrants to the United States?" *Demography* 16 (1979): 417-23.

Hornby, W. F., and M. Jones. *An Introduction to Population Geography*. Cambridge: Cambridge UP, 1980.

Howe, G. M. *National Atlas of Disease Mortality in the United Kingdom*. London: Nelson, 1970.

——, ed. *A World Geography of Human Diseases*. London: Academic P, 1977.

Hudson, T. "Health Migration in America's Nineteenth Century Westward Movement." *Ecumene* 13 (1981): 50-58.

Johnston, R. J. *Geography and Geographers: Anglo-American Human Geography since 1945*. London: Arnold, 1979.

Jones, E. F. "Fertility Decline in Australia and New Zealand 1861-1936." *Population Index* 37 (1971): 301-38.

Jones, Huw R. *Population Geography*. 2nd ed. London: Paul Chapman. 1990. Bibliography 286-317.

Jones, R. C. "Undocumented Migration from Mexico: Some Geographical Questions." *Annals, Association of American Geographers* 72 (1982): 77-87.

Kahn, H., et al. *The Next 200 Years: A Scenario for America and the World*. New York: Morrow, 1976.

Kellerman, Aharon. *Time, Space, and Society: Geographical Societal Perspectives*. Dordrecht: Kluwer, 1989.

Knights, P. *The Plain People of Boston 1830-60*. New York: Oxford UP, 1971.

Knodel, J. "Law, Marriage and Illegitimacy in Nineteenth-Century Germany." *Population Studies* 20 (1966-67): 279-94.

Kosinski, L., and R. M. Prothero, eds. *People on the Move: Studies on Internal Migration*. London: Methuen, 1975.

Krause, J. "Changes in English Fertility and Mortality 1781-1850." *Economic History Review* 11 (1958): 52-70.

Larson, A. *Patterns of Contraceptive Use Around the World*. Washington: Population Reference Bureau, 1981.

Lee, R. D., ed. *Population Patterns in the Past*. New York: Academic P, 1977.

Lenski, G. *The Religious Factor: A Sociological Study of Religion's Impact on Politics, Economics and Family Life*. New York: Doubleday, 1961.

McGlashan, N., ed. *Medical Geography, Techniques and Analysis*. London: Methuen, 1972.

McKeown, T. *The Modern Rise of Population*. London: Arnold, 1976.

McKeown, T., and R. Brown. "Medical Evidence Related to English Population Changes in the Eighteenth Century." *Population Studies* 9 (1955): 119-41.

Madigan, F. "Are Sex Mortality Differentials Biologically Caused?" *Milbank Memorial Fund Quarterly* 35 (1957): 202-23.

Matras, J. "The Social Strategy of Family Formation: Some Variations in Time and Space." *Demography* 2 (1965): 349-62.

Meadows, D. H., et al. *The Limits to Growth, a Report for the Club of Rome's Project on the Predicament of Mankind*. New York: Universe Books, 1972.

Myrdal, G. *Population: A Problem for Democracy*. Cambridge: Harvard UP, 1940.

Newman, James L., and Gordon E. Matzke. *Population: Patterns, Dynamics, and Prospects*. Englewood Cliffs: Prentice-Hall, 1984. Bibliography 287-97.

Page, H., and R. Lesthaeghe. *Child-Spacing in Tropical Africa*. London: Academic P, 1982.

Peters, Gary L., and Robert P. Larkin. *Population Geography: Problems, Concepts, and Prospects*. 3rd ed. Dubuque: Kendal/Hunt, 1990.

Petras, E. M. "Towards a Theory of International Migration: The New Division of Labor." *Sourcebook on the New Immigration*. Ed. R. L. Bryce-LaPorte. New Brunswick: Transaction Books, 1980. 439-49.

Razzell, P. *The Conquest of Smallpox: The Impact of Inoculation on Smallpox Mortality in Eighteenth Century Britain*. Firle, Sussex: Caliban Books, 1977.

Reddaway, W. *The Economics of Declining Population*. London: Macmillan, 1939.

Rich, W. *Smaller Families through Social and Economic Progress*. Washington: Overseas Development Council, 1973.

Romaniuk, A. "Increase in Natural Fertility during the Early Stages of Modernization: Evidence from an African Case Study, Zaire." *Population Studies* 34 (1980): 293-310.

Sauer, R. "Infanticide and Abortion in Nineteenth Century Britain." *Population Studies* 32 (1978): 81-93.

Sauvy, A. *Zero Growth*. Oxford: Blackwell, 1975.

Shorter, E., J. Knodel, and E. Van de Walle. "The Decline of Non-Marital Fertility in Europe 1880-1940." *Population Studies* 25 (1971): 375-93.

Shrewsbury, J. *A History of Bubonic Plague in the British Isles*. Cambridge: Cambridge UP, 1970.

Shryock, H., and J. Siegel. *The Methods and Materials of Demography*. New York: Academic P, 1976.

Sigerist, H. *Civilization and Disease*. Ithaca: Cornell UP, 1943.

Spengler, J. *Population and America's Future*. San Francisco: W. H. Freeman, 1975.

Stewart, J. Q. "Empirical Mathematical Rules Concerning the Distribution and Equilibrium of Population." *Geographical Review* 37 (1947): 461-85.

Wrigley, E. A., ed. *An Introduction to English Historical Demography.* London: Weidenfeld and Nicolson, 1966.

Wynne-Edwards, V. "Self-Regulating Systems in Populations of Animals." *Science* 147 (1965): 1543-48.

9. GEOLOGY AND GEOPHYSICS

A. GENERAL (INCLUDING GEOCHRONOLOGY)

Acock, M. "Age of the Universe." *Philosophy of Science* 50 (March 1983): 130-45.

Allaart, J. H. "The Pre-3760 M.Y. Old Supracrustal Rocks of the Isua Area, Central West Greenland, and the Associated Occurrence of Quartz-Banded Ironstone." *The Early History of the Earth*. Ed. B. F. Windley. London: Wiley, 1976. 177-89.

Bailey, E., Sr. *James Hutton: Founder of Modern Geology*. New York: Elsevier, 1971.

Bailey, S. W. "Cation Ordering and Pseudosymmetry in Layer Silicates." *American Mineralogist* 60 (1975): 175-87.

Barnes, Charles W. *Earth, Time, and Life: An Introduction to Physical and Historical Geology*. 2nd ed. New York: Wiley, 1988.

Bayer, U. *Pattern Recognition Problems in Geology and Paleontology*. New York: Springer, 1985.

Buckland, William. *Geology and Mineralogy Considered with Reference to Natural Theology*. Ed. Stephen J. Gould. 2 vols. 1836; reprint, Salem: Ayer, 1980.

Buffon, George Louis Leclerc, Comte de. *Théorie de la terre*. In his *Oeuvres complètes de M. le Cte de Buffon*. 24 vols. Paris, 1774.

Burnet, Thomas. *Telluris Theoria Sacra*. London, 1681. The English translation of the full work did not appear until 1690. It reads in many parts like a work of fiction.

Cairns-Smith, A. G. "Precambrian Solution Photochemistry, Inverse Segregation and Banded Iron Formations." *Nature* 276 (1978): 807-08.

Cook, Melvin Alonzo. *Geological Chronometry*. Salt Lake City: U of Utah, Department of Metallurgy, 1956.

Cradwick, P. D. G., et al. "Imogolite, a Hydrated Aluminium Silicate of Tubular Structure." *Nature (Physical Science)* 240 (1972): 187-89.

Cuvier, Georges. *Essay on the Theory of the Earth: Mineralogical Notes, and an Account of Cuvier's Geological Discoveries*. Trans. Robert Kerr. Ed. Claude C. Albritton, Jr. 1817; reprint, Salem: Ayer, 1978.

Davies, G. L. *The Earth in Decay: A History of British Geomorphology 1578-1878*. New York: Elsevier, 1969.

Eicher, Don L. *Geologic Time*. 2nd ed. Englewood Cliffs: Prentice-Hall, 1976.

Franke, Herbert W. *Methoden der Geochronologie: Die Suche nach den*

Daten der Erdgeschichte. Berlin: Springer, 1969.

Goldich, Samuel S., et al. *Precambrian Geology and Geochronology of Minnesota*. Saint Paul: Minnesota Geological Survey, 1961.

Gould, Stephen J. *Time's Arrow—Time's Cycle: Myth and Metaphor in the Discovery of Geological Time*. Cambridge: Harvard UP, 1988.

Hallam, Anthony. *Great Geological Controversies*. Oxford: Oxford UP, 1983.

Harper, Christopher T., ed. *Geochronology: Radiometric Dating of Rocks and Minerals*. Stroudsburg: Dowden, Hutchinson, 1973.

Hooykaas, R. *The Principle of Uniformity in Geology, Biology, and Theology*. Leiden: E. J. Brill, 1963.

Hutton, J. *Theory of the Earth with Proofs and Illustrations*. Edinburgh: William Creech, 1795.

Johnson, L. T. "Clay Minerals in Pennsylvania Soils in Relation to Lithology of the Parent Rock and Other Factors—1." *Clays and Clay Minerals* 18 (1970): 247-60.

Keller, W. D. "Classification of Kaolins Exemplified by Their Texture in Scan Electron Micrographs." *Clays and Clay Minerals* 26 (1978): 1-20.

Keller, W. D., and R. F. Hanson. "Dissimilar Fabrics by Scan Electron Microscopy of Sedimentary Versus Hydrothermal Kaolins in Mexico." *Clays and Clay Minerals* 23 (1975): 201-04.

Kirkaldy, John F. *Geological Time*. Edinburgh: Oliver and Boyd, 1971.

Lahav, N., D. White, and S. Chang. "Peptide Formation in the Prebiotic Era: Thermal Condensation of Glycine in Fluctuating Clay Environments." *Science* 201 (1978): 67-69.

Lyell, Charles. *Principles of Geology*. 3 vols. London: Murray, 1830-33.

Mansfield, C. F., and S. W. Bailey. "Twin and Pseudotwin Intergrowths in Kaolinite." *American Mineralogist* 57 (1972): 411-25.

Marvin, U. *Continental Drift*. Washington: Smithsonian Institution P, 1973.

Millot, G. "Clay." *Scientific American* 240.4 (1979): 76-84.

Moorbath, S. "The Oldest Rocks and the Growth of Continents." *Scientific American* 236.3 (1977): 92-104.

Ozima, M. *Geohistory: Global Evolution of the Earth*. Trans. J. Wakabayashi. New York: Springer, 1987.

Porter, Roy. *The Making of Geology: Earth Science in Britain, 1660-1815*. Cambridge: Cambridge UP, 1977.

Rosenberg, G. D., and S. K. Runcorn, eds. *Growth Rhythms and the History of the Earth's Rotation*. New York: Wiley, 1975.

Sarjeant, William A. *Geologists and the History of Geology: An International Bibliography from the Origins to 1978*. 5 vols. Salem: Ayer,

n.d.
——. *Geologists and the History of Geology: An International Bibliography, Supplement 1979-1984, and Additions*. Vols. 6 and 7. Melbourne: Krieger, 1987.

Seyfert, C. K., and L. A. Sirkin. *Earth History and Plate Tectonics*. New York: Harper and Row, 1973.

Steno, Nicolaus (Stensen). *The Earliest Geological Treatise (1667)*. New York: Macmillan, 1958.

Thornes, John B., and D. Brundsen. *Geomorphology and Time*. New York: Wiley, 1977.

Toulmin, Stephen, and June Goodfield. *The Discovery of Time*. 1965; reprint, Chicago: U of Chicago P, 1982.

Wada, S., and K. Wada. "Density and Structure of Allophane." *Clay Minerals* 12 (1977): 289-98.

Walker, J. C. G. "Implications for Atmospheric Evolution of the Inhomogeneous Accretion Model of the Origin of the Earth." *The Early History of the Earth*. Ed. B. F. Windley. London: Wiley, 1976. 537-46.

York, Derek. *The Earth's Age and Geochronology*. Oxford: Pergamon, 1972.

Zeuner, Frederic E. *Dating the Past: An Introduction to Geochronology*. 4th ed. 1946; reprint, New York: Hafner, 1970.

B. PALEONTOLOGY

Allsworth-Jones, P. *The Szeletian: And the Transition from Middle to Upper Paleolithic in Central Europe*. New York: Oxford UP, 1986.

Ameur, R. C., J.-J. Jaeger, and J. Michaux. "Radiometric Age of Early Hipparion Fauna in North-West Africa." *Nature* 261 (1976): 38-39.

Beerbower, James R. *Search for the Past: An Introduction to Paleontology*. 2nd ed. Englewood Cliffs: Prentice-Hall, 1968.

Berry, William B. N. *Growth of a Prehistoric Time Scale, Based on Organic Evolution*. Revised ed. Palo Alto: Blackwell Scientific, 1987.

Black, Rhona M. *The Elements of Paleontology*. 2nd ed. Cambridge: Cambridge UP, 1988. Includes bibliography.

Boucot, A. J. *Evolution and Extinction Rate Controls*. New York: Elsevier, 1975.

Bowen, Robert. *Paleotemperature Analysis*. New York: Elsevier, 1966.

Brenchley, P. J., ed. *Fossils and Climate*. New York: Wiley, 1984.

Buffetaut, Eric. *A Short History of Vertebrate Paleontology*. New York:

Routledge, Chapman and Hall, 1986.
Carroll, Robert L. *Vertebrate Paleontology and Evolution*. New York: W. H. Freeman, 1987.
Clarkson, E. N. K. *Invertebrate Paleontology and Evolution*. London: Allen and Unwin, 1979.
Cline, R. M., and J. D. Hays, eds. *Investigation of Late Quaternary Paleoceanography and Paleoclimatology*. Boulder: Geological Society of America, 1976.
Cockburn, Aiden, and Eve Cockburn. *Mummies, Diseases and Ancient Cultures*. Abridged ed. New York: Cambridge UP, 1983.
Colbert, Edwin H. *The Great Dinosaur Hunters and Their Discoveries*. New York: Dover, 1984.
Cope, Edward D. *The Origin of the Fittest: Essays on Evolution and the Primary Factors of Organic Evolution*. Salem: Ayer, 1974.
Cracraft, Joel, and Niles Eldredge, eds. *Phylogenetic Analysis and Paleontology*. New York: Columbia UP, 1979.
Croucher, Ronald, and Alan R. Woolley. *Fossils, Minerals and Rocks: Collection and Preservation*. New York: Cambridge UP, 1982.
Crux, Jason A. *Calcareous Nannofossils and Their Applications*. New York: Wiley, 1988.
Dawson, John William. *The Chain of Life in Geological Time: The Origin and Succession of Animals and Plants*. London: Religious Tract Society, [1880].
Day, William. *Genesis on Planet Earth: The Search for Life's Beginning*. 2nd ed. New Haven: Yale UP, 1984. Includes bibliography.
Desmond, Adrian John. *The Hot-Blooded Dinosaurs: A Revolution in Paleontology*. London: Blond and Briggs, 1975.
Downie, Charles, and William Anthony S. Sarjeant. *Bibliography and Index of Fossil Dinoflagellates and Acritarchs*. New York: Geological Society of America, 1964.
Driesch, Angela von den. *A Guide to the Measurement of Animal Bones from Archaeological Sites*. Cambridge: Peabody Museum, Harvard U, 1976.
Dyer, Judith, and Frederick Schram. *A Manual of Invertebrate Paleontology*. Albany: SUNY P, 1983.
Edwards, W. N. *The Early History of Paleontology*. Woodland Hills: Sabbot, Natural History Books, 1976.
Fortey, Richard. *Fossils: The Key to the Past*. Woodland Hills: Sabbot, Natural History Books, 1983.
Gould, Stephen Jay. *Wonderful Life: The Burgess Shale and the Nature of History*. Harmondsworth: Penguin, 1989.
Hallam, A. *Atlas of Palaeobiogeography*. New York: Elsevier, 1972.

Hitchcock, Edward. *A Report on the Sandstone of the Connecticut Valley, Especially Its Fossil Footmarks*. Salem: Ayer, 1974.

Hsu, Kenneth J., and Helmut J. Weissert, eds. *South Atlantic Palaeoceanography*. New York: Cambridge UP, 1985.

Klemm, D. D., and H. J. Schneider, eds. *Time- and Strata-Bound Ore Deposits*. New York: Springer, 1978.

Kurten, Bjorn. *On Evolution and Fossil Mammals*. New York: Columbia UP, 1987.

Larwood, G. P., ed. *Extinction and Survival in the Fossil Record*. New York: Oxford UP, 1988.

Levinton, Jeffrey. *Genetics, Paleontology, and Macroevolution*. New York: Cambridge UP, 1988.

Lyell, C. *On the Geological Evidences of the Antiquity of Man*. London, 1862.

McKinney, M. L., ed. *Heterochrony in Evolution: A Multidisciplinary Approach*. New York: Plenum, 1988.

Mahaney, W. C. *Quaternary Dating Methods*. New York: Elsevier, 1984.

Odin, Gilles S. *Numerical Dating in Stratigraphy*. 2 vols. New York: Wiley, 1982.

Rudwick, M. J. S. "Charles Lyell's Dream of a Statistical Paleontology." *Paleontology* 21 (1978): 225-44.

——. *The Great Devonian Controversy*. Chicago: U of Chicago P, 1985.

——. *The Meaning of Fossils*. London: Macdonald, 1972.

Schoch, Robert M. *Phylogeny Reconstruction in Paleontology*. New York: Van Nostrand Reinhold, 1986.

Schroeder, J., and B. H. Purser, eds. *Reef Diagenesis*. New York: Springer, 1986.

Seposki, J. John, Jr. *A Compendium of Fossil Marine Families*. Milwaukee: Milwaukee Public Museum, 1982.

Shipman, Pat. *Life History of a Fossil: Introduction to Taphonomy and Paleoecology*. Cambridge: Harvard UP, 1981.

Simpson, George G. *Fossils and the History of Life*. New York: W. H. Freeman, 1984.

Smith, Andrew. *Echinoid Palaeobiology*. Winchester: Unwin Hyman, 1984.

Smith, Fred H., and Frank Spencer. *The Origins of Modern Humans: A World Survey of Fossil Evidence*. New York: Alan R. Liss, 1984.

Spoczynska, Joy O. *Fossils: A Study in Evolution*. Lanham: Rowman and Littlefield, 1971.

Stirton, Reuben Arthur. *Time, Life, and Man: The Fossil Record*. New York: Wiley, 1959.

Sunderland, Luther D. *Darwin's Enigma: Fossils and Other Problems*.

San Diego: Master Book Publishers, 1984. Includes bibliography.
Swinnerton, H. H. *Outlines of Palaeontology*. London: Arnold, 1923.
Swinton, W. E. *Fossil Birds*. 3rd ed. Woodland Hills: Sabbot, Natural History Books, 1975.
Webby, B. D., ed. *Fossil Collections of the World: An International Guide*. Washington: International Palaeontological, 1989.
Weishampel, D. B. *Evolution of Jaw Mechanisms in Ornithopod Dinosaurs*. New York: Springer, 1984.
Whittington, H. B. *The Burgess Shale*. New Haven: Yale UP, 1985.
Wilson, M. D., and E. D. Pittman. "Authigenic Clays in Sandstones: Recognition and Influence on Reservoir Properties and Paleo-environmental Analysis." *J of Sedimentology and Petrology* 47 (1977): 3-31.

10. HISTORY AND HISTORIOGRAPHY

A. GENERAL

Aron, R. *Introduction à la philosophie de l'histoire*. Paris: Gallimard, 1948.

Bernheim, E. *Lehrbuch der historischen Methode und der Geschichtsphilosophie*. Leipzig, 1903.

Bloch, Marc. *The Historian's Craft*. Trans. Peter Putman. Manchester: Manchester UP, 1954.

Bodin, Jean. *Methodus ad facilem historiarum cognitionem*. Paris, 1566.

——. *Methodus ad facilem historiarum cognitionem*. 1650; reprint, Aalen: Scientia Verlag, 1967.

Bossuet, J. *Discours sur l'histoire universelle*. [Paris], 1681.

Braudel, Fernand. *On History*. Trans. Sarah Matthews. Chicago: U of Chicago P, 1980.

Brown, Norman O. *Life Against Death: The Psychoanalytical Meaning of History*. Middletown: Wesleyan UP, 1959.

Carr, David. *Time, Narrative, and History*. Bloomington: Indiana UP, 1986.

Childe, V. G. *History*. London: Watts, 1947.

Collingwood, R. G. *Essays in the Philosophy of History*. Austin: U of Texas P, 1965.

——. *The Idea of History*. Oxford: Clarendon P, 1947.

Condorcet. *Esquisse d'un tableau historique des progrès de l'esprit humain*. [Paris], 1795.

Croce, B. *History: Its Theory and Practice*. Trans. Douglas Ainslie. New York: Russell and Russell, 1960.

——. *Theory and History of Historiography*. Trans. Douglas Ainslie. London: Harrap, 1921.

Davillé, L. *Leibniz historien*. Paris, 1909.

Dawson, H. Christopher. *The Dynamics of World History*. Ed. J. J. Mulloy. London: Sheed and Ward, 1957.

Dilthey, Wilhelm. *Pattern and Meaning in History*. Trans. and ed. H. P. Rickman. New York: Harper and Row, 1961.

Dinkler, Erich, ed. *Zeit und Geschichte*. Tübingen: Mohr, 1964.

Dux, Günter. *Die Zeit in der Geschichte. Ihre Entwicklungslogik vom Mythos zur Weltzeit*. Frankfurt: Suhrkamp, 1989.

Erikson, Erik. *Young Man Luther: A Study in Psychoanalysis and History*. New York: Norton, 1962.

Ettlinger, M. *Leibniz als Geschichtsphilosoph*. Munich, 1921.

Foucault, Michel. *The Order of Things: An Archaeology of the Human Sciences*. New York: Vintage Books, 1973.

Franklin, Julian. *Jean Bodin and the Sixteenth Century Revolution in Law and History*. New York: Columbia UP, 1963.

Haber, F. C. "The Darwinian Revolution in the Concept of Time." *The Study of Time I*. Ed. J. T. Fraser, F. C. Haber, and G. H. Müller. New York: Springer, 1972. 383-401.

Hegel, G. W. F. *The Philosophy of History*. Trans. J. B. Sibree. New York: Dover, 1956.

Hempel, C. G. "The Functions of General Laws in History" (1942). Reprinted in *Theories of History*. Ed. P. Gardiner. New York: Free P, 1969. 344-56.

Herder, J. G. von. *Reflections on the Philosophy of History*. Trans. T. O. Churchill. Chicago: U of Chicago P, 1968.

Jaspers, K. *The Origin and Goal of History*. Trans. M. Bullock. London: Routledge and Kegan Paul, 1953.

Kant, I. *On History*. Ed. L. W. Beck. New York: Bobbs Merrill, 1963.

Kelley, Donald. *The Foundations of Modern Historical Scholarship*. New York: Columbia UP, 1970.

Koselleck, R. *Vergangene Zukunft: Zur Semantik geschichtlicher Zeiten*. Frankfurt: Suhrkamp, 1985.

Krieger, Leonard. *Time's Reasons: Philosophies of History Old and New*. Chicago: U of Chicago P, 1989.

La Popelinière, Henri Voisin de. *L'histoire des histoires*. Paris, 1599.

Leijonhufvud, A. "Time in Theory and History or Why I Am Not a Historian." *Agricultural History* 60 (Winter 1986): 84-88.

Major-Poetzl, Pamela. *Michael Foucault's Archaeology of Western Culture: Toward a New Science of History*. Chapel Hill: U of North Carolina P, 1979.

Manuel, Frank. *Isaac Newton Historian*. Cambridge: Harvard UP, 1963.

Meyerhoff, Hans, ed. *The Philosophy of History in Our Time*. Garden City: Doubleday, 1959.

Newton, Isaac. *The Chronology of Ancient Kingdoms Amended, to which is prefixed a Short Chronicle from the First Memory of Things in Europe to the Conquest of Persia by Alexander the Great*. London, 1728.

Niceron, J. P., ed. *Mémoires pour servir à l'histoire des hommes illustres dans la république des lettres*. Paris, 1737.

Oakeley, H. *History and Self*. London: Williams and Norgate, 1934.

Oakeshott, M. *On History and Other Essays*. Oxford: Blackwell, 1985.

Poole, Reginald Lane. *Chronicles and Annals*. Oxford: Clarendon P,

1926.

Ranke, Leopold von. *Das Briefwerk*. Ed. Walter Fuchs. Hamburg: Hoffmann und Campe, 1949.

——. *Tagebücher*. Ed. Walter Fuchs. Munich: R. Oldenbourg, 1964.

Rotenstreich, Nathan. *Time and Meaning in History*. Boston: Reidel, 1987.

Rule, J. C., and B. F. Crosby. "Bibliography of Works on Arnold J. Toynbee, 1946-1960." *History and Theory* 4 (1965): 212-33.

Santamaria, U., and A. M. Bailey. "A Note on Braudel's Structure as Duration." *History and Theory* 23.1 (1984): 78-83.

Schaeffler, Richard. *Die Struktur der Geschichtszeit*. Frankfurt: Klostermann, 1963.

Shuter, William. "History as Palingenesis in Pater and Hegel." *Publications of the Modern Language Association* 86 (1971): 411-21.

Somerwell, D. C. *A Study of History*. 2 vols. New York: Dell, 1965.

Sorokin, P. A. "Toynbee's Philosophy of History." *J of Modern History* 12 (1940): 374-87.

Stern, Fritz, ed. *Varieties of History*. Cleveland: World, 1956.

Thompson, J. W. *A History of Historical Writing*. 2 vols. New York: Macmillan, 1942.

Toynbee, A. J. *An Historian's Approach to Religion*. Oxford: Oxford UP, 1956.

——. *A Study of History*. 12 vols. New York: Oxford UP, 1954-61.

Voisé, W. "On Historical Time in the Works of Leibniz." *The Study of Time II*. Ed. J. T. Fraser, and N. Lawrence. New York: Springer, 1975. 114-21.

White, Hayden. *Metahistory: The Historical Imagination in Nineteenth-Century Europe*. Baltimore: Johns Hopkins UP, 1973.

Whitrow, Gerald James. *Time in History: The Evolution of Our General Awareness of Time and Temporal Perspective*. Oxford: Oxford UP, 1988.

B. EASTERN

Denta, R. C., ed. *The Idea of History in the Ancient Near East*. New Haven: Yale UP, 1955.

Elvin, Mark. *The Pattern of the Chinese Past*. Stanford: Stanford UP, 1973.

Gardner, C. S. *Chinese Traditional Historiography*. 1938; reprint, Cambridge: Harvard UP, 1961.

Hulsewé, A. F. "Notes on the Historiography of the Han Period."

Historians of China and Japan. Ed. W. G. Beasley and E. G. Pulleyblank. London: Oxford UP, 1961.

Kierman, F. A. *Ssuma Chhien's Historiographical Attitude as Reflected in Four Late Warring States Biographies*. Wiesbaden: Harrassowitz, 1962.

Ling-Shêng, Yang. "The Organisation of Chinese Official Historiography: Principles and Methods of the Standard Histories from the Thang through the Ming Dynasty." *Historians of China and Japan*. Ed. W. G. Beasley, and E. G. Pulleyblank. London: Oxford UP, 1961.

Loon, P. van der. "The Ancient Chinese Chronicles and the Growth of Historical Ideals." *Historians of China and Japan*. Ed. W. G. Beasley, and E. G. Pulleyblank. London: Oxford UP, 1961.

Mandal, Kumar K. *A Comparative Study of the Concepts of Space and Time in Indian Thought*. Varanasi, India: Chowkhambe Sanskrit Series Office, 1968.

Meskill, J., ed. *The Pattern of Chinese History: Cycles, Development, or Stagnation?* Westport: Greenwood, 1965.

Moule, A. C., and W. P. Yetts. *The Rulers of China, 221 B.C. to A.D. 1949*. London: Routledge and Kegan Paul, 1957.

Reischauer, E. O. "The Dynastic Cycle." *The Pattern of Chinese History*. Ed. J. Meskill. Westport: Greenwood, 1965.

Sivin, Nathan. "The Chinese Conception of Time." *Earlham Review* 1 (1966): 82-91.

Smith, V. A. *The Early History of India*. Oxford: Clarendon P, 1904.

Thomas, E. J. *History of Buddhist Thought*. London, 1933.

Watson, B. *Ssuma Chhien, Grand Historian of China*. New York: Columbia UP, 1958.

Wilson, G. M. "Time and History in Japan." *American Historical Review* 85 (June 1980): 557-71.

Yü-Fu, Chin. *Chung-Kuo Shih-Hsüeh Shih (A History of Chinese History-Writing)*. Peking: Chung-hua, 1962.

Yu-Shan, Han. *Elements of Chinese Historiography*. Hollywood: Hawley, 1955.

Xin, He. "Academic Disparity between China and the West: A Trial Research on Comparative Cultural History." *J of Dialectics of Nature* 5.2 (1983): 38-47.

C. THE IDEA OF PROGRESS

Baillie, J. *The Belief in Progress*. Oxford: Oxford UP, 1950.

Bury, J. B. *The Idea of Progress*. London: Macmillan, 1920.

Crombie, A. C. "Some Attitudes to Scientific Progress: Ancient, Medieval and Early Modern." *History of Science* 13 (1975): 213-30.

Dawson, C. *Progress and Religion*. London: Sheed and Ward, 1931.

Dodds, E. R. *The Ancient Concept of Progress and Other Essays*. Oxford: Oxford UP, 1973.

Ferrarotti, F. *The Myth of Inevitable Progress*. Westport: Greenwood, 1985.

Macey, Samuel L. "Literary Images of Progress: The Rise and Fall of a Western Ideal." *Time, Science, and Society in China and the West: The Study of Time V*. Ed. J. T. Fraser, N. Lawrence, and F. C. Haber. Amherst: U of Massachusetts P, 1986. 93-103.

D. SCIENCE AND TECHNOLOGY

Ashton, T. S. "The Technical Innovations." *The Industrial Revolution 1760-1830*. Revised ed. London: Oxford UP, 1962. 58-93.

Babbage, Charles. *On the Economy of Machinery and Manufactures*. London: Charles Knight, 1832.

——. *On the Economy of Machinery and Manufactures*. 4th ed., enlarged. 1835; reprint, New York: Kelley, 1963.

Benjamin, F., and G. J. Toomer. *Campanus of Novara and Medieval Planetary Theory*. Madison: U of Wisconsin P, 1971.

Berriman, A. E. *Historical Metrology*. New York: Greenwood, 1969.

Brody, D. "Time and Work During Early American Industrialism." *Labor History* 30 (Winter 1989): 5-46.

Cipolla, Carlo M. *European Culture and Overseas Expansion*. Harmondsworth: Penguin, 1970.

Crowther, J. G. *Scientists of the Industrial Revolution*. London: Cresset P, 1962.

Dexiu, Zou. "On the Causes of Backwardness of China's Agricultural Science and Technology in Modern Times." *J of Dialectics of Nature* 5.2 (1983): 48-51.

Dijksterhuis, E. J. *The Mechanization of the World Picture*. Oxford: Oxford UP, 1961.

Drachman, A. G. *The Mechanical Technology of Greek and Roman Antiquity*. Madison: U of Wisconsin P, 1963.

Duncan, C. S. "The Scientist as a Comic Type." *Modern Philology* 14 (1916): 281-91.

Farrington, Benjamin. *Greek Science*. Baltimore: Penguin, 1969.

Flinn, M. W. *Origins of the Industrial Revolution*. [London]: Longmans,

1966.

Fontenelle, Bernard Le Boivier de. *A Plurality of Worlds*. Trans. John Glanvill. 1688; reprint, London: Nonsuch P, 1929.

Fraser, J. T. "The Problems of Exporting Faust." *Time, Science, and Society in China and the West: The Study of Time V*. Ed. J. T. Fraser, N. Lawrence, and F. C. Haber. Amherst: U of Massachusetts P, 1986. 1-20.

Gauquelin, Michel. *The Cosmic Clocks: From Astrology to a Modern Science*. Chicago: H. Regnery, 1967.

Gimpel, Jean. *The Medieval Machine: The Industrial Revolution of the Middle Ages*. New York: Penguin, 1977.

Goodman, G. K. *The Dutch Impact on Japan (1640-1853)*. Leiden: E. J. Brill, 1967.

Guantao, Jin, et al. "The Evolution of Chinese Science and Technology." *Time, Science, and Society in China and the West: The Study of Time V*. Ed. J. T. Fraser, N. Lawrence, and F. C. Haber. Amherst: U of Massachusetts P, 1986. 170-80.

———. "Historical Change in the Structure of Science and Technology." *J of Dialectics of Nature* 5.1 (1983): 14-24.

———. "The Structure of Science and Technology—On the Factors Delaying the Development of Science and Technology in China in Comparison with the West since the 17th Century." *J of Dialectics of Nature* 4.5 (1983): 7-24.

Guye, Samuel, and Henry Michel. *Time and Space*. Trans. Diana Dolan, and Samuel W. Mitchell. London: Pall Mall P, 1970.

Hall, Hubert, and Frieda J. Nicholas, eds. *Select Tracts and Table Books Relating to English Weights and Measures (1100-1742)*. In *Camden Third Series*. Vol. 41. London: Offices of the Society, 1929.

Henderson, John B. *The Development and Decline of Chinese Cosmology*. New York: Columbia UP, 1984.

Hooke, Robert. *The Posthumous Works*. 1705; reprint, New York: Johnson Reprint, 1969.

Howse, Derek. *Greenwich Time and the Discovery of the Longitude*. Oxford: Oxford UP, 1980.

Hunter, Michael. *Science and Society in Restoration England*. Cambridge: Cambridge UP, 1981.

Johnstone, William D. *For Good Measure*. New York: Holt Rinehart, 1975. Deals with weights and measures.

Kahn, Charles H. *Anaximander and the Origins of Greek Cosmology*. New York: Columbia UP, 1960.

Kirk, G. S. *Heraclitus, the Cosmic Fragments*. Cambridge: Cambridge UP, 1954.

Lianghui, Qui. "On the Causes for the Lagging Behind of China's Metallurgical Technology in Modern Times." *J of Dialectics of Nature* 5.2 (1983): 52-57.

Mantoux, Paul. *The Industrial Revolution in the Eighteenth Century.* London: Cape, 1961.

Musson, A. E., and Eric Robinson. *Science and Technology in the Industrial Revolution.* Manchester: Manchester UP, 1969.

Needham, Joseph. *The Grand Titration: Science and Society in East and West.* London: Allen and Unwin, 1969.

——. "Science and China's Influence on the World." *The Legacy of China.* Ed. Raymond Dawson. Oxford: Oxford UP, 1964.

——. *Science and Civilisation in China, Vol. 2: History of Scientific Thought.* Cambridge: Cambridge UP, 1956.

Owen, Robert. *Observations on the Effect of the Manufacturing System.* London: Richard and Arthur Taylor, 1815.

Peide, Diao. "Notes on the Symposium Probing the Causes of Backwardness of Science and Technology in China in Modern Times." *J of Dialectics of Nature* 5.1 (1983): 32-34.

Plato. *Plato's Cosmology.* Trans. F. M. Cornford. New York: Harcourt, Brace, 1937.

Power, Henry. *Experimental Philosophy.* London: John Martin and James Allestry, 1663.

Price, Derek J. de Solla, ed. *The Equatorie of the Planetis.* Cambridge: Cambridge UP, 1955. Price claims the work for Chaucer.

Renzong, Qui. "Cultural and Intellectual Attitudes That Prevented the Spontaneous Emergence of Modern Science in China." *Time, Science and Society in China and the West: The Study of Time V.* Ed. J. T. Fraser, N. Lawrence, and F. C. Haber. Amherst: U of Massachusetts P, 1986. 181-84.

Rush, Philip, and John O'Keefe. *Weights and Measures.* London: Methuen, 1962.

Sexton, James. "Brave New World and the Rationalization of Industry." *English Studies in Canada* 12 (December 1986): 429-38.

Singer, Charles, et al., eds. *A History of Technology.* 8 vols. Oxford: Clarendon P, 1954-84.

Sivin, Nathan. "Chinese Alchemy and the Manipulation of Time." *Isis* 67 (1967): 513-27.

——. *Cosmos and Computation in Early Chinese Mathematical Astronomy.* Leiden: E. J. Brill, 1969.

——. "On the Limits of Empirical Knowledge in the Traditional Chinese Sciences." *Time, Science, and Society in China and the West: The Study of Time V.* Ed. J. T. Fraser, N. Lawrence, and F.

C. Haber. Amherst: U of Massachusetts P, 1986. 151-69.

Sprat, Thomas. *The History of the Royal Society*. Ed. Jackson I. Cope, and Harold Whitmore Jones. 1667; reprint, St. Louis: Washington U Studies, 1959.

Stephens, C. "'The Most Reliable Time': William Bond, the New England Railroads, and Time Awareness in 19th-Century America." *Technology and Culture* 30 (January 1989): 1-24.

Stevens, W. M. "Compotistica et Astronomica in the Fulda School." *Saints, Scholars and Heroes* 2 (1979): 27-63.

——. "The Figure of the Earth in Isidore's *De natura rerum*." *Isis* 71 (1980): 268-77.

Szamosi, Géza. *The Twin Dimensions: Inventing Time and Space*. New York: McGraw-Hill, 1986.

Wenzhao, Lin. "An Analysis of the Causes of the Failure of Science [in China] to Develop in Modern Times." *J of Dialectics of Nature* 5.1 (1983): 7-14.

Yalan, Chen. "The Effect of Qing Dynasty Autocratic Monarchy on Science and Technology." *J of Dialectics of Nature* 5.3 (1983): 62-68.

Zhenghai, Song, et al. "Why Did Zheng He's Voyage Fail to Lead the Chinese to Make the 'Great Geographical Discovery'?" *J of Dialectics of Nature* 5.4 (1983): 25-31.

E. WESTERN

Ariotti, P. E. "The Concept of Time in Western Antiquity." *The Study of Time II*. Ed. J. T. Fraser, and N. Lawrence. New York: Springer, 1975. 69-80.

Gibbon, Edward. *The History of the Decline and Fall of the Roman Empire*. Ed. J. B. Bury. 7 vols. London: Methuen, 1896.

Haber, Francis C. "Time, Technology, Religion, and Productivity Values in Early Modern Europe." *Time, Science, and Society in China and the West: The Study of Time V*. Ed. J. T. Fraser, N. Lawrence, and F. C. Haber. Amherst: U of Massachusetts P, 1986. 79-92.

Hartog, C. "Time and Metaphor in Gibbon's History." *Clio* 12 (Winter 1983): 153-68.

Herodotus. *The Histories*. Trans. A. de Sélincourt. Harmondsworth: Penguin, 1972.

Hunter, Virginia. *Past and Process in Herodotus and Thucydides*. Princeton: Princeton UP, 1982.

Huppert, George. *The Idea of Perfect History: Historical Erudition and*

Historical Philosophy in Renaissance France. Urbana: U of Illinois P, 1970.

Macey, Samuel L. "Cronus in the Eternal City: Time in Rome and England." *Social Science* 53 (Summer 1978): 139-46.

Phillips, Mark. *Francesco Guicciardini: The Historian's Craft*. Toronto: U of Toronto P, 1977.

Schoeps, H. J. *Vorläufer Spenglers (Studien zum Geschichtspessimismus im 19 Jahrhundert)*. Leiden: E. J. Brill, 1953.

Spengler, Oswald. *The Decline of the West*. 2 vols. New York: Alfred A. Knopf, 1945.

Voegelin, E. "On Hegel—A Study in Sorcery." *The Study of Time I*. Ed. J. T. Fraser, F. C. Haber, and G. H. Müller. New York: Springer, 1972. 418-51.

Wilcox, Donald J. *The Development of Florentine Humanist Historiography during the Fifteenth Century*. Cambridge: Harvard UP, 1969.

——. "Guicciardini and the Humanist Historians." *Annali d'Italianistica* 2 (1984): 19-33.

——. "The Sense of Time in Western Historical Narratives from Eusebius to Machiavelli." *Classical Rhetoric and Medieval Historiography*. Ed. Ernst Breisach. Kalamazoo: Western Michigan UP, 1985. 200-15.

11. HOROLOGY

A. GENERAL WORKS AND HISTORIES

Aked, Charles Kenneth. *Horology in Provincial and Rural Museums.* London: Antiquarian Horological Society, 1974.

Allix, Charles. *Carriage Clocks, Their History and Development.* New York: Apollo, 1974.

Ariotti, P. E. "Celestial Reductionism of Time." *Studi Internazionali di Filosofia* (Turin) (Autumn 1972): 91-120.

——. "The Conception of Time in Late Antiquity." *International Philosophical Quarterly* 12 (December 1972): 526-53.

——. "Toward Absolute Time." *Annals of Science* 30 (1973): 31-50.

Attali, Jacques. *Histoires du temps.* Paris: Fayard, 1982.

Baillie, Granville H. *Clocks and Watches: An Historical Bibliography.* London: NAG P, 1951.

——. *Watchmakers and Clockmakers of the World.* London: NAG P, 1947. See Loomes for Vol. 2.

Basserman-Jordan, Ernst von. *The Book of Old Clocks and Watches.* Trans. H. Alan Lloyd. Ed. Hans von Bertele. 4th ed. London: Allen and Unwin, 1964.

Bedini, Silvio A. *The Scent of Time: A Study of the Use of Fire and Incense for Time Measurement in Oriental Countries.* Philadelphia: American Philosophical Society, 1963.

Bertele, Hans von. "Precision Timekeeping in the Pre-Huygens Era." *Horological Journal* 95 (December 1953): 794-816.

Bolli, Jean-Jacques. *L'aspect horloger des relations commerciales américano-suisses de 1929 à 1950.* La Chaux-de-Fonds, Éditions de la Suisse Horlogère, 1956.

Briatore, Luigi. *Cronologia e tecniche della misura del tempo.* Florence: G. Barbèra, 1976.

Britten, Frederick J. *Britten's Watch and Clock Maker's Handbook, Dictionary and Guide.* 16th ed., revised and edited by Richard Good. London: Eyre Methuen, 1978. The standard guide.

Brusa, Giuseppe. *L'arte dell'orologeria in Europa: Sette secoli di orologi meccanici.* Busto Arsizio: Bramante, 1978.

Bruton, Eric. *Clocks and Watches.* Feltham: Hamlyn, 1968.

——. *Dictionary of Clocks and Watches.* New York: Arco Publishing, 1962.

——. *The History of Clocks and Watches.* London: Orbis, 1979. Includes bibliography.

Burlingame, Roger. *Dictator Clock: 5,000 Years of Telling Time*. New York: Macmillan, 1966.

Cardinal, Catherine. *L'horlogerie dans l'histoire, les arts et les sciences*. Lausanne: Scriptar, 1983.

——. *La montre*. Fribourg: Office du livre, 1985.

Carrera, R. *Les heures de l'amour*. Lausanne: Scriptar, n.d. About erotic watches.

Chamberlain, Paul M. *It's About Time*. New York: Richard R. Smith, 1941.

Chapuis, Alfred. *De horogiis in arte*. Lausanne: Scriptar, 1954.

Church, R. A. "Nineteenth Century Clock Technology in Britain, U.S. and Switzerland." *Economic History Review* 28.4 (November 1975): 616-30.

Cipolla, Carlo. *Clocks and Culture, 1300-1700*. New York: Walker, 1967.

Clutton, Cecil, and George Daniels. *Watches*. 2nd ed. London: Sotheby Parke Bernet, 1979.

Cunynghame, Sir Henry Hardinge Samuel. *Time and Clocks: A Description of Ancient and Modern Methods of Measuring Time*. Detroit: Singing Tree P, 1970.

Cuss, T. P. Camerer. *The Camerer Cuss Book of Antique Watches*. Ed. T. A. Camerer Cuss. Woodbridge, England: Antique Collectors' Club, 1976.

Cutmore, M. *Watches 1850-1980*. Newton Abbot: David and Charles, 1989.

De Carle, Donald. *Complicated Watches and Their Repair*. New York: Bonanza, 1979.

——. *Watch and Clock Encyclopedia*. 2nd ed. London: NAG P, 1959.

Defossez, L. *Les savants du XVIIième siècle et la mesure du temps*. Lausanne: Éditions du Journal Suisse d'Horlogerie et de Bijouterie, 1946.

De Lalande, J.-J. Lefrançois, and Ferdinand Berthoud. *Essai sur l'horlogerie*. 2 vols. Paris, 1763.

Edwardes, Ernest L. *Weight-Driven Chamber Clocks of the Middle Ages and Renaissance*. Altrincham: J. Sherratt, 1965.

Erbrich, Klaus. *Präzisionspendeluhren: von Graham bis Reifler*. Munich: Callwey, 1978.

Fanelli, Joseph. *A Century of Fine Carriage Clocks*. Bronxville: Clock Trade Enterprises, 1987.

Fleet, Simon. *Clocks*. New York: Putnam's, 1961.

G[atty], A[lfred]. *The Bell: Its Origins, History and Uses*. London, 1847.

Gélis, Edouard. *L'horlogerie ancienne: histoire, décor et technique*.

Paris: Grund, 1949.
Good, Richard. *Watches in Colour*. Poole: Blandford P, 1978.
Goolkasian, P. "Picture Differences in Processing Clock Times." *American J of Psychology* 97.2 (Summer 1984): 259-83.
Goudsmit, Samuel A., et al. *Time*. New York: Time-Life Books, 1966.
Guyot, Edmond. *Histoire de la détermination de l'heure*. La Chaux-de-Fonds: La Chambre Suisse de l'Horlogerie, 1968.
Haber, F. C. "The Cathedral Clock and the Cosmological Clock Metaphor." *The Study of Time II*. Ed. J. T. Fraser, and N. Lawrence. New York: Springer, 1975. 399-416.
Haswell, J. Eric. *Horology: The Science of Time Measurement and the Construction of Clocks, Watches and Chronometers*. London: Chapman and Hall, 1928.
Hering, Daniel W. *The Lure of the Clock*. New York: New York UP, 1932.
Hooke, Robert. *Diary*. Ed. H. W. Robinson, and Walter Adams. London: Wykeham Publications, 1968.
Huygens, Christiaan of Zulichem. "Horologium" (1658). Trans. Ernest L. Edwardes. *Antiquarian Horology* 7 (December 1970): 43ff.
Jespersen, James, and Jane Fitz-Randolph. *From Sundials to Atomic Clocks: Understanding Time and Frequency*. New York: Dover, 1982.
Johnson, Chester. *Clocks and Watches*. New York: Odyssey, 1964.
Landes, David S. *Revolution in Time: Clocks and the Making of the Modern World*. Cambridge: Belknap P of Harvard UP, 1983.
Laviolette, Jean G. "De la vente en horlogerie." *Horlogerie Ancienne* 12 (1982): 55-74.
Le Lionnais, François. *Le temps*. Paris: Robert Delpire, 1959.
Lepaute, J.-A. *Traité d'horlogerie*. Paris, 1755.
Lloyd, H. Alan. *The Collector's Dictionary of Clocks*. New York: Barnes, 1964.
——. *Old Clocks*. 4th ed. London: Ernest Benn, 1970.
——. *Some Outstanding Clocks over Seven Hundred Years 1250-1950*. London: Leonard Hill, 1958.
——. "Timekeepers—An Historical Sketch." *The Voices of Time*. Ed. J. T. Fraser. 2nd ed. Amherst: U of Massachusetts P, 1981. 388-400.
Loomes, Brian. *Watchmakers and Clockmakers of the World*. Vol. 2. London: NAG P, 1976. See also under Baillie.
Macey, Samuel L. "The Influence of Horology on Literature." *Expression, Communication and Experience in Literature and Language*. Ed. Ronald G. Popperwell. London: Modern Humanities Research

Association, 1973. 209-12.

Maurice, Klaus, and Peter Heuer. *European Pendulum Clocks*. West Chester: Schiffer, 1988.

Meis, Reinhard. *Pocket Watches*. West Chester: Schiffer, 1987.

——. *Taschenuhren: Von der Halsuhr zum Tourbillon*. Munich: Callwey, 1979.

Milham, Willis I. *Time and Timekeepers*. 2nd ed. New York: Macmillan, 1945.

Morpurgo, Enrico. *L'origine dell'orologio tascabile (The Origin of the Watch)*. Rome: Edizioni "La Clessidra," 1954. Text in Italian and English.

North, J. D. "Monasticism and the First Mechanical Clocks." *The Study of Time II*. Ed. J. T. Fraser, and N. Lawrence. New York: Springer, 1975. 381-98.

Patrizzi, Oswaldo, and Fabienne-X. Sturm. *Montres de fantaisie, 1790-1850*. Geneva: Tribune Editions, 1979.

Planchon, Mathieu. *L'horloge: son histoire rétrospective, pittoresque et artistique*. Paris: Henri Laurens, 1898.

Pomella, Fulgido. *L'orologio da portare addosso*. Ivrea: Priuli & Verlucca, 1978.

Price, D. de Solla. "Clockwork Before the Clock and Timekeepers Before Timekeeping." *The Study of Time II*. Ed. J. T. Fraser, and N. Lawrence. New York: Springer, 1975. 367-80.

Quercetani, Roberto L. *A World History of Track and Field Athletics 1864-1964*. London: Oxford UP, 1964.

Rawlings, A. L. *The Science of Clocks and Watches*. Luling: Caldwell Industries, 1974.

Reid, Thomas. *Treatise on Watch and Clockmaking*. Edinburgh, 1826.

Robertson, J. Drummond. *The Evolution of Clockwork*. London: Cassell, 1931.

Royer, Collard F. B. *Skeleton Clocks*. London: NAG P, 1969.

Saunier, Claudius. *Treatise on Modern Horology in Theory and Practice*. Trans. Julieu Tripplin, and Edward Rigg. London: W. and G. Foyle, 1952.

Smith, Alan, ed. *The Country Life International Dictionary of Clocks*. New York: Putnam's, 1979. Includes bibliography.

——. *The Guinness Book of Clocks*. Enfield: Guinness Superlatives, 1984.

Tait, Hugh. *Clocks and Watches*. Cambridge: Harvard UP, 1983.

Torrens, D. S. "Some Notes on the History of Machine Watchmaking." *Horological J* 89 (April 1947): 177-84.

Tyler, E. J. "The Earliest Known Clock to Show Hours, Minutes and

Seconds [c. 1540]." *Horological J* 115 (December 1972): 19.
Ullyett, Kenneth. *Clocks and Watches*. London: Hamlyn, 1971.
——. *In Quest of Clocks*. London: Hamlyn, 1968.
Ungerer, Alfred. *Les horloges astronomiques et monumentales les plus remarquables de l'antiquité jusqu'à nos jours*. Strasbourg, 1931.
Ward, F. A. B. "The Earliest Illustration of a Watch." *Antiquarian Horology* 4 (December 1964): 278, and front cover.
——. *Time Measurement*. London: Science Museum, 1947.
Whitney, Marvin E. *The Ships Chronometer*. Cincinnati: American Watchmakers Institute, 1984.
Wins, Alphonse. *L'horloge à travers les âges*. Mons: Léon Dequesne, 1924.
Wood, Edward J. *Curiosities of Clocks and Watches*. 1866; reprint, Wakefield: EP Publishing, 1973.
"World Watch Production Increases: Total Market Value Declines." *Horological J* 129 (May 1987): 28-29.
Young, Edward. *Labor in Europe and America: A Special Report on the Rates of Wages, the Cost of Subsistence, and the Condition of the Working Classes in Great Britain, Germany, France, Belgium, and Other Countries of Europe; also in the United States and British America*. Washington, 1875. On p. 608, Young estimates current watch production and values.

B. AUTOMATA

Bedini, Silvio A. "The Role of Automata in the History of Technology." *Technology and Culture* 5 (1964): 24-42.
Chapuis, Alfred, and Edmond Droz. *Automata*. Trans. Alec Reid. Neuchâtel: Griffon, 1958.
Chapuis, Alfred, and Edouard Gélis. *Le monde des automates: étude historique et technique*. 2 vols. Paris: privately printed, 1928.
Cohen, J. *Human Robots in Myth and Science*. London: Allen and Unwin, 1966.
A Loan Exhibition of Antique Automatons... November 3-December 5, 1950. New York: À La Vieille Russie, Inc., n.d.
Neumann, John von. *Theory of Self-Reproducing Automata*. Ed. and completed by A. W. Burns. Urbana: U of Illinois P, 1969.
Perregaux, Charles, and F.-Louis Perrot. *Les Jaquet-Droz et Leschot*. Neuchâtel: Attinger Frères, 1916.
Perrot, Diane. "*The Grotto* the Long Lost Automata by Jaquet Droz." *Antiquarian Horology* 5 (December 1966): 170-72.

Price, Derek de Solla. "Automata and the Origins of Mechanism and Mechanistic Philosophy." *Technology and Culture* 24 (1964): 9-23.

Vaucanson. *An Account of the Mechanism of an Automaton....* Trans. J. T. Desaguliers. London: Stephen Varillon, 1742.

C. FRANCE

Bassermann-Jordan, Ernst von. *Die Standuhr Philipps des Guten von Burgund.* Leipzig: W. Diebener, 1927.

Beeson, C. F. C. "Perpignan 1356 and the Earliest Clocks." *Antiquarian Horology* 7 (June 1970): 408-14.

——. *Perpignan 1356: The Making of a Tower Clock and Bell for the King's Castle.* London: Antiquarian Horological Society, 1983.

Chapiro, Adolphe. "Jean-Antoine Lépine, 1720-1814, an 'Unknown' Maker." *Antiquarian Horology* 9 (September 1975): 443-54.

Chapuis, Alfred. *A.-L. Breguet pendant la Révolution française à Paris, en Angleterre et en Suisse.* Neuchâtel: Éditions du Griffon, 1953.

Conservatoire National des Arts et Métiers (Paris museum). *Catalogue, Section JB: Horlogerie.* Paris, 1947.

——. *Horloges et automates, septembre-octobre, 1954.* Paris: Les Presses Artistiques, 1954.

Daniels, George. *The Art of Breguet.* New York: Sotheby Parke Bernet, 1975.

Develle, E. *Les horlogers blésois au XVIième et au XVIIième siècle.* 3rd ed. Nogent-le-Roi: Librairie des Arts et Métiers, 1978.

Edey, Winthrop. *French Clocks.* New York: Walker, 1967.

Girod, François. "Un précurseur: Frédéric Japy (1749-1812)." *Association Nationale des Collectionneurs et Amateurs d'Horlogerie Ancienne* 14 (May 1975): 37.

Isambert-Jamati, Viviane. *L'industrie horlogère dans la région de Besançon: étude sociologique.* Paris: Presses Universitaires, 1955.

Lebon, E. *Études historiques, morales et statistiques sur l'horlogerie en Franche-Comté.* Besançon, 1860.

Marshall, W. "Voltaire: Politician, Statesman, Financier, Author." *Horological J* 113 (July 1971): 19-20.

Maurice, Klaus. *Die französische Pendule des 18. Jahrhunderts: Ein Beitrag zu ihrer Ikonologie.* Berlin: Gruyter, 1967.

Megnin, Georges. *Naissance, développement et situation actuelle de l'industrie horlogère à Besançon.* Besançon, 1909.

Panicali, Roberto. *Watch Dials of the French Revolution 1789-1800.* Lausanne: Scriptar, 1972.

Roblot, Charles. *Collection de cadrans de montres*. Paris, 1900.

Sabrier, J. D. "Histoire de l'échappement à ancre: les montres à ancre françaises de la première génèration." *Association Nationale des Collectionneurs et Amateurs d'Horlogerie Ancienne* 23 (December 1978): 45-60.

Sandoz, Charles. *Origine et développement de l'industrie horlogère à Besançon et en Franche Comté*. Besançon, 1893.

Savoye, Charles. *Notice sur l'origine de la fabrique d'horlogerie de Besançon*. Paris, 1878.

Schwilguè, Charles. *Kurze Beschreibung der astronomischen Uhr des Strassburger Münsters*. Strasbourg: L. F. LeRoux, 1842.

Sully, Henry. *Règle artificielle du temps*. New ed., revised and enlarged by Julien Le Roy. Paris: Grégoire Dupuis, 1737.

Ungerer, Alfred. *L'horloge astronomique de la Cathédrale de Strasbourg*. Paris, 1922.

Ungerer, Theodore, and l'abbé André Glory. "L'astrologue au cadran solaire de la Cathédrale de Strasbourg (1493)." *Archives Alsaciennes d'Histoire de l'Art* 12 (1933): 73-108.

Vial, Eugène, and Claudius Côte. *Les horlogers lyonnais de 1550 à 1650*. Mâcon: Protat Frères, 1927.

Walter, Joseph. "Le Mystère 'Stella' des trois mages joué à la Cathédrale de Strasbourg au XIIième siècle." *Archives Alsaciennes d'Histoire de l'Art* 8 (1929): 39-50.

D. GERMANY, AUSTRIA, AND THE NETHERLANDS

Coole, P. G., and E. Neumann. *The Orpheus Clocks*. London: Hutchinson Educational, 1972.

Friedrich, Carl. "Die Augsburger Uhrmacherei während des 18. Jahrhunderts." *Allgemeines J der Uhrmacherkunst* (Leipzig) 11 (1886): nos. 24 ff.

Groiss, Eva. "Das Augsburger Uhrmacher-Handwerk." *Die Welt als Uhr: Deutsche Uhren und Automaten, 1550-1650*. Ed. Klaus Maurice, and Otto Mayr. Munich: Bayerisches Nationalmuseum, 1980.

Kahlert, Helmut. *Bibliographie zur Schwarzwalduhr*. Furtwangen: Deutsches Uhrenmuseum Furtwangen, 1984.

——. *300 Jahre Schwarzwälder Uhrenindustrie*. Gernsbach: C. Katz, 1986.

Kochmann, Karl. *The Black Forest Cuckoo Clock: European Clockmaking*. 4th ed. Concord: Antique Clocks Publishing, 1983.

Mackensen, Ludolf von. *Die erste Sternwarte Europas mit ihren Instrumenten und Uhren: 400 Jahre Jost Bürgi in Kassel*. Munich: Callwey,

1979. Includes contributions by Hans von Bertele and John H. Leopold.

Maurice, Klaus. *Die deutsche Räderuhr: Zur Kunst und Technik des mechanischen Zeitmessers im deutschen Sprachraum*. 2 vols. Munich: Beck, 1976.

——. *Zeit von den Gestirnen: Sonnen-, Mond- und Sternuhren aus drei Jahrhunderten*. Munich: Bayerische Versicherungskammer und Nationalmuseum, 1978.

Maurice, Klaus, and Otto Mayr, eds. *The Clockwork Universe: German Clocks and Automata 1550-1650*. New York: Neal Watson, 1980.

Neher, F. L. *Ein Jahrhundert Junghans: Ein Beitrag zur Technik- und Kulturgeschichte der Uhr*. Schramberg, [1961].

Spierdijk, C. *Klokken en Klokkenmakers, 1300-1900*. Amsterdam: De-Bussy, 1962.

Walther, Susanne, and Franz Sharinger. *Horological Museum, Vienna*. Vienna: Museum's Own Publications, 1980.

Wieschebrink, Theodor. *Die astronomische Uhr im Dom zu Münster*. Ed. Erich Hüttenhain. Münster: Aschendorff, 1968.

Zinner, E. *Deutsche und niederländische astronomische Instrumente des 11.-18. Jahrhunderts*. 2nd ed. 1967; reprint, Munich: Beck, 1979.

E. GREAT BRITAIN

Aked, Charles K. *Complete List of English Horological Patents up to 1853*. Ashford: Brant Wright, 1975.

——. "William Derham and *The Artificial Clockmaker*." *Antiquarian Horology* 6 (March, June, and September 1970): 362-69, 416-27, and 495-505.

Allix, Charles. "Mudge Milestones: Watch Dates." *Antiquarian Horology* 12 (Summer 1981): 627-34.

Atkins, Samuel E., and William Henry Overall. *Some Account of the Worshipful Company of Clockmakers of the City of London*. [London]: privately printed, 1881.

Beeson, C. F. C. *Clockmaking in Oxfordshire, 1400-1850*. Ticehurst: Antiquarian Horological Society, 1962.

——. *English Church Clocks, 1280-1850*. London: Antiquarian Horological Society, 1971.

Bird, Anthony. *English House Clocks 1600-1850*. New York: Arco Publishing, 1973.

Bromley, John, ed. *The Clockmakers' Library: The Catalogue of the Books and Manuscripts in the Library of the Worshipful Company of*

Clockmakers. London: Sotheby Parke Bernet, 1977.

Bruton, Eric. *The Longcase Clock*. New York: Arco Publishing, 1964.

Burchall, A. "Pitt's Clock and Watch Tax: A Reassessment." *Antiquarian Horology* 13 (September 1982): 454-61.

Cliborne, A. F. "Mudge Milestones: Addenda." *Antiquarian Horology* 13 (December 1981): 144-45.

Daniell, J. A. *The Making of Clocks and Watches in Leicestershire and Rutland*. Leicester: Leicestershire Archaeological Society, 1951.

Daniels, George. "Thomas Mudge, the Complete Horologist." *Antiquarian Horology* 13 (December 1981): 150-73.

Dawson, Percy G. *The Iden Clock Collection*. Woodbridge, Suffolk: Antique Collector's Club, 1987.

D[erham], W[illiam]. *The Artificial Clockmaker: A Treatise of Watch and Clock-Work*. 2nd ed., enlarged. London: James Knapton, 1700. The first edition was published in 1696.

Ellmers, Chris. "The Impact of the 1797 Tax on Clocks and Watches on the London Trade." *Collectanea Londiniensia: Studies in London Archaeology and History Presented to Ralph Merrifield*. Ed. Joanna Bird, Hugh Chapman, and John Clark. [London]: London and Middlesex Archaeological Society, 1978.

Epinasse, Margaret. *Robert Hooke*. London: Heinemann, 1956.

Ferriday, Peter. *Lord Grimthorpe, 1816-1905*. London: Murray, 1957.

Good, Richard. "The First Lever Watch Made by Thomas Mudge." *Pioneers of Precision Timekeeping*. Ed. Cecil Clutton and Humphrey Quill. N.p.: Antiquarian Horological Society, monograph no. 3, n.d. 31-45.

——. "Watch by Thomas Mudge, London No. 574 with Perpetual Calendar Mechanism." *Antiquarian Horology* 13 (December 1981): 178-87.

Hall, A. R. "Robert Hooke and Horology." *Notes and Records* (The Royal Society, London) 7 (1950): 166-77.

Howse, Derek. "The Tompion Clocks at Greenwich and the Dead-Beat Escapement." *Antiquarian Horology* 7 (December 1970, and March 1971): 18-34, and 114-33.

Jagger, Cedric. *The Artistry of the English Watch*. Rutland: C. E. Tuttle, 1988.

——. *Royal Clocks: British Monarchy and Its Timekeepers 1300-1900*. Philadelphia: Trans-Atlantic Publications, 1983.

Lloyd, H. Alan. "Samuel Watson." *Antiquarian Horology* 1 (December 1954): 60-61.

Loomes, Brian. *Complete British Clocks*. Newton Abbot: David and Charles, 1978.

——. *The Early Clockmakers of Great Britain*. London: NAG P, 1981.

——. *Grandfather Clocks and Their Cases*. Newton Abbot: David and Charles, 1985.

——. *Lancashire Clocks and Clockmakers*. Newton Abbot: David and Charles, 1975.

——. *Yorkshire Clockmakers*. Clapham, Yorkshire: Dalesman Books, 1972.

Mercer, Vaudrey. *The Frodshams*. London: Antiquarian Horological Society, 1981.

——. *The Life and Letters of Edward John Dent, Chronometer Maker, and Some of His Successors*. London: Antiquarian Horological Society, 1977.

Millburn, John R. "Some Horological Extracts from Stukeley's Diaries." *Antiquarian Horology* 6 (September 1969): 206-11.

Newman, Bernard. *Clockmaking and Watchmaking in Colchester*. Feltham: Country Life Books, 1969.

North, John D. *Richard of Wallingford*. 3 vols. Oxford: Clarendon P, 1974.

Ponsford, Clive N. *Devon Clocks and Clockmakers*. Newton Abbot: David and Charles, 1985.

Quare, Daniel. "For to Keep a Pendulum Watch in Good Order." *Horological J* 121 (July 1978): 4-7. First published about 1700.

Roberts, Derek. *British Skeleton Clocks*. Ithaca: Antique Collectors' Club, 1987.

Robinson, T. *The Longcase Clock*. Ithaca: Antique Collectors' Club, 1981.

Rose, Ronald E. *English Dial Clocks*. 2nd ed. Ithaca: Antique Collectors' Club, 1988.

Shenton, Rita. *Christopher Pinchbeck and His Family*. Ashford: Brant Wright, 1976.

Smith, Alan, ed. *A Catalogue of Tools for Watch and Clock Makers by John Wyke of Liverpool*. Charlottesville: U of Virginia P, 1978.

Smith, John. *Horological Dialogues*. London, 1675.

Symonds, R. W. *Thomas Tompion: His Life and Work*. London: Spring Books, 1969.

Turner, Anthony J. "Christopher Wren and the Wadham Clock." *Antiquarian Horology* 7 (June 1971): 229-30.

Ward, F. A. B. *Time Measurement: Descriptive Catalogue of the Collection, Science Museum*. London: Her Majesty's Stationery Office, 1966.

Watson, E. "The St. Alban's Clock." *Antiquarian Horology* 1 (Winter 1979): 576-84.

W[ay], A. "Original Documents: Agreement between the Dean and Chapter of St. Paul's, London, and Walter the Orgoner...Communicated by Sir Frederic Madden, K. H." *Archaeological J* 12 (1855): 173-77. About expenses related to the Norwich clock.

Weiss, Leonard. *Watchmaking in England 1760-1820*. London: Robert Hale, 1982.

White, George. *English Lantern Clocks*. Ithaca: Antique Collectors' Club, 1989.

F. ITALY AND SPAIN

Aked, Charles K. "Catalan Clocks and Clockmakers to the End of the Middle Ages." *Antiquarian Horology* 10 (Spring 1978): 722-27.

Bedini, Silvio A. *XIVth and XVth Century Public Clocks of the Papal Marches*. Florence: L. S. Olschki, 1962.

——. "The Instruments of Galileo Galilei." *Galileo, Man of Science*. Ed. E. McMullin. New York: Basic Books, 1967.

Bedini, Silvio A., and Francis R. Maddison. "Mechanical Universe: The Astrarium of Giovanni de Dondi." *Transactions of the American Philosophical Society* n.s., 56.5 (1966): 5-69.

Dondi, Giovanni dall'Orologio. *Tractatus astrarii...* (Biblioteca capitolare de Padova, Cod. D 39). Ed. Antonio Barzon, et al. Vatican City: The Vatican, 1960.

Galilei, Galileo. *L'usage du cadran ou de l'horloge physique universal*. Paris, 1639.

Lloyd, H. Alan. "Giovanni de Dondi's Horological Masterpiece." *La Suisse Horlogère* (International Edition) 70 (July 1955): 49-71.

Simoni, Antonio. *Orologi italiani dal cinquecento all'ottocento*. N.p.: Antonio Vallardi Editore, 1965.

Turner, Anthony J. "The Tragicall History of Giovanni de' Dondi." *J of the History of Astronomy* 6 (1975): 126-31.

G. NEAR AND FAR EAST

Bedini, Silvio A. "Oriental Concepts of the Measure of Time." *The Study of Time II*. Ed. J. T. Fraser, and N. Lawrence. New York: Springer, 1975. 451-84.

Bernard, Henri. "Les adaptations chinoises d'ouvrages européens 1514-1688." *Momumenta Serica*. Peking, 1945.

Bonnant, Georges. "The Introduction of Western Horology in China."

La Suisse Horlogère (International Edition) 74.3 (1959): 28-38.

Borchardt, L., ed. *Altaegyptische Zeitmessung*. Berlin: W. de Gruyter, 1920.

Chapuis, Alfred. *La montre chinois*. Neuchâtel: Attinger Frères, 1919.

——. "Montres et pendules 'chinoises' et chinoises." *J Suisse d'Horlogerie et de Bijouterie* 12 (December 1932): 248-49.

——. *Relations de l'horlogerie suisse avec la Chine: la montre "chinoise."* Neuchâtel: Attinger Frères, [1919].

Combridge, John H. "Chinese Steelyard Clepsydras." *Antiquarian Horology* 12 (Spring 1981): 530-35.

D'Elia, Pasquale M. *Storia dell'introduzione del Cristianesimo in Cina, scritta da Matteo Ricci*. In *Fonte ricciane*. Ed. Pascale M. D'Elia. Vol. 1. Rome: Libreria dello Stato, 1942.

Dennis, Faith. "Some Jewelled 'Toys' of Georgian London." *Bulletin of the Metropolitan Museum of Art* 5.6 (February 1947): 164-68.

Enshoff, D. "Ricci's Uhren." *Die Katolischen Missionen* 65 (1937): 190-94.

Greenberg, Michael. *British Trade and the Opening of China, 1800-42*. Cambridge: Cambridge UP, 1952.

Harcourt-Smith, Simon. *A Catalogue of Various Clocks, Watches, Automata...in the Palace Museum and the Wu Ying Tien, Peiping*. Peking: Palace Museum, 1933.

Hayward, John. "English Clocks for China." *Antiques* (July 1959): 46-49.

Hill, Donald R. *Arabic Water Clocks* (Sources and Studies in the History of Arabic-Islamic Science: History of Technology Series, number 4). Aleppo: U of Aleppo, 1981.

Jaeger, Fritz. *Das Buch von den wonderlichen Maschinen in Asia Major*. Leipzig, 1944.

Kurz, O. *European Clocks and Watches in the Near East*. London: The Warburg Institute, 1975.

Lloyd, H. Alan. "English Clocks for the Chinese Market." *Antique Collector* (February 1951): 25-29.

MacGowan, D. J. "Modes of Timekeeping Known Among the Chinese." *Patent Office Report*. Washington: Government Printing Office, 1851.

Mody, N. H. N. *Japanese Clocks*. London: Routledge and Kegan Paul, 1968.

Needham, Joseph, Wang Ling, and Derek J. de Solla Price. *Heavenly Clockwork: The Great Astronomical Clocks of Medieval China*. Cambridge: Cambridge UP, 1960.

Pfister, Louis. *Notices biographiques et bibliographiques sur les Jesuites*

des anciens missions de Chine, 1553-1773. Shanghai, 1932-34.
Read, John. "Daimyo Clocks." *Horological J* 130 (June 1988): 12-15, 20-21.
Schoy, Karl. *Sonnenuhren der spätarabischen Astronomie*. Brussels: M. Weissenbruch, 1924.
——. *Über den Gnomonschatten und die Schattentafeln der arabischen Astronomie*. Hanover: H. LaFaire, 1923.
White, William C., and P. M. Millman. *An Ancient Chinese Sundial*. Toronto: U of Toronto P, 1938.
Wiedemann, Eilhard E. G. *Über die Uhren im Bereich der islamischen Kultur*. Halle: E. Karras, 1915.
Yamaguchi, Ryuji. *The Clocks of Japan*. Tokyo: Nippon Hyoron-Sha Publishing, 1950.
Yang, Yuming. *L'industrie horlogère suisse et ses relations avec la Chine de 1949 à 1982*. Geneva: Institut universitaire de hautes études internationales, 1984.

H. NORTH AMERICA (UNITED STATES)

Abbott, Henry G. *The Watch Factories of America Past and Present*. Chicago, 1888.
Andrews, William, and Seth Atwood. *The Time Museum: An Introduction*. Rockford: The Time Museum, 1983.
Bailey, Chris H. *Two Hundred Years of American Clocks and Watches*. Englewood Cliffs: Prentice-Hall, 1975.
Battison, Edwin A. *The American Clock 1725-1865*. Greenwich: New York Graphic Society, 1973.
Bolino, August C. *The Watchmakers of Massachusetts*. Washington: Kensington Historical P, 1987.
Distin, William. *The American Clock*. New York: Bonanza Books, 1983.
Drepperd, Carl W. *American Clocks and Clockmakers*. New York: Doubleday, 1947.
Eckhardt, George H. *Pennsylvania Clocks and Clockmakers: An Epic of Early American Science, Industry and Craftsmanship*. New York: Devin-Adair, 1955.
——. *United States Clock and Watch Patents 1790-1890: The Record of a Century of American Horology and Enterprise*. New York: privately printed, 1960.
Faber, Edward, et al. *American Wristwatches*. West Chester: Schiffer, 1988.

Gibbs, James W. "George Washington's Watch." *Bulletin of the National Association of Watch and Clock Collectors* 15 (December 1971): 12-13.

——. *Pennsylvania Clocks and Watches*. University Park: Pennsylvania State UP, 1984.

Harris, H. R. *Nineteenth Century American Clocks*. White Plains: Emerson Books, 1981.

Hering, Daniel Webster. *The Lure of the Clock: An Account of the James Arthur Collection of Clocks and Watches at New York University*. New York: New York UP, 1932.

Hoopes, Penrose R. *Connecticut Clockmakers of the Eighteenth Century*. Hartford: E. V. Mitchell, 1930.

Howard, Robert A. "Interchangeable Parts Reexamined: The Private Sector of the American Arms Industry on the Eve of the Civil War." *Technology and Culture* 19.4 (1978): 633-49.

Husher, R. W., and W. W. Welch. *A Study of Simon Willard's Clocks*. Nahant: Husher and Welch, 1980.

Jerome, Chauncey. *History of the American Clock Business for the Past Sixty Years and Life of Chauncey Jerome, Written by Himself*. New Haven, 1860.

Ly, Tran Duy. *Seth Thomas Clocks and Movements*. Arlington: Arlington Book Company, 1985.

Marsh, E. A. *Watches by Automatic Machinery at Waltham*. Chicago, 1896.

Maust, Don. *Early American Clocks: A Collection of Essays on Early American Clocks and Their Makers*. Uniontown: E. G. Warman, 1971.

Moore, Charles W. *Timing a Century: History of the Waltham Watch Company*. Cambridge: Harvard UP, 1945.

Mussey, Barrows. *Young Father Time (Eli Terry)*. New York: Newcomer Society in North America, 1950.

New Haven Clock Company. *Illustrated Catalogue of Clocks Manufactured by the New Haven Clock Company*. Bristol: American Clock and Watch Museum, 1976.

Palmer, Brooks. *The Book of American Clocks*. New York: Macmillan, 1959.

——. *A Treasury of American Clocks*. New York: Macmillan, 1967.

Pertuch, Walter A. R., and Emerson W. Hilker. *Horological Books and Pamphlets in the Franklin Institute Library*. 2nd ed. Philadelphia: Franklin Institute, 1968.

Sanderson, Edmund L. *Waltham Industries: A Collection of Sketches of Early Firms and Founders*. Waltham: Waltham Historical Society,

1957.
Shaffer, Douglas H. *Clocks*. Ed. Brenda Gilchrist. New York: Cooper-Hewitt Museum, 1980.
Sketch, Carolyn Wood. "Early Colonial Clockmakers in Philadelphia." *Pennsylvania Magazine of History and Biography* 56 (1932): 225-35.
Sposato, Kenneth A. *The Dictionary of American Clock and Watch Makers*. White Plains: K. A. Sposato, 1983.
Swedberg, Robert W., and Hariett Swedberg. *American Clocks and Clockmakers*. Radnor: Wallace-Homestead, 1989.
Townsend, George E. *Almost Everything You Wanted to Know about American Watches and Didn't Know Who to Ask*. Privately published, 1971.
——. *American Railroad Watches*. Privately published, 1977.
Treiman, Lawrence W. "Railroad Watches and Time Service." *Bulletin of the National Association of Watch and Clock Collectors, Inc.* 15 (October 1972): 151-75.
Turner, Anthony. *Time Museum Catalogue of the Collection*. Vol. 1, *Time Measuring Instruments*. Part 1, *Astrolabes and Related Devices*. Ed. Bruce Chandler. Rockford: Time Museum, 1985.
——. *Time Museum Catalogue of the Collection*. Vol. 1, *Time Measuring Devices*. Part 3, *Water-Clocks, Sand-Glasses, Fire-Clocks*. Ed. Bruce Chandler. Rockford: Time Museum, 1984.
United States Tariff Commission. *Watches* (War Changes in Industry Series, report no. 20). Washington: Government Printing Office, 1947.
Varkaris, Jane. *The Canada and Hamilton Clock Companies*. Erin, Ontario: Boston Mills P, 1986.
Waldo, L. "The Mechanical Art of American Watchmaking." *J. Society of Arts* 34 (1886): 740-51.
Whitton, Blair. *American Clockwork Toys*. West Chester: Schiffer, 1981.
Willard, John Ware. *Simon Willard and His Clocks*. New York: Dover, 1968.
Woodbury, R. S. "The Legend of Eli Whitney and Interchangeable Parts." *Technology and Culture* 1 (Summer 1960): 235-53.

I. SWITZERLAND

Babel, Antony. *Les métiers dans l'ancienne Genève: histoire corporative de l'horlogerie, de l'orfèvrerie et des industries annexes*. Geneva: A. Kündig, 1916.
Bachelin, A. *L'horlogerie neuchâteloise*. Neuchâtel: Attinger Frères,

1888.

Chapuis, Alfred. *Grands artisans de la chronométrie: histoire de l'horlogerie au Locle*. Neuchâtel: Éditions du Griffon, 1958.

——, ed. *L'horlogerie: une tradition helvétique*. Neuchâtel: Éditions de la Bourgade, 1948.

——. *Montres et émaux de Genève: Louis XIV, Louis XV, Louis XVI et Empire*. Geneva: Rolex, 1944. Based on the Wilsdorf collection.

Collection de montres et automates Maurice et Edouard M. Sandoz. In *Histoire de l'horlogerie*. Le Locle: Musée d'Horlogerie, 1976. Fascicle 3.

Fallet-Scheurer, M. *Le travail à domicile dans l'horlogerie suisse et ses industries annexes*. Bern: Imprimerie de l'Union, 1912.

Françillon, André. *Histoire de la fabrique de Longines*. Saint-Imier: privately published, 1947.

Guye, S. "The Vital Necessity and the History of Mechanization in the Swiss Watch Industry." *La Suisse Horlogère et Revue Internationale de l'Horlogerie* 80 (December 1965): 12-18.

Guyot, Edmond. *Dictionnaire des termes utilisés dans la mesure du temps*. La Chaux-de-Fonds: La Chambre Suisse de l'Horlogerie, 1953.

Huber, Martin. *Patek Philippe, Genève*. Zürich: Verlag P. Ineichen, 1982.

Jaquet, Eugène, and Alfred Chapuis. *Technique and History of the Swiss Watch*. 2nd ed. London: Spring Books, 1970.

Jequier, François. *Une enterprise horlogère du Val-de-Travers: Fleurier Watch Co SA*. Neuchâtel: La Baconnière, 1972.

——. *De la forge à la manufacture horlogère*. Lausanne: Bibliothèque historique vaudoise, 1983.

Pfleghart, A. *Die schweizerischen Uhrenindustrie: Ihre geschichtliche Entwicklung und Organisation*. Leipzig: Duncker und Humblot, 1908.

Pinot, Robert. *Paysans et horlogers jurassiens*. Geneva: Gronauer, 1979.

Sabrier, Jean-Claude, and Andre Imbert. "An Early Anonymous Swiss Self-Winding Watch." *Antiquarian Horology* 7 (March 1972): 526-27.

Scheurer, F. *Les crises de l'industrie horlogère dans le canton de Neuchâtel*. Neuveville: Éditions Beerstecher, 1914.

Stetter, Hilmar. *L'exportation du Swiss Made: transfert de l'appareil de production industrielle dans le Tiers Monde*. Geneva: Éditions Pierre-Marcel Favre, 1981.

Thomann, Charles. *Les dignitaires de l'horlogerie*. Neuchâtel: Éditions du Griffon, 1981.

Tissot, André. *Voyage de Pierre Jaquet-Droz à la cour du roi d'Espagne, 1758-1759*. Neuchâtel: Éditions de la Baconnière, 1982.

Weiss, Rolf. *Uhrensammlung Kellenberger, Winterthur*. Basel: Gesellschaft für Schweizerische Kunstgeschichte, 1974.

J. TIMEKEEPERS: SPECIFIC INVENTIONS AND TRADES

Ahrens, Gerd. "Frühe Lochsteine aus Rubin." *Chronométrophilia* (La Chaux-de-Fonds) 6 (May 1979): 53-56.

Ariotti, P. E. "Aspects of the Conception and Development of the Pendulum in the Seventeenth Century." *Archive for History of Exact Sciences* 8 (1972): 330-407.

Atwood, S. G. "The Development of the Pendulum as a Device for Regulating Clocks Prior to the 18th Century." *The Study of Time II*. Ed. J. T. Fraser, and N. Lawrence. New York: Springer, 1975. 417-50.

Bertele, Hans von. "The Origin of the Differential Gear." *Antiquarian Horology* 2 (December 1956): 14-15.

——. *Zur Geschichte der Äquationsuhren Entwicklung*. Vienna: Springer, 1956.

Caillard, Bernard. "The History of the Pendulum Watch." *Antiquarian Horology* 3 (March 1960): 41-43.

Chapuis, Alfred, and Eugène Jaquet. *The History of the Self-Winding Watch, 1770-1931*. Neuchâtel: Éditions du Griffon, 1956.

Daumas, Maurice. *Scientific Instruments of the Seventeenth and Eighteenth Centuries and their Makers*. Trans. Mary Holbrook. London: B. T. Batsford, 1972.

Davies, W. O. *Gears for Small Mechanisms*. London: NAG P, 1970.

Dobson, R. D. "The Development of the Pendulum Clock." *Antiquarian Horology* 13 (March 1982): 270-79.

——. "Huygens, the Secret in the Coster-Fromanteel 'Contract': The Thirty-Hour Clock." *Antiquarian Horology* 12 (Summer 1980): 192-96.

Drover, C. B. "The Brussels Miniature." *Antiquarian Horology* 3 (September 1962): 357-61.

Les ébauches: ouvrage publié à l'occasion du 25ième anniversaire d'Ébauches S. A. Neuchâtel: Éditions de la Baconnière, [1951].

Edwardes, Ernest L. *The Story of the Pendulum Clock*. Altrincham: J. Sherratt, 1977.

——. "The Suspended Foliot and New Light on Early Pendulum Clocks." *Antiquarian Horology* 12 (June 1981): 614-26, and 634.

Fried, Henry B. *Bench Practices for Clockmakers*. Cincinnati: American Watchmakers Institute, 1984.

Hurst, Michael. "The First Twelve Years of the Pendulum Clock." *Antiquarian Horology* 6 (June 1969): 146.

Huygens, Christiaan. *Horologium Oscillatorium*. Paris, 1673.

Leiter, Alfred, and Alma Helfrich Dorner. *Email-Malerei: Kostbarkeiten unter den Taschenuhren: Zur Technik und Geschichte der Email-Uhren*. Kornwestheim: Werbe-Verlag, 1977.

Michel, Henry. "Some New Documents in the History of Horology." *Antiquarian Horology* 3 (March 1962): 288-91.

Pellaton, James C. *Watch Escapements*. London: NAG P, 1949.

Thorndike, Lynn. "Invention of the Mechanical Clock about 1271 A.D." *Speculum* 16 (April 1941): 242-43.

Wadsworth, Francis. "A History of Repeating Watches." *Antiquarian Horology* 6 (September 1965): 364-67; and 5 (December 1965, March 1966, and June 1966): 24-26, 48-52, and 90-92.

The Watch Escapement. Cincinnati: American Watchmakers Institute, 1984.

Wenzel, J. "Equation Clocks." *Antiquarian Horology* 13 (September 1981): 24-43.

K. TIMEKEEPERS (NON-MECHANICAL): SUNDIALS, SANDGLASSES, AND CLEPSYDRAE

Bedini, Silvio A. "The Compartmentalized Cylindrical Clepsydrae." *Technology and Culture* 3 (1962): 115-41.

The Book of Old Sundials and Their Mottoes. Edinburgh: T. N. Foulis, 1922.

Boursier, Charles. *800 devises de cadrans solaires*. Paris: Berger-Levrault, 1936.

Cole, T. W. *Origin and Use of Church Scratch-Dials*. Wimbledon: Hill Bookshop, 1935.

Combridge, John H. "The Celestial Balance: A Practical Reconstruction." *Horological J* 104 (February 1962): 82-86.

Cousins, Frank W. *Sundials: A Simplified Approach by Means of the Equatorial Dial*. London: J. Baker, 1969.

Drover, C. B. "A Medieval Monastic Water-Clock." *Antiquarian Horology* 1 (December 1954): 56.

——. "Sand-Glass 'Sand.'" *Antiquarian Horology* 3 (June 1960): 62-67.

Earle, Alice Morse. *Sundials and Roses of Yesterday*. 1902; reprint, Detroit: Singing Tree P, 1969.

Gouk, Penelope. *The Ivory Sundials of Nurenberg, 1500-1700*. Cambridge: Whipple Museum of the History of Science, 1988.

Green, Arthur R. *Sundials, Incised Dials or Mass-Clocks*. London: Society for Promoting Christian Knowledge, 1926.

Hill, Donald R. *On the Construction of Water Clocks*. London: Turner and Devereux, 1976.

Jünger, Ernst. *Das Sanduhrbuch*. Frankfurt: Klostermann, 1957.

Körber, Hans G. *Zur Geschichte der Konstruktion von Sonnenuhren und Kompassen des 16. bis 18. Jahrhunderts*. Berlin: VEB Deutscher Verlag der Wissenschaften, 1965.

Loske, Lothar M. *Die Sonnenuhren: Kunstwerke der Zeitmessung und ihre Geheimnisse*. 2nd ed. Berlin: Springer, 1970.

Maddison, F., B. Scott, and A. Kent. "An Early Medieval Water-Clock." *Antiquarian Horology* 3 (1962): 348-53.

Marshall, Roy K. *Sundials*. New York: Macmillan, 1963.

Noble, Joseph V., and Derek de Solla Price. "The Water Clock in the Tower of Winds." *American J of Archaeology* 72 (1968): 345-55.

Sonnen- und Turmuhren. Dresden: Sachsenverlag, 1959.

Spackman, Henry S. *The Timepiece of Shadows: A History of the Sun Dial*. New York: W. T. Comstock, 1895.

Waugh, Albert E. *Sundials: Their Theory and Construction*. New York: Dover, 1973.

Zinner, Ernst. *Alte Sonnenuhren an europäischen Gebäuden*. Wiesbaden: F. Steiner, 1964.

L. TIME'S MEASUREMENT: PRECISE NON-MECHANICAL (INCLUDING ATOMIC, QUARTZ, AND ELECTRIC)

Aked, Charles K. "Electricity, Magnetism and Clocks." *Antiquarian Horology* 7 (December 1971): 398-451.

Bain, Alexander. *A Short History of the Electric Clock* (1852). Ed. W. D. Hackmann. London: Turner and Devereux, 1973.

Bay, Z., and J. A. White. "The Speed of Light and the New Meter." *Acta Physiologica Academiae Scientiarum Hungaricae* 36.1 (1974): 91-106.

Beehler, R. E., R. C. Mockler, and J. M. Richardson. "Caesium Beam Atomic Time and Frequency Standards." *Metrologia* 1.3 (1965): 114-31.

Bovay, J.-P. *Omega Sports Timing: Photosprint History*. Bienne: Omega, 1980.

Carroll, J. A. "An Absolute Scale of Time." *Nature* 184 (1959): 260-61. About atomic clocks and radioactive decay.

Clemence, G. M. "Time and Its Measurement." *American Scientist* 40.2

(1952): 260-69.

———. "Time Measurement for Scientific Use." *The Voices of Time*. Ed. J. T. Fraser. 2nd ed. Amherst: U of Massachusetts P, 1981. 401-14.

Ditisheim, Paul. *Précision des garde-temps piézo-électriques et des pendules astronomiques....* Besançon, 1937.

Essen, L. "The National Physical Laboratory's Caesium Standard." *Proceedings of the Royal Society*. London, 1947.

Explanatory Supplement to the Astronomical Ephemeris and the American Ephemeris and Nautical Almanac. 4th ed. London: Her Majesty's Stationery Office, 1977. This is the major reference work for scientific time reckoning.

"First All Electronic LCD Watch." *Horological J* 114 (April 1972): 16-18.

Gerharz, R. "Long-Period Time Interval Markers from Mercury Transits." *J. British Astronomical Association* 85.3 (1975): 252-54.

Good, Richard. "An Analysis of the Electric Watch." *Horological J* 99 (July 1957): 406-12.

———. "An Analysis of the LIP Electric Watch." *Horological J* 102 (January 1960, and February 1960): 18-23, 35, and 90-93.

———. "An Analysis of the Swiss Electric Watch." *Horological J* 102 (November 1960): 696-708.

———. "Year of the Quartz Crystal Watch." *Horological J* 112 (May 1970): 5-12.

Hope-Jones, F. *Electrical Timekeeping*. London: NAG P, 1940.

Il'in, V. G., et al. "High Precision Time and Frequency Measurements in the National Economy and Scientific Investigations." *Measurement Techniques* 17.7 (1974): 1033-37.

Janich, Peter. *Protophysics of Time: Constructive Foundation and History of Time Measurement*. Boston: Reidel, 1985.

Kartaschoff, P. *Frequency and Time*. New York: Academic P, 1978.

Kawasumi, K., and K. Ushikoshi. "Analogue Quartz Watches Consuming Low Electric Power." *Proceedings: Tenth International Congress of Chronometry*. Ed. Société Suisse de Chronométrie. Büren, 1979. 2: 313-20.

Kurita, H., et al. "Multifunctional Digital Quartz Watches." *Proceedings: Tenth International Congress of Chronometry*. Ed. Société Suisse de Chronométrie. Büren, 1979. 2:301-307.

Markowitz, William. "Time Measurement Techniques in the Microsecond Region." *Engineer's Digest* 135 (July-August 1962): 9-18.

Martin, D. "Measure Time Interval Precisely." *Electronic Design* 22.24 (1974): 162-67.

"Piezo Electric Quartz Resonators." *Horological J* 109 (April 1967, and

May 1967): 8-10, and 14-15.

Prokhovnik, S. J. "The Nature of Time and Its Measurement." *J and Proceedings. Royal Society of New South Wales* 103.3-4 (1970): 123-26. About arrow of time, cosmic time, relativity, and space-time.

"Quartz Crystal Wristwatch." *Horological J* 110 (March 1968): 14-18.

Ramsey, N. F. "History of Atomic and Molecular Standards of Frequency and Time." *IEEE Transactions on Instrumentation and Measurement* 21.2 (1972): 90-99.

Rondeau, René. *The Watch of the Future: The Story of the Hamilton Electric Watch*. Corte Madera: R. Rondeau, 1989.

Sen, A. K., and B. Saha. "Worldwide Standards of Time and Frequency." *J of Scientific and Industrial Research* 32.6 (1973): 300-11.

Thomson, Malcolm M. *The Beginning of the Long Dash: A History of Timekeeping in Canada*. Toronto: U of Toronto P, 1978.

Ward, F. A. B. "Physics in Time Measurement." *J of Scientific Instruments* 43.8 (1966): 536-40.

Weaver, J. D. *Electrical and Electronic Clocks and Watches*. Feltham: Newnes, 1982.

Wolsky, A. M. "The Scales of Length and Time in Classical and Modern Physics." *American J of Physics* 39.5 (1971): 529-33.

12. LAW

A. GENERAL

Braithwaite, Thomas Wolfe. *A Manual of Times of Procedure in Chancery, Embracing Chiefly the Provisions of the General Rules and Orders of the Court*. London: E. Cox, 1864.

Engel, D. M. "Law, Time, and Community." *Law and Society Review* 21.4 (1987): 605-37. Includes bibliography.

Explanatory Report on the European Convention on the Calculation of Time-Limits. Strasbourg: Council of Europe, 1973.

Lindgren, Kevin E. *Time in the Performance of Contracts: Especially for the Sale of Land*. 2nd ed. Sydney, Australia: Butterworths, 1982.

Meigs, Joseph Vincent. *Time: The Essence of Patent Law*. New York: Baker, Voorhis, 1940.

Shetreet, S. "Time Standards for Justice." *Dalhousie Law J* 5 (November 1979): 729-36.

Singer, R. G., and R. C. Hand. "Sentencing Computation: Laws and Practices." *Criminal Law Bulletin* 10 (May 1974): 318-47.

Squillante, A. M. "Time for Performance or Payment of Contract Terms." *Criminal Law J* 84 (January 1979): 21-25.

Steinberg, M. I. "Dismissal with or without Prejudice Under the Speedy Trial Act: A Proposed Interpretation." *J of Criminal Law* 68 (March 1977): 1-14.

——. "1974 Amendments to the Freedom of Information Act: The Safety Valve Provision of 552 (a) (6) (C) Excusing Agency Compliance with Statutory Time Limits—A Proposed Interpretation." *The Notre Dame Law Review* 52 (December 1976): 235-60.

——. "Right to Speedy Trial: The Constitutional Right and Its Applicability to the Speedy Trial Act of 1974." *The J of Criminal Law and Criminology* 66 (September 1975): 229-39.

Towers, L. A. "Time Constraints on Trials." *The Review of Litigation* 6 (Spring 1987): 175-89.

Townsend, R. "Pre-accusation Delay." *South Texas Law J* 24 (1983): 69-123.

B. LIMITATION OF ACTIONS

Abney, D. L. "For Whom the Statute Tolls: Medical Malpractice Under the Federal Tort Claims Act." *Notre Dame Law Review* 61

(1986): 696-721.

Allen, D. K. "Contracts—Torts—Statute of Limitations—When Does Time Begin to Run?" *The Canadian Bar Review* 61 (December 1983): 876-82.

Andre, S. J. "California Personal Injury Statutes of Limitations: The Modern Tort and the Judicial Abandonment of an Archaic Doctrine." *Santa Clara Law Review* 27 (Fall 1987): 657-92.

Andrews, T. "Infant Tolling Statutes in Medical Malpractice Cases: State Constitutional Challenges." *J of Legal Medicine* 5 (September 1984): 469-87.

Angell, Joseph Kinnicut. *A Treatise on the Limitations of Actions at Law and Suits in Equity and Admiralty: With an Appendix Containing the American and English Statutes of Limitations*. 6th ed. Revised by John Wilder May. Boston: Little, Brown, 1876.

Anglin, Francis A. *Limitations of Actions Against Trustees and Relief from Liability for Technical Breaches of Trust*. Toronto: Canada Law Book Company, 1900.

Bintliff, B. "Statutes of Limitations for Court Actions in Colorado." *The Colorado Lawyer* 12 (June 1983): 895-902.

Birnbaum, S. I. "'First Breath's' Last Gasp: The Discovery Rule in Products Liability Cases." *Forum* 13 (Fall 1977): 279-92.

———. "Statutes of Limitations in Environmental Suits: The Discovery Rule Approach." *Trial* 16 (April 1980): 38-41.

Burch, F. B., Jr. "Practitioner's Guide to the Statutes of Limitations in Product Liability Suits." *U of Baltimore Law Review* 5 (Fall 1975): 23-46.

Buswell, Henry Foster. *The Statute of Limitations and Adverse Possession: With an Appendix Containing the English Acts of Limitation*. Boston: Little, Brown, 1889.

Campbell, R. G. "Medical Malpractice Under the Tort Claims Act: Limitations Problems." *The Air Force Law Review* 17 (Spring 1975): 42-54.

Carlucci, P. P. "The Continuous Treatment Doctrine and the Statute of Limitations." *New York State Bar J* 56 (February 1984): 21-27.

Carter, P. B. "The Foreign Limitation Periods Act 1984." *The Law Quarterly Review* 101 (January 1985): 68-78.

Cervera, Alejo de. *The Statute of Limitations in American Conflicts of Laws*. Rio Piedras: U of Puerto Rico P, 1966.

Classen, H. W. "An Investigation into the Statute of Limitations and Product Identification in Asbestos Litigation." *Howard Law J* 30 (1987): 1-26.

Clausnitzer, M. "Statute of Limitations for Murder in the Federal

Republic of Germany." *The International and Comparative Law Quarterly* 29 (April-July 1980): 473-79.

Collins, C. H. "Limitation of Action Statutes for Architects and Builders—An Examination of Constitutionality." *Federation of Insurance Counsel Quarterly* 29 (Fall 1978): 41-68.

Côté, L. "The Operation in Time of the Statute of Frauds and of the Statute of Limitations." *Revue de Droit, Université de Sherbrooke* 16 (1985): 315-49.

Creswell, R. W. "Statutes of Limitation: Counterproductive Complexities." *Mercer Law Review* 37 (Fall 1985): 1-77.

David, H. "Postponement of Limitation Periods: The Innocent Defendant." *The Advocates' Quarterly* 6 (February 1986): 385-408.

Davis, M. F. "Solving Statute of Limitations Problems Under the Fair Credit Reporting Act." *Indiana Law Review* 18 (Spring 1985): 507-20.

Davis, P. J. "Limitations of the Law of Limitation." *The Law Quarterly Review* 98 (April 1982): 249-78.

Doane, D. G. "Claims by or against Dissolved Companies Under the Australian Companies Acts." *Monash U Law Review* 9 (June 1974): 370-80.

Doyle, M. A., and M. P. Kenny. "The Statute of Limitations Applicable to Criminal Enforcement of the Sherman Act: Restraint of Trade or Enjoyment of the Spoils?" *Arizona State Law J* (1986): 183-205.

Drill, J. A., and L. M. Hambleton. "Latent Injury Cases Applying Statutes of Limitations and Statutes of Repose." *Trial* 19 (November 1983): 106-13.

Dworkin, T. M. "Product Liability of the 1980s: 'Repose Is Not the Destiny' of Manufacturers." *North Carolina Law Review* 61 (October 1982): 33-66.

Eichengrun, J. "The Statute of Limitation for Constructive Trusts in North Carolina." *Wake Forest Law Review* 21 (1986): 613-31.

Ellis, N. S. "Applicability of the Statute of Limitations to Truth-in-Landing Counterclaims." *American Business Law J* 19 (Winter 1982): 471-91.

Erwin, G. D. "Securities Fraud and the Statute of Limitations: The Strange Case of the 'Modified Uniform' Securities Act." *Creighton Law Review* 10 (December 1976): 324-50.

Eyre, P. "Limitation: House Owners v. Local Authorities." *The Solicitors' J* 121 (December 23-30, 1977): 839-40.

Ferris, R. L. "Notice of Knowledge of Injury and Limitations of Actions Under the Iowa Workers' Compensation Act." *Drake Law*

Review 25 (1979-80): 775-92.

Fischer, J. M. "California Statutes of Limitation: Special Project. The Limits of Statutes of Limitation." *Southern U Law Review* 16 (1986): 1-435. Annotated table of California statutes of limitation.

Ford, M. J. W. "Pension Benefits: Statutory Claims, Jury Trials and Limitations Periods." *Columbia Human Rights Law Review* 14 (Spring-Summer 1982): 75-99.

Freeman, M. "A Question of Time." *California Lawyer* 4 (January 1984): 32-35.

Ghiardi, J. D. "Computing Time in Tort Statutes of Limitation." *Marquette Law Review* 64 (Summer 1981): 575-606.

Giancola, P. J. "Erroneous Medical Reports: Extending Limitations." *Trial Lawyers Quarterly* 15 (1983): 64-70.

Grasso, C. T. "The Statute of Limitations as Applied to Medical Malpractice Actions Brought Under the Federal Tort Claims Act." *Military Law Review* 117 (Summer 1987): 1-152.

Gravelis, N. P. "Time Limit Clauses and Judicial Review—The Relevance of Context." *Modern Law Review* 41 (July 1978): 383-403.

Grossman, M. R. "Statutes of Limitations and the Conflict of Laws: Modern Analysis." *Arizona State Law J* (1980): 1-65.

Haddad, W. A. "Delayed Appeals in Civil and Criminal Cases." *Florida Bar J* 57 (October 1983): 525-28.

Harlan, M. C. "Circumstances That Extend the Statute of Limitations Under the Willfulness Provisions of the Age Discrimination in Employment Act of 1967." *Labor Law J* 36 (November 1985): 842-49.

Heller, G. W. "The District of Columbia's Architects' and Builders' Statute of Repose: Its Application and Need for Amendment." *Catholic U Law Review* 34 (Summer 1985): 919-77.

Houlihan, D. L. "Limitations of Action: Strict Liability in Tort—The Legislature Has Intervened." *Illinois Bar J* 67 (December 1978): 214-21.

Jacobs, N. "English Proceedings and Foreign Limitation Periods." *New Law J* 136 (May 16 and 23, 1986): 471-73, 499-500.

Jones, M. A. "Limitation: The Manifestation of Latent Defects." *Northern Ireland Legal Quarterly* 34 (Autumn 1983): 241-47.

Josling, J. F. *Periods of Limitation: With Practical Notes*. 5th ed. London: Oyez, 1981.

Kanovitz, J. R. "The Seller Fiddles and the Clock Ticks: Seller's Cure and the U.C.C. Statute of Limitations." *Notre Dame Law Review* 60 (1985): 318-44.

Kelley, P. J. "Discovery Rule for Personal Injury Statutes of Limitations: Reflections on the British Experience." *Wayne Law Review* 24 (September 1978): 1641-82.

——. "Statutes of Limitations in the Era of Compensation Systems: Workmen's Compensation Limitations Provisions for Accidental Injury Claims." *Washington U Law Quarterly* (1974): 541-631.

Kendal, G. H. "Computing Limits of Time." *UBC Law Review* 17 (1983): 275-97.

Kenny, P. H. "Limitation and Registered Leases." *The Conveyancer* (May-June 1982): 201-07.

Kerr, T. "Tort—Defective Premises—Statute of Limitations." *Dublin U Law J* n.s., 5 (June 1983): 133-39.

Khan, A. N. "Limitation Limits." Parts 1-2. *The Solicitor's J* 127 (May 20 and 27, 1983): 331-32, 352-53.

Kibble-Smith, B. "Statutes of Limitation and Section 1983: Implications for Illinois Civil Rights Law." *The John Marshall Law Review* 20 (Spring 1987): 415-31.

Koffler, J. H. "Legal Malpractice Statutes of Limitations: A Critical Analysis of a Burgeoning Crisis." *Akron Law Review* 20 (Fall 1986): 209-60.

Leibman, J. H., and T. M. Dworkin. "Time Limitations Under State Occupational Disease Acts." *The Hastings Law J* 36 (January 1985): 287-376.

Letvin, D. J. "In Search of the Spirit of Lipsey: Discovery of Malpractice and the Statute of Limitations." *Southern Illinois U Library J* (November 1978): 345-64.

Levy, S. J. "Scientific Developments and Judicial Change." *The Trial Lawyer's Guide* 26 (Spring 1982): 31-50.

Lewis, C. B. "Judicial Review, Time Limits and Retrospectivity." *Public Law* (Spring 1987): 21-27.

McGovern, F. E. "Status of Statutes of Limitations and Statutes of Repose in Product Liability Actions: Present and Future." *Forum* 16 (Winter 1981): 416-50.

——. "Variety, Policy and Constitutionality of Product Liability Statutes of Repose." *American U Law Review* 30 (Spring 1981): 579-641.

Mallen, R. E. "Panacea or Pandora's Box? A Statute of Limitations for Lawyers." *California State Bar J* 52 (January-February 1977): 22-24.

Martin, J. A. "Statutes of Limitations and Rationality in the Conflict of Laws." *Washburn Law J* 19 (Spring 1980): 405-22.

Martin, M. M. "Statute of Repose for Product Liability Claims." *Fordham Law Review* 50 (April 1980): 745-80.

Mastoris, J. F. "Statutes of Limitation in Workers' Compensation Proceedings." *California Western Law Review* 15 (Fall 1979): 32-92.

Milhollin, G. L. "Interest Analysis and Conflicts Between Statutes of Limitation." *Hastings Law J* 27 (September 1975): 1-53.

Miller, M. E. "Barbed Wire in the Borderland: Statute of Limitations Choices for Wrongful Discharge Claims." *San Diego Law Review* 23 (July-August 1986): 833-74.

Nackley, J. V. "Initial Filing Period in Ohio Workers' Compensation Law." *North Kentucky Law Review* 7 (1980): 33-48.

Nock, G. R. "Pleading the Statute of Limitations in Criminal Cases." *Brigham Young U Law Review* (1977): 77-96.

Partington, Martin. *Claim in Time: A Study of the Time Limit Rules for Claiming Social Security Benefits*. London: F. Pinter, 1978.

Phillips, J. G. "Limitations and the Malpractice Syndrome: The Patient Knows All and Nothing at All—Simultaneously." *Medical Trial Technique Quarterly* 34 (1987): 185-201.

Pickering, M. H. "Should Time Run from the Date of Damage or from the Date of Its Discovery?" *Law Institute J* 57 (May 1983): 443-44.

Pineus, Kaj. *Time-Barred Actions*. London: Lloyds of London P, 1984.

Rafferty, N. "The Impact of Concurrent Liability in Contract and Tortious Negligence Upon the Running of Limitation Periods." *U of New Brunswick Law J* 32 (1983): 189-206.

——. "Limitation and Latent Damage—The Canadian Experience." *Oxford J of Legal Studies* 6 (Winter 1986): 464-71.

——. "Limitation Periods in Negligence and the Hidden Cause of Action." *Canadian Business Law J* 12 (February 1987): 338-65.

Reader, H. H., and H. P. Polk. "The One-Year Suit Limitation in Fire Insurance Policies: Challenges and Counterpunches." *Forum* 19 (Fall 1983): 24-47.

Reeves, M. E., and H. L. Hirsh. "For Whom the Bell Tolls: The Statute of Limitations and Medical Malpractice." *Medical Trial Technique Quarterly* 32 (Spring 1986): 414-29.

Reynolds, O. M. "Statute of Limitations Problems in Products Liability Cases—Exercises in Privity, Symmetry, and Repose." *Oklahoma Law Review* 38 (Winter 1985): 667-96.

Rossiter, C. J., and M. Stone. "Latent Defects in Buildings: When Does the Cause of Action Arise?" *Australian Law J* 59 (October 1985): 606-15.

Segalla, T. F., and R. A. Galbo. "Asbestos: New York's Approach to the Statute of Limitations." *New York State Bar J* 57 (November 1985): 28-32.

Shaw, B., P. Cihon, and M. Myers. "The Discovery Rule: Fairness in

Toxic Tort Statutes of Limitations." *Cleveland State Law Review* 33 (1984-85): 491-504.

Sisson, L. F., III, and J. B. Kelley. "Statutes of Limitations for the Design and Building Professions—Will They Survive Constitutional Attack?" *Insurance Counsel J* 49 (April 1982): 243-50.

Smith, P. F. "Limitation and the Land Registration Act 1925." *New Law J* 131 (July 9, 1981): 718-20.

Stewart, C. E. "The Government Suspension Provision of the Clayton Act's Statute of Limitations: For Whom Does it Toll?" *St. John's Law Review* 60 (Fall 1985): 70-92.

Stimson, D. C. "Statutes of Limitations in Trademark Actions." *The Trademark Reporter* 71 (November-December 1981): 605-19.

Stone, P. A. "Time Limitation in the English Conflict of Laws." *Lloyd's Maritime and Commercial Law Quarterly* (1985): 497-512.

Thomas, D. R. "Power of Court to Extend Time for Commencing Arbitration Proceedings." *Lloyd's Maritime and Commercial Law Quarterly* (November 1981): 529-49.

Todd, S. "Latent Defects in Property and the Limitation Act: A Defence of the 'Discoverability' Test." *New Zealand U Law Review* 10 (December 1983): 311-31.

Todd, Stephen. *Limitation Periods in Personal Injury Claims*. London: Sweet and Maxwell, 1982.

Torrey, D. B. "Time Limitations in the Pennsylvania Workmen's Compensation and Occupational Disease Acts: Theoretical Doctrine and Current Applications." *Duquesne Law Review* 24 (Summer 1986): 975-1127.

Uelmen, G. F. "Making Sense Out of the California Criminal Statute of Limitations." *Pacific Law J* 15 (October 1983): 35-82.

Vickers, J. "Overriding the Time Bar in Personal Injury Actions." *New Law J* 130 (April 17, 1980): 380-82.

Wanger, B. R. "Medical Products Liability and the Statute of Limitations." *Medical Trial Technique Quarterly* 32 (Fall 1985): 192-206.

Watson, G. D. "Amendment of Proceedings After Limitation Periods." *The Canadian Bar Review* 53 (May 1975): 237-301.

Wheeler, M. E., and R. J. Jones. "Statute of Limitations for Antitrust Damage Actions: Four Years or Forty?" *The U of Chicago Law Review* 41 (Fall 1973): 72-90.

Williams, C. "The Statute of Limitations: Prospective Warranties, and Problems of Interpretation in Article Two of the UCC." *The George Washington Law Review* 52 (November 1983): 67-112.

Williams, Jeremy S. *Limitations of Actions in Canada*. 2nd ed. Toronto: Butterworths, 1980.

Wilson, J. R. "Dormant Judgments, Limitations and the Touch of Fraud." *Baylor Law Review* 32 (Summer 1980): 327-45.

Wood, Horace Day. *A Treatise on the Limitations of Actions at Law and in Equity*. 4th ed. Revised by Dewitt C. Moore. 2 vols. Albany: M. Bender, 1916.

Wurfel, S. W. "Statutes of Limitations in the Conflict of Laws." *North Carolina Law Review* 52 (February 1974): 489-574.

13. LITERATURE AND LANGUAGE

A. GENERAL

Alkon, Paul K. "Boswellian Time." *Studies in Burke and His Time* 14 (1973): 239-56.

——. "Johnson and Time Criticism." *Modern Philology* 85 (May 1988): 543-57.

Backscheider, Paula R., ed. *Probability, Time, and Space in Eighteenth-Century Literature*. New York: AMS P, 1979.

Bender, John, and David E. Wellbery, eds. *Chronotypes: The Construction of Time*. Stanford: Stanford UP, 1991.

Bergson, Henri. *Creative Evolution*. Trans. Arthur Mitchell. New York: Holt, 1911.

Burke, Peter. *The Renaissance Sense of the Past*. London: Edward Arnold, 1969.

Chatman, Seymour. *Story and Discourse: Narrative Structure in Fiction and Film*. Ithaca: Cornell UP, 1980.

Cohn, Dorrit. "Narrated Monologue: Definition of a Fictional Style." *Comparative Literature* 28 (1966): 97-112.

——. *Transparent Minds: Narrative Modes for Presenting Consciousness in Fiction*. Princeton: Princeton UP, 1978.

Crane, R. S. "Critical and Historical Principles of Literary History." *The Idea of the Humanities and Other Essays Critical and Historical*. Chicago: U of Chicago P, 1967. 2:45-156.

De Man, Paul. "The Rhetoric of Temporality." *Interpretation: Theory and Practice*. Ed. Charles S. Singleton. Baltimore: Johns Hopkins P, 1969. Ch. 6.

Ebeling, Herman J. "The Word *Anachronism*." *Modern Language Notes* 52 (February 1937): 120-21.

Fasel, Ida. "Spatial Form and Spatial Time." *Western Humanities Review* 16 (1962): 223-34.

Fjelde, Rolf. "Time, Space, and Wyndham Lewis." *Western Review* 15 (1950-51): 201-12.

Ford, George H. "Death, Literature, and Its Consolations." *The Study of Time IV*. Ed. J. T. Fraser, N. Lawrence, and D. Park. New York: Springer, 1981. 46-58.

——. "Humanities and the Experiences of Time." *Time, Science and Society in China and the West: The Study of Time V*. Ed. J. T. Fraser, N. Lawrence, and F. C. Haber. Amherst: U of Massachusetts P, 1986. 141-48.

Fox, A. A. O., ed. *An Anthology of Clocks and Watches*. Swansea: published by the editor, n.d.

Frank, Manfred. *Das Problem "Zeit" in der deutschen Romantik*. Munich: Winkler, 1972.

Furst, L. R. "Realism and Its 'Code of Accreditation.'" *Comparative Literature Studies* 25.2 (1988): 101-26. With bibliography.

Grivelet, Michel. "Shakespeare's 'War with Time': The Sonnets and *Richard II*." *Shakespeare Survey* 23 (1970): 69-78.

Gross, Harvey. *The Contrived Corridor: History and Fatality in Modern Literature*. Ann Arbor: U of Michigan P, 1971.

Grubbs, Henry A. "Sartre's Recapturing of Lost Time." *Modern Language Notes* 73 (1958): 515-22.

Hayles, N. Katherine. *Chaos Bound: Orderly Disorder in Contemporary Literature and Science*. Ithaca: Cornell UP, 1990.

——. *The Cosmic Web: Scientific Field Models and Literary Strategies in the Twentieth Century*. Ithaca: Cornell UP, 1984.

Holbrook, David. *Evolution and the Humanities*. New York: St. Martin's P, 1987.

Holdheim, William Wolfgang. *The Hermeneutic Mode: Essays on Time in Literature and Literary Theory*. Ithaca: Cornell UP, 1984.

Humes, Jay. *Two Against Time: A Study of the Very Present Worlds of Paul Claudel and Charles Péguy*. Chapel Hill: U of North Carolina Department of Languages, 1978.

Ianni, L. A. "Lawrence Ferlinghetti's Fourth Person Singular and the Theory of Relativity." *Wisconsin Studies in Contemporary Literature* 8 (1967): 392-406.

Iser, Wolfgang. *The Act of Reading: A Theory of Aesthetic Response*. Baltimore: Johns Hopkins UP, 1978.

——. "Perception, Temporality, and Action as Modes of Subjectivity, W. Faulkner: *The Sound and the Fury*." In his *The Implied Reader*. Baltimore: Johns Hopkins UP, 1974. 136-52.

Jauss, Hans Robert. "Literary History as a Challenge to Literary Theory." *New Literary History* 2 (1970): 7-37.

Jordan, Richard D. *The Temple of Eternity: Thomas Traherne's Philosophy of Time*. Port Washington: Kennikat P, 1972.

Kinser, Samuel. "Chronotopes and Catastrophes: The Cultural History of Mikhail Bakhtin." *J of Modern History* 56.2 (1984): 301-10.

Lewis, Wyndham. *Time and Western Man*. Boston: Beacon P, 1957.

Lynen, John F. *The Design of the Present: Essays on Time and Form in American Literature*. New Haven: Yale UP, 1969.

Macey, Samuel L. *Clocks and the Cosmos: Time in Western Life and Thought*. Hamden: Archon Books, 1980.

——. "A Review of Quinones' *Renaissance Discovery of Time* and Iwasaki's *Sword and the Word: Shakespeare's Tragic Sense of Time.*" *Shakespeare Quarterly* 26 (Summer 1975): 307-10.

Mack, Sara. *Patterns of Time in Vergil*. Hamden: Archon Books, 1978.

Meyerhoff, Hans. *Time in Literature*. Berkeley: U of California P, 1960.

——. "The Time of Life." *Forum* 3.9 (1962): 13-17.

Miller, J. Hillis. "Geneva or Paris? The Recent Work of Georges Poulet." *U of Toronto Quarterly* 39 (1969-70): 212-28.

——. "The Literary Criticism of Georges Poulet." *Modern Language Notes* 78.5 (December 1963): 471-88.

Miner, Earl. "Time, Sequence, and Plot in Restoration Literature." *Studies in Eighteenth-Century Culture V*. Ed. Ronald C. Rosbottom. Madison: U of Wisconsin P, 1976. 67-85.

Müller, Gunther. "Erzählzeit und erzählte Zeit." *Festschrift für P. Kluckhorn*. Tübigen, 1948.

Noon, William T. "Modern Literature and the Sense of Time." *Thought* 33 (1958): 571-603.

Norton, Glyn P. "Retrospective Time and the Musical Experience in Rousseau." *Modern Language Quarterly* 34 (1973): 131-45.

Ong, Walter J. "Evolution, Myth, and Poetic Vision." *Comparative Literature Studies* 3 (1966): 1-20.

Passler, David L. *Time, Form and Style in Boswell's Life of Johnson*. New Haven: Yale UP, 1971.

Patrides, C. A., ed. *Aspects of Time*. Manchester: Manchester UP, 1976.

——. "The Renascence of the Renaissance: T. S. Eliot and the Pattern of Time." *Michigan Quarterly Review* 12 (1973): 172-96.

——. "The Renaissance View of Time: A Bibliographical Note." *Notes and Queries* n.s., 10 (1963): 409-10.

Paul, T. "The Worm Ouroboros: Time Travel, Imagination, and Entropy." *Extrapolation* 24 (Fall 1983): 272-79.

Pike, Burton. "Time in Autobiography." *Comparative Literature* (1976): 326-42.

Poulet, Georges. *Studies in Human Time*. Trans. Elliott Coleman. Baltimore: Johns Hopkins P, 1956.

——. "Timelessness and Romanticism." *J of the History of Ideas* 15 (January 1954): 3-22.

Priestley, J. B. *Man and Time*. Garden City: Doubleday, 1964.

Prior, S. P. "Patience—Beyond Apocalypse." *Modern Philology* 83 (May 1986): 337-48.

Quinones, Ricardo J. "Four Phases of Time and Literary Modernism." *The Study of Time II*. Ed. J. T. Fraser, and N. Lawrence. New

York: Springer, 1975. 122-35.
——. *Mapping Literary Modernism: Time and Development.* Princeton: Princeton UP, 1985.
——. *The Renaissance Discovery of Time.* Cambridge: Harvard UP, 1972.
Richter, David. *Fable's End.* Chicago: U of Chicago P, 1978.
Ritter, Alexander, ed. *Zeitgestaltung in der Erzählkunst.* Darmstadt: Wissenschaftliche Buchgesellschaft, 1978.
Rubino, C. A. "'Smitten by a Desire to See Again What I Had Once Loved': Homecomings in Homer, Virgil, and Proust." *Modern Language Notes* 99.4 (September 1984): 941-50.
Salm, Peter. *Pinpoint of Eternity: European Literature in Search of the All-Encompassing Moment.* Lanham: UP of America, 1986.
Shattuck, Roger. "Making Time: A Study of Stravinsky, Proust, and Sartre." *Kenyon Review* 25 (1963): 248-63.
Sherwood, Merriam. "Magi and Mechanics in Medieval Fiction." *Studies in Philology* 44 (October 1947): 567-92.
Soulsby, Marlene Pilarcik [Marlene A. Pilarcik]. "Mind and Time: A Comparative Reading of Haiku, Kafka, and Le Guin." *Time and Mind: The Study of Time VI.* Ed. J. T. Fraser. Madison: International Universities P, 1989. 261-76.
Spanos, William V. "Modern Literary Criticism and the Spatialization of Time: An Existential Critique." *J of Aesthetics and Art Criticism* 29 (1970): 87-104.
Stern, Madeleine B. "Counterclockwise: Flux of Time in Literature." *Sewanee Review* 44 (1936): 338-65.
Sternberg, Meir. *Expositional Modes and Temporal Ordering in Fiction.* Baltimore: Johns Hopkins UP, 1978.
——. "What is Exposition? An Essay in Temporal Delimitation." *The Theory of the Novel: New Essays.* Ed. John Halperin. New York: Oxford UP, 1974. 25-70.
Sturzl, Erwin A. *The Lackey of Eternity: The Concept of Time in Elizabethan Literature.* 2 vols. Wolfeboro: Longwood Publishing, 1986.
Suda, David. *The Moving Image: Immutability, Metaphors, and the Time Clocks Tell.* New York: UP of America, 1989.
Toyokuni, Takashi. "A Modern Man Obsessed by Time: A Note on 'The Man Who Loved Islands.'" *D. H. Lawrence Review* 7 (1974): 78-82.
Uhlig, C. "Forms of Time and Varieties of Change in Literary Texts." *Comparative Literature* 37 (Fall 1985): 289-300.
Vonalt, Larry P. "Of Time and Literature." *Sewanee Review* 77 (1969): 164-70.

Walker, G. *The Strong Necessity of Time: The Philosophy of Time in Shakespeare and in Elizabethan Literature*. The Hague: Mouton, 1976.

Walter, G. F. "John Donne's Changing Attitudes to Time." *Studies in English Literature* 14 (1974): 79-89.

Weber, Jean-Paul. "Edgar Poe or the Theme of the Clock." *Poe: A Collection of Critical Essays*. Ed. Robert Regan. Englewood Cliffs: Prentice-Hall, 1967. 79-97.

Weitz, Morris. "T. S. Eliot: Time as a Mode of Salvation." *Sewanee Review* 60 (1952): 48-64.

Wood, David. *The Deconstruction of Time*. Atlantic Highlands: Humanities P, 1988.

B. CLOCKWORK DIABOLISM

Baldensperger, Fernand. "Une nouvelle française peu connue sur le machinisme menaçant." *Modern Language Notes* (May 1945): 321-23.

Baudelaire, Charles. "L'horloge." *Les fleurs du mal*. Ed. Antoine Adam. Paris: Éditions Garnier Frères, 1959. 87, 374.

Brentano, Clemens. *Gockel, Hinkel und Gackeleia*. In *Die deutschen Romantiker*. Ed. Gerhard Stenzel. Salzburg: Verlag "Das Bergland-Buch," n.d. 2:560-631.

Butler, Samuel. *Erewhon or Over the Range*. London: Jonathan Cape, 1921.

Hoffmann, E. T. A. *Selected Writings*. Ed. and trans. Leonard J. Kent, and Elizabeth C. Knight. Chicago: U of Chicago P, 1969. See "The Sandman."

Macey, Samuel L. "Richardson's Grandison and Austen's General Tilney: Contrasting Authorial Views on Clockworklike Efficiency." *Transactions. Johnson Society of the Northwest* 14 (1983): 1-16.

——. "Time, Clockwork, and the Devil in Robertson Davies' Deptford Trilogy." *Studies in Robertson Davies' Deptford Trilogy*. Ed. Robert G. Lawrence and Samuel L. Macey. Victoria: ELS Monograph Series, 1980. 32-44.

Poe, Edgar Allan. *The Complete Tales and Poems*. New York: Modern Library, 1938. See "Devil in the Belfry," "A Predicament," "The Pit and the Pendulum," and "The Bells."

Tennyson, Alfred. *The Devil and the Lady*. Ed. Charles Tennyson. London: Macmillan, 1930.

C. DRAMA

Anders, G. "Being Without Time: On Beckett's Play *Waiting for Godot.*" *Samuel Beckett*. Ed. M. Esslin. Englewood Cliffs: Prentice-Hall, 1965. 140-51.

Barnet, Sylvan. "Prodigality and Time in *The Merchant of Venice.*" *Publications of the Modern Language Association* 87 (1972): 26-30.

Bennett, B. K. "Strindberg and Ibsen: Toward a Cubism of Time in Drama." *Modern Drama* 26 (September 1983): 262-81.

Bennett, R. B. "Four Stages of Time: The Shape of History in Shakespeare's Second Tetralogy." *Shakespeare Studies* 19 (1987): 61-85.

Blissett, William. "The Wide Gap of Time: *The Winter's Tale.*" *English Literary Renaissance* 1 (1971): 52-70.

Brennan, A. "Helena Versus Time's Winged Chariot in *All's Well That Ends Well.*" *Midwest Quarterly* 21 (Summer 1980): 391-411.

Brinkley, E. S. "Proustian Time and Modern Drama: Beckett, Brecht, and Fugard." *Comparative Literature Studies* 25.4 (1988): 352-66.

Broude, Ronald. "Time, Truth, and Right in *The Spanish Tragedy.*" *Studies in Philology* 68 (1971): 130-45.

Buland, Mable. *The Presentation of Time in the Elizabethan Drama*. New York: Holt, 1912.

Burkman, K. H. "Harold Pinter's *Betrayal*: Life before Death—and after." *Theatre J* 34 (December 1982): 505-18.

Cirillo, Albert R. "Time, Light, and the Phoenix: The Design of *Samson Agonistes.*" *Calm of Mind*. Ed. Joseph A. Wittreich. Cleveland: Case Western Reserve UP, 1971. 209-33.

Davidson, Clifford. "The Triumph of Time." *Dalhousie Review* 50 (1970): 170-81. About *Hamlet*.

Driver, Tom F. *The Sense of History in Greek and Shakespearean Drama*. New York: Columbia UP, 1960.

——. "The Shakespearean Clock: Time and the Vision of Reality in *Romeo and Juliet* and *The Tempest.*" *Shakespeare Quarterly* 15 (1964): 363-70.

Foreman, Walter C., Jr. *The Music of the Close*. Lexington: UP of Kentucky, 1978.

Freytag, Gustav. *Die Technik des Dramas*. 10th ed. Leipzig: S. Hirzel, 1905.

Frye, Northrop. *Fools of Time: Studies in Shakespearean Tragedy*. Toronto: U of Toronto P, 1967.

Garner, S. B., Jr. "Time and Presence in *The Winter's Tale.*" *Modern Language Quarterly* 46 (December 1985): 347-67.

Groat, Cegienas de. *Zeitgestaltung im Drama Max Frischs*. Amsterdam:

Rodopie, 1977.

Hamel, G. "Time in *Richard III.*" *Shakespeare Survey* 40 (1988): 41-49.

Hunter, Frederick J. "The Value of Time in Modern Drama." *J of Aesthetics and Art Criticism* 16 (1957-58): 194-201.

Iwasaki, Soji. *The Sword and the Word: Shakespeare's Tragic Sense of Time*. Tokyo: Shinozaki Shorin P, 1973.

Jorgensen, Paul A. "'Redeeming Time' in *Henry IV.*" *Tennessee Studies in Literature* 5 (1960): 101-09.

Kastan, David Scott. *Shakespeare and the Shapes of Time*. Hanover: UP of New England, 1982.

Kaula, David. "The Time Sense of *Antony and Cleopatra.*" *Shakespeare Quarterly* 15 (1964): 211-23.

Lapp, John C. "The Unities of Time and Place." In his *Aspects of Racinian Tragedy*. Toronto: U of Toronto P, 1955. Ch. 2.

Lindenbaum, Peter. "Time, Sexual Love, and the Uses of Pastoral in *The Winter's Tale.*" *Modern Language Quarterly* 33 (1972): 3-22.

Link, Franz H. *Dramaturgie der Zeit*. Freiburg: Verlag Rombach, 1977.

Macey, Samuel L. "The Relationship Between Our New Sense of Time and Our Sense of an Ending in Tragedy." *The Study of Time IV*. Ed. J. T. Fraser, N. Lawrence, and D. Park. New York: Springer, 1981. 94-102.

Maclean, H. "'Looking Before and After': Hal and Hamlet Once More." *Papers on Language and Literature* 23 (Summer 1987): 273-89.

McNeir, Waldo F. "The Tempest: Space-Time and Spectacle-Theme." *Arlington Quarterly* 2.4 (1970): 29-58.

Montgomery, Robert L. "The Dimensions of Time in *Richard II.*" *Shakespeare Studies* 4 (1968): 73-85.

Morgenstern, Irvin. *The Dimensional Structure of Time, Together with the Drama and Its Timing*. New York: Philosophical Library, 1960.

Mueller, Martin. "Time and Redemption in *Samson Agonistes* and *Iphigenie auf Tauris.*" *U of Toronto Quarterly* 41 (1972): 227-45.

Nosworthy, J. M. *Shakespeare Puts the Clock Back*. Cardiff: U of Wales P, 1978.

Peterson, Douglas L. *Time, Tide and Tempest: A Study of Shakespeare's Romances*. San Marino: Huntington Library, 1973.

Pike, L. J. "Liturgy and Time in Counterpoint: A View of T. S. Eliot's *Murder in the Cathedral.*" *Modern Drama* 23 (September 1980): 277-91.

Priestley, J. B. *Three Time-Plays*: Dangerous Corner, Time and the Conways, *and* I Have Been Here Before. London: Pan Books, 1952.

Quinones, Ricardo J. "Views of Time in Shakespeare." *J of the History of Ideas* 26 (1965): 327-52.

Radke, Judith J. "The Theater of Samuel Beckett: 'Une durée à animer.'" *Yale French Studies* 29 (1962): 57-64.

Rayner, A. "Harold Pinter: Narrative and Presence." *Theatre J* 40 (December 1988): 482-97.

Rechtien, Brother John, S. M. "Time and Eternity Meet in the Present." *Texas Studies in Literature and Language* (1964): 5-21. About Beckett's *Godot*.

Robinson, James E. "Time and *The Tempest*." *J of English and German Philology* 63 (1964): 255-67.

Romilly, Jacqueline de. *Time in Greek Tragedy*. Ithaca: Cornell UP, 1968.

Seng, Peter J. "Songs, Time, and the Rejection of Falstaff." *Shakespeare Survey* 15 (1962): 31-40.

Spanos, William V. *The Christian Tradition in Modern British Verse Drama: The Poetics of Sacramental Time*. New Brunswick: Rutgers UP, 1967.

——. "Modern Drama and the Aristotelian Tradition: The Formal Imperatives of Absurd Time." *Contemporary Literature* 12 (1971): 345-72.

Spencer, Benjamin T. "*2 Henry IV* and the Theme of Time." *U of Toronto Quarterly* 13 (1944): 394-99.

Stirling, Brent. *Unity in Shakespearean Tragedy*. New York: Columbia UP, 1956.

Stroud, Ronald. "The Bastard to the Time in *King John*." *Comparative Drama* 6 (1972): 154-66.

Sypher, Wylie. *The Ethic of Time: Structures of Experience in Shakespeare*. New York: Seabury P, 1976.

Tanselle, G. Thomas. "Time in *Romeo and Juliet*." *Shakespeare Quarterly* 15 (1964): 349-61.

Tembeck, Robert. "Dialectic and Time in [Sartre's] *The Condemned of Altona*." *Modern Drama* 12 (1969): 10-17.

Toliver, Harold E. "Falstaff, the Prince, and the History Play." *Shakespeare Quarterly* 16 (1965): 63-80.

——. "Shakespeare and the Abyss of Time." *J of English and German Philology* 64 (1965): 234-54.

Torrance, Robert M. "Modes of Being and Time in the World of *Godot*." *Modern Language Quarterly* 28 (1967): 77-95.

Turner, Frederick. *Shakespeare and the Nature of Time*. Oxford: Clarendon P, 1971.

Ungvàri, T. "Time and the Modern Self: A Change in Dramatic

Form." *The Study of Time I*. Ed. J. T. Fraser, F. C. Haber, and G. H. Müller. New York: Springer, 1972. 470-78.

Vincent, Deirdre. *The Eternity of Being: On the Experience of Time in Goethe's* Faust. Bonn: Bouvier, 1987.

Waller, G. F. *The Strong Necessity of Time: The Philosophy of Time in Shakespearian and Elizabethan Literature*. The Hague: Mouton, 1976.

Ward, Anne. "Speculations on Eliot's Time-World: An Analysis of *The Family Reunion* in Relation to Hulme and Bergson." *American Literature* 21 (1949): 18-34.

D. FICTION

Alexandrov, V. E. "Relative Time in *Anna Karenina*." *Russian Review* 41 (April 1982): 159-68.

Alkon, Paul K. *Defoe and Fictional Time*. Athens: U of Georgia P, 1979.

———. "The Odds Against Friday: Defoe, Bayes, and Inverse Probability." *Probability, Time, and Space in Eighteenth-Century Literature*. Ed. Paula R. Backscheider. New York: AMS P, 1979. 29-61.

Baine, Rodney M. "Roxana's Georgian Setting." *Studies in English Literature* 15 (1975): 459-71.

Baird, Theodore. "The Time-Scheme of *Tristram Shandy* and a Source." *Publications of the Modern Language Association* 51 (1936): 803-20.

Bakhtin, Mikhail Mikhailovitch. "Forms of Time and of the Chronotype in the Novel." *The Dialogic Imagination: Four Essays by M. M. Bakhtin*. Trans. Caryl Emerson and Michael Holquist. Ed. Michael Holquist. Austin: U of Texas P, 1981.

Barthold, Bonnie J. *Black Time: Fiction of Africa, the Caribbean, and the United States*. New Haven: Yale UP, 1981.

Beznos, Maurice J. "Aspects of Time According to the Theories of Relativity in Marcel Proust's *À la recherche du temps perdu*." *Ohio U Review* 10 (1968): 74-102.

Biemel, Walter. *Zeitigung und Romanstruktur: Philosophische Analysen zur Deutung des modernen Romans*. Freiburg: K. Alber, 1985.

Bisson, L. A. "Proust, Bergson, and George Eliot." *Modern Language Review* 40 (1945): 104-14.

Blewett, David. "*Roxana* and the Masquerades." *Modern Language Review* 65 (1970): 499-502.

Blodgett, Harriet. "Joyce's Time Mind in *Ulysses*: A New Emphasis."

James Joyce Quarterly 5 (1967): 22-29.

Bloom, E. A., and L. D. Bloom. "Hostage to Fortune: Time, Chance, and Laurence Sterne." *Modern Philology* 85 (May 1988): 499-513.

Bork, Alfred M. "Durrell and Relativity." *Centennial Review* 7 (1963): 191-203.

Bowling, Lawrence E. "What is the Stream of Consciousness Technique?" *Publications of the Modern Language Association* 65 (1950): 333-45.

Brand, Gerd. *Welt, Ich, und Zeit*. The Hague: Martinus Nijhoff, 1955.

Brée, Germaine. *Du temps perdu au temps retrouvé*. Paris: Société d'Éditions, 1950.

——. *Marcel Proust and Deliverance from Time*. Trans. C. J. Richards and A. D. Truitt. 2nd ed. New Brunswick: Rutgers UP, 1969.

Brivic, Sheldon R. "Time, Sexuality, and Identity in Joyce's *Ulysses*." *James Joyce Quarterly* 7 (1969): 30-51.

Brown, Robert C. "Laurence Sterne and Virginia Woolf: A Study in Literary Continuity." *U of Kansas City Review* 26 (1959): 153-59.

Bryant, E. V. "Tree-Clock in Bernard Malamud's *Idiots First*." *Studies in Short Fiction* 20 (Winter 1983): 52-54.

Butler, Colin. "Borges and Time: With Particular Reference to 'A Refutation of Time.'" *Orbis Literarum* 28 (1973): 148-61.

Butler, Eliza M. "The Element of Time in Goethe's *Werther* and Kafka's *Prozess*." *German Life and Letters* 12 (1958-59): 248-58.

Carpenter, Mary Wilson. *George Eliot and the Landscape of Time: Narrative Form and Protestant Apocalyptic History*. Chapel Hill: U of North Carolina P, 1986.

Champigny, Robert. "Proust, Bergson and Other Philosophers." *Publications of the Modern Language Association* 73 (1958): 129-35.

Cheng, V. T. "A Chronology of *The Good Soldier*." *English Language Notes* 24 (September 1986): 91-97.

Church, Margaret. "Fiction: The Language of Time—Thomas Mann and James Joyce." *The Study of Time III*. Ed. J. T. Fraser, N. Lawrence, and D. Park. New York: Springer, 1978. 500-11.

——. "Thomas Mann: Time." *The Hopkins Review* 3 (1950): 20-29.

——. *Time and Reality: Studies in Contemporary Fiction*. Chapel Hill: U of North Carolina P, 1963.

The Clockmakers Outcry against the Author of the Life and Opinions of Tristram Shandy. London: J. Burd, 1760. Four editions in 1760 in response to Sterne's *Tristram Shandy*.

Coetzee, J. M. "Time, Tense and Aspect in Kafka's *The Burrow*." *Modern Language Notes* 96 (April 1981): 556-79.

Cohn, Dorrit. "Kafka's Eternal Present: Narrative Tense in 'Ein Landarzt' and Other First-Person Stories." *Publications of the Modern Language Association* 83 (1968): 144-50.

Cook, Albert. "Proust: The Invisible Stilts of Time." *Modern Fiction Studies* 4 (1958): 118-26.

"Coventry's Watch." *The Dickensian* 47 (June 1951): 17.

Curry, S. S. "Into the Shadow of Hesitation: Time and Identity in Gide's Middle Fiction." *Twentieth-Century Literature* 28 (Fall 1982): 233-51.

Daiches, David. "Time and Sensibility." *Modern Language Quarterly* 25 (1964): 486-92.

Dassin, Joan. "The Dialectics of Recurrence: The Relation of the Individual to Myth and Legend in Thomas Mann's *Joseph and His Brothers*." *Centennial Review* 15 (1971): 362-90.

Davis, J. K. "Time and the Demonic in William Faulkner and Flannery O'Connor." *Studies in the Literary Imagination* 20 (Fall 1987): 123-43.

Dickens, Charles. Master Humphrey's Clock *and* A Child's History of England. London: Oxford UP, 1958.

Diemer, Alwin. *Edmund Husserl*. Meisenheim am Glan: Verlag Anton Hain, 1965.

Dobie, Ann B. "Early Stream-of-Consciousness Writing: *Great Expectations*." *Nineteenth-Century Fiction* 25 (1971): 405-16.

Donaldson, Ian. "The Clockwork Novel: Three Notes on an Eighteenth-Century Analogy." *Review of English Studies* 21 (1970): 14-22.

Dunn, Albert A. "The Articulation of Time in *The Ambassadors*." *Criticism* 14 (1972): 137-50.

Dyck, Martin. "Relativity in Physics and in Fiction." *Studies in German Literature of the 19th and 20th Centuries*. Ed. Siegfried Mews. Chapel Hill: U of North Carolina P, 1970. 174-85.

Erickson, John D. "The Proust-Einstein Relation: A Study in Relative Point of View." *Marcel Proust: A Critical Panorama*. Ed. Larkin B. Price. Urbana: U of Illinois P, 1973. 247-76.

Ewton, Ralph W., Jr. "The Chronological Structure of Thomas Mann's *Die Geschichten Jaakobs*." *Rice U Studies* 1.4 (1964): 27-40.

Fagan, Edward R. "Disjointed Time and the Contemporary Novel." *J of General Education* 23 (1971): 151-60.

Fawkner, Harold William. *The Timescapes of John Fowles*. Rutherford: Fairleigh Dickinson UP, 1984.

Fetz, Howard W. "Of Time and the Novel." *Xavier U Studies* 8.2 (1969): 1-17.

Fitzgerald, G. D. "Chronology of F. M. Dostoevskij's *The Possessed*."

Slavic and East European J 27 (Spring 1983): 19-46.

Fleischauer, John F. "Simultaneity in Nabokov's Prose Style." *Style* 5 (1971): 57-69.

Follett, W. "Time and Thomas Mann." *Atlantic Monthly* (June 1938): 792-94.

Ford, George H. "The Eternal Moment: D. H. Lawrence's *The Rainbow* and *Women in Love.*" *The Study of Time III*. Ed. J. T. Fraser, N. Lawrence, and D. Park. New York: Springer, 1978. 512-36.

——. "Dickens and the Voices of Time." *Nineteenth-Century Fiction* 24 (1970): 428-48.

Friedman, Melvin. "The Novels of Samuel Beckett: An Amalgam of Joyce and Proust." *Comparative Literature* 12 (1960): 47-58.

——. *Stream of Consciousness: A Study in Literary Method.* New Haven: Yale UP, 1955.

Frohock, W. M. "Thomas Wolfe: Time and the National Neurosis." In his *The Novel of Violence in America*. 2nd ed. Dallas: Southern Methodist UP, 1957. Ch. 3.

Ganzel, Dewey. "Chronology in *Robinson Crusoe.*" *Philological Quarterly* 40 (1961): 495-512.

Garber, Frederick. "Time and the City in Rilke's *Malte Laurids Brigge.*" *Contemporary Literature* 11 (1970): 324-39.

Genette, Gérard. "Time and Narrative in *À la recherche du temps perdu.*" Trans. Paul De Man. *Aspects of Narrative*. Ed. J. Hillis Miller. New York: Columbia UP, 1971. 93-118.

Gerhardi, Gerhard C. "Psychological Time and Revolutionary Action in *Le rouge et le noir.*" *Publications of the Modern Language Association* 88 (1973): 1115-25.

German, Howard, and Sharon Kaehele. "The Dialectic of Time in [Virginia Woolf's] *Orlando.*" *College English* 24 (1962): 35-41.

Grenander, M. E., B. J. Rahn, and F. Valvo. "The Time-Scheme in [James's] *The Portrait of a Lady.*" *American Literature* 32 (1960): 127-35.

Gross, Beverly. "Narrative Time and the Open-Ended Novel." *Criticism* 8 (1966): 362-76.

Guiguet, Jean. "Time and Space." In his *Virginia Woolf and Her Works*. Trans. Jean Stewart. London: Hogarth P, 1965. 382-98.

Hagemann, E. R. "Only Let the Story End as Soon as Possible: Time-and-History in Ernest Hemingway's *In Our Time.*" *Modern Fiction Studies* 26 (Summer 1980): 255-62.

Hamilton, Gary D. "The Past in the Present: A Reading of [Faulkner's] *Go Down, Moses.*" *Southern Humanities Review* 5 (1971): 171-81.

Harkness, Bruce. "Conrad on Galsworthy: The Time Scheme of *Fraternity*." *Modern Fiction Studies* 1.2 (1955): 12-18.

Harper, Ralph. "Remembering Eternity: St. Augustine and Proust." *Thought* 34 (1959): 569-606.

Hartley, Lodwick. "Of Time and Mrs Woolf." *Sewanee Review* 47 (1939): 235-41. About *The Years*.

Higdon, David Leon. "The Chronology of *Moll Flanders*." *English Studies* 56 (1975): 316-19.

——. *Time and English Fiction*. London: Macmillan, 1977.

Holl, Oskar. *Der Roman als Funktion und Überwindung der Zeit und Gleichzeitigkeit im deutschen Roman des 20. Jahrhunderts*. Bonn: H. Bouvier, 1968.

Holtz, William. "Time's Chariot and *Tristram Shandy*." *Michigan Quarterly Review* 5 (1966): 197-203.

Horowitz, Floyd R. "The Christian Time Sequence in Henry James's *The American*." *CLA J* 9 (1966): 234-45.

Hudspeth, Robert N. "Conrad's Use of Time in *Chance*." *Nineteenth-Century Fiction* 21 (1966-67): 283-89.

Huertas-Jourda, J. "Structures of the 'Living Present': Husserl and Proust." *The Study of Time II*. Ed. J. T. Fraser, and N. Lawrence. New York: Springer, 1975. 163-95.

Humphrey, Robert. *Stream of Consciousness in the Modern Novel*. Berkeley: U of California P, 1954.

Hungerford, H. "Past and Present in [Faulkner's] *Light in August*." *American Literature* 55 (May 1983): 183-98.

Hutchens, Eleanor N. "The Novel as Chronomorph." *Novel* 5 (1972): 215-22.

Hutchinson, James D. "Time: The Fourth Dimension in Faulkner." *Southern Dakota Review* 6.2-3 (1968): 91-103.

Ireland, K. R. "Future Recollections of Immortality: Temporal Articulation in Jane Austen's *Persuasion*." *Novel* 13 (Winter 1980): 204-20.

Izevbaye, D. S. "Time in the African Novel." *J of Commonwealth Literature* 17.1 (1982): 74-89.

Jacobs, Robert G. "H. G. Wells, Joseph Conrad, and the Relative Universe." *Conradiana* 1 (1968): 51-55.

Karst, Roman. "Franz Kafka: Word-Space-Time." *Mosaic* 3.4 (1970): 1-13.

Kermode, Frank. *The Sense of an Ending: Studies in the Theory of Fiction*. New York: Oxford UP, 1967.

Klawitter, Robert. "Henri Bergson and Joyce's Fictional World." *Comparative Literature Studies* 3 (1966): 429-37.

Klotz, Marvin. "The Triumph Over Time: Narrative Form in William Faulkner and William Styron." *Mississippi Quarterly* 17 (1963-64): 10-20.

Kohler, David. "Time in the Modern Novel." *College English* 10 (19-48): 15-24.

Kort, Wesley A. "Human Time in Hemingway's Fiction." *Modern Fiction Studies* 26 (Winter 1980-81): 579-96.

——. *Modern Fiction and Human Time: A Study in Narrative and Belief.* Tampa: U of Florida P, 1985.

Krojino, Anto. *Die Zeitauffassung bei Thomas Mann, gesehen im Lichte der Goetheschen Phänomenallehre.* Las Vegas: Peter Lang, 1978.

Kumar, Shiv K. *Bergson and the Stream of Consciousness Novel.* London: Blackie, 1962.

——. "Space-Time Polarity in *Finnegans Wake.*" *Modern Philology* 54 (1957): 230-33.

Lankford, W. T. "Deep of Time: Narrative Order in *David Copperfield.*" *English Literary History* 46 (Fall 1979): 452-67.

Lay, M. M. "Temporal Ordering in the Fiction of Margaret Drabble." *Critique* 21.3 (1980): 73-84.

Lee, D.H. "Denial of Time and the Failure of Moral Choice: Camus's *The Stranger*, Faulkner's *Old Man*, Wright's *The Man Who Lived Underground.*" *College Language Association J* 23 (March 1980): 364-71.

Leech, Clifford. "The Shaping of Time: *Nostromo* and *Under the Volcano.*" *Imagined Worlds.* Ed. Maynard Mack and Ian Gregor. London: Methuen, 1968. 233-42.

Lehman, B. H. "Of Time, Personality, and the Author: A Study of *Tristram Shandy.*" *Studies in the Comic, U of California Publications in English* 8 (1941): 233-50.

Le Poidevin, R. "Time and Truth in Fiction." *British J of Aesthetics* 28 (Summer 1988): 248-58.

Lodge, David. "Thomas Hardy and Cinematographic Form." *Novel* 7 (1974): 246-54.

Macey, Samuel L. "Clocks and Chronology in the Eighteenth-Century Novel." *Eighteenth-Century Life* 7.2 (January 1982): 96-104.

——. "The Linear and Circular Time Schemes in Sterne's *Tristram Shandy.*" *Notes and Queries* n.s., 36.4 (December 1989): 477-79.

——. "A Review of Alkon's *Defoe and Fictional Time.*" *Modern Philology* 78.4 (May 1981): 435-38.

——. "The Time Schemes in *Moll Flanders.*" *Notes and Queries* n.s., 16.9 (September 1969): 336-37.

Macksey, Richard. "Architecture of Time: Dialectics and Structure."

Proust. Ed. René Girard. Englewood Cliffs: Prentice-Hall, 1962. 104-21.

Maclay, Joanna H. "The Aesthetics of Time in Narrative Fiction." *Speech Teacher* 18 (1969): 194-96.

Manierre, Virginia. "How Long Is a Minute?" *New Mexico Quarterly* 26 (1956): 238-48. About Thomas Mann.

Mein, Margaret. *Proust's Challenge to Time*. Manchester: Manchester UP, 1962.

Mendilow, Adam A. *Time and the Novel*. 1952; reprint, New York: Humanities P, 1972.

Mizruchi, S. L. "The Politics of Temporality in *The Bostonians*." *Nineteenth-Century Fiction* 40 (September 1985): 187-215.

Moloney, Michael F. "The Enigma of Time: Proust, Virginia Woolf, and Faulkner." *Thought* 32 (1957): 69-85.

Monteiro, George. "Bankruptcy in Time: A Reading of Faulkner's *Pylon*." *Twentieth-Century Literature* 4 (1958): 9-20.

Moore, S. "Chronological Difficulties in the Novels of William Gaddis." *Critique* 22.1 (1980): 79-91.

Morris, Jill. *Time and Timelessness in Virginia Woolf*. Hicksville: Exposition P, 1977.

Motz, Reighard. *Time As Joyce Tells It*. Mifflinburg: Mulford Colebrook, 1978.

Moyer, Patricia. "Time and the Artist in Kafka and Hawthorne." *Modern Fiction Studies* 4 (1958): 295-306.

Murillo, Louis Andrew. *The Golden Dial: Temporal Configuration in* Don Quijote. Oxford: Dolphin Book Company, 1975.

Murray, H. "The Birthday of Bloom the Apostle." *James Joyce Quarterly* 21 (Spring 1984): 282-85.

Nettels, Elsa. "*The Ambassadors* and the Sense of the Past." *Modern Language Quarterly* 31 (1970): 220-35.

Neufeldt, Leonard N. "Time and Man's Possibilities in [Faulkner's] *Light in August*." *Georgia Review* 25 (1971): 27-40.

Newton-De Molina, David. "'Tempus edax rerum': A Note on an Aspect of Frank Kermode's Theory of Fiction." *Critical Quarterly* 12 (1970): 352-81.

Nodelman, P. "Interpretation and the Apparent Sameness of Children's Novels." *Studies in the Literary Imagination* 18 (Fall 1985): 5-20.

Norton, Roger C. *Hermann Hesse's Futuristic Idealism*. Frankfurt: Peter Lang, 1973.

O'Brien, R. A. "Time, Space, and Language in Lawrence Durrell." *Waterloo Review* 6 (1961): 16-24.

O'Donnell, P. "Journeying to the Center: Time, Pattern, and Transcendence in William Golding's *Free Fall.*" *Ariel* 11 (July 1980): 83-98.

O'Rourke, D. "Exiles in Time: Gallant's *My Heart is Broken.*" *Canadian Literature* 93 (Summer 1982): 98-107.

Osborne, L. MacKenzie. "The 'Chronological Frontier' in Thomas Hardy's Novels." *Studies in the Novel* 4 (1972): 543-55.

Park, S. "Time and Ruins in John Fowles's *Daniel Martin.*" *Modern Fiction Studies* 31 (Spring 1985): 157-63.

Parker, Dorothy. "The Time Scheme of *Pamela* and the Character of B." *Texas Studies in Literature and Language* 1 (1969): 695-704.

Paul, David. "Time and the Novelist." *Partisan Review* 21 (1954): 636-49.

Perlis, Alan D. "[Faulkner's] *As I Lay Dying* as a Study of Time." *South Dakota Review* 10.1 (1972): 103-10.

Petersen, C. R. "Time and Stress: *Alice in Wonderland.*" *J of the History of Ideas* 46 (July-September 1985): 427-33.

Pouillon, Jean. *Temps et roman*. Paris: Gallimard, 1946.

Pritchett, V. S. "Time Frozen: [Faulkner's] *A Fable.*" *Partisan Review* 21 (1954): 557-61.

Proust, Marcel. *À la recherche du temps perdu*. 3 vols. Paris: Gallimard, 1954.

——. *Remembrance of Things Past*. Trans. C. K. Scott Moncrieff, et al. 3 vols. New York: Random House, 1981.

Prusok, Rudi. "Science in Mann's *Zauberberg*: The Concept of Space." *Publications of the Modern Language Association* 88 (1973): 52-61.

Radandt, Friedhelm. "Transitional Time in Keller's *Züricher Novellen.*" *Publications of the Modern Language Association* 89 (1974): 77-84.

Radloff, B. "Time and Timepieces: A Note on Quentin's Section in [Faulkner's] *The Sound and the Fury.*" *English Language Notes* 23 (March 1986): 51-57.

Raleigh, John Henry. *Time, Place, and Idea: Essays on the Novel.* Carbondale: Southern Illinois UP, 1968.

Ramérey i Molas, Pere. *Tiempo y naeración*. Madrid: Gredos, 1978.

Read, Phyllis J. "The Illusion of Personality: Cyclical Time in Durrell's *Alexandria Quartet.*" *Modern Fiction Studies* 13 (1967): 389-99.

Renza, L. A. "Killing Time with Mark Twain's Autobiographies." *English Literary History* 54 (Spring 1987): 157-82.

Richter, Harvena. "Canonical Hours in [Woolf's] *Mrs Dalloway.*" *Modern Fiction Studies* 28 (Summer 1982): 236-40.

——. *Virginia Woolf: The Inward Voyage*. Princeton: Princeton UP, 1970.

Robbe-Grillet, A. "Time and Description in Fiction Today." *For a New Novel*. Trans. R. Howard. New York: Grove P, 1965. 143-56.

Robson, Vincent. "The Psychosocial Conflict and the Distortion of Time: A Study of Diver's Disintegration in [Fitzgerald's] *Tender Is the Night*." *Language and Literature* 1.2 (1972): 55-64.

Rogers, Frederick R. S. "Charles Dickens and Horology." *Antiquarian Horology* 7 (December 1970): 60-65.

Rubin, Louis D., Jr. "Joyce and Sterne: A Study in Affinity." *Hopkins Review* 3.2 (1950): 14-22.

Sallé, Jean Claude. "Sterne's Distinction of Time and Duration." *Review of English Studies* n.s., 6 (1955): 180-82.

Schwarz, Peter Paul. *Aurora. Zur romantischen Zeitstruktur bei Eichendorff*. Berlin: Gehlen, 1970.

Shattuck, Roger. *Proust's Binoculars*. New York: Random House, 1963.

Shaw, Harry E. *The Forms of Historical Fiction: Sir Walter Scott and His Successors*. Ithaca: Cornell UP, 1983.

Sherbo, Arthur. "Time and Place in Richardson's *Clarissa*." *Boston U Studies in English* 3 (1957): 139-46.

——. "The Time-Scheme in [Fielding's] *Amelia*." *Boston U Studies in English* 4 (1960): 223-28.

Slade, Joseph W. "The Functions of Eternal Recurrence in Thomas Mann's *Joseph and His Brothers*." *Symposium* 25 (1971): 180-97.

Slochower, Harry. "Marcel Proust: Revolt Against the Tyranny of Time." *Sewanee Review* 51 (1943): 370-81.

Smedman, M. S. "Springs of Hope: Recovery of Primordial Time in 'Mythic' Novels for Young Readers." *Children's Literature* 16 (1988): 91-107. Includes bibliography.

Smith, M. "Chronology of [Faulkner's] *Go Down, Moses*." *Mississippi Quarterly* 36 (Summer 1983): 319-28.

Smyth, W. F. "Lawrence Durrell: Modern Love in Chamber Pots and Space-Time." *Edge* 2 (Spring 1964): 105-16.

Soulsby, Marlene Pilarcik [Marlene A. Pilarcik]. "Beginnings and Endings: Hesse and Kawabata." *The Study of Time IV*. Ed. J. T. Fraser, N. Lawrence, and D. Park. New York: Springer, 1981. 119-35.

——. "Dialectics and Change in Kawabata's *The Master of Go*." *Modern Language Studies* 16.4 (Fall 1986): 9-21.

——. "Zen and the Way of Reading Hermann Hesse's *Das Glasperlenspiel*." *Studia Mystica* 1.1 (Spring 1988): 3-17.

Spencer, Sharon. *Space, Time and Structure in the Modern Novel.* New York: New York UP, 1971.

Spiegel, Alan. "Flaubert to Joyce: Evolution of a Cinematographic Form." *Novel* 6 (1973): 229-43.

Stallman, R. W. "Time and [Conrad's] *The Secret Agent,*" and "Gatsby and the Hole in Time." In his *The Houses that James Built*. East Lansing: Michigan State UP, 1961. 111-50.

———. "Time and the Unnamed Article in *The Ambassadors.*" *Modern Language Notes* 72.1 (January 1957): 27-32.

Steel, Gareth H. *Chronology and Time in* À la recherche du temps perdu. Geneva: Droz, 1979.

Stein, William Bysshe. "*Almayer's Folly*: The Terrors of Time." *Conradiana* 1.1 (1968): 27-34.

———. "Conrad's East: Time, History, Action, and *Maya.*" *Texas Studies in Literature and Language* 7 (1965): 265-83.

Steinberg, Erwin R. *The Stream of Consciousness and Beyond in* Ulysses. Pittsburgh: U of Pittsburgh P, 1973.

Steinmetz-Schönemann, Helga. *Die Bedeutung der Zeit in den Romanen von Marguerite Duras*. Amsterdam: Rodopi, 1976.

Sterne, Laurence. *The Life and Opinions of Tristram Shandy, Gentleman*. Ed. James Aitken Work. Indianapolis: Odyssey P, 1940.

Stevick, Philip. "Fielding and the Meaning of History." *Publications of the Modern Language Association* 86 (1971): 561-68.

Stone, Edward. "From James to Balderston: Relativity and the 20s." *Modern Fiction Studies* 1.2 (1955): 2-11.

Stone, Harry. "Dickens and the Interior Monologue." *Philological Quarterly* 38 (1959): 52-65.

Struc, Roman S. "*The Magic Mountain*: Time and Timelessness." *Research Studies* 39 (1971): 83-95.

Struve, Gleb. "*Monologue intérieur*: The Origins of the Formula and the First Statement of Its Possibilities." *Publications of the Modern Language Association* 69 (1954): 1101-11.

Sutherland, J. A. "The Handling of Time in *Vanity Fair.*" *Anglia* 89 (1971): 349-56.

Swann, C. "Dating the Action of [Melville's] *Bartleby.*" *Notes and Queries* n.s., 32 (September 1985): 357-58.

Swiggart, Peter. "Moral and Temporal Order in [Faulkner's] *The Sound and the Fury.*" *Sewanee Review* 61 (1953): 221-37.

———. "Time in Faulkner's Novels." *Modern Fiction Studies* 1.2 (1955): 25-29.

Thieberger, Richard. *Der Begriff der Zeit bei Thomas Mann, vom* Zauberberg *zum* Joseph. Baden-Baden: Verlag für Kunst und

Wissenschaft, 1952.

Tintner, A. R. "The Great Condition: Henry James and Bergsonian Time." *Studies in Short Fiction* 21 (Spring 1984): 111-15.

Tintner, Sylvie. *Colette et le temps surmonté*. Geneva: Slatkine, 1980.

Tobin, Patricia D. *Time and the Novel: The Genealogical Imperative*. Princeton: Princeton UP, 1978.

Tompkins, Jane P. "The Redemption of Time in [James's] *Notes of a Son and Brother*." *Texas Studies in Literature and Language* 14 (1973): 681-90.

Torgovnick, M. "Closure and the Shape of Fictions: The Example of *Women in Love*." *The Study of Time IV*. Ed. J. T. Fraser, N. Lawrence, and D. Park. New York: Springer, 1981. 147-58.

Vande Kieft, Ruth M. "Faulkner's Defeat of Time in *Absalom, Absalom!*" *Southern Review* 6 (1970): 1100-09.

Vickery, Olga W. "Faulkner and the Contours of Time." *Georgia Review* 12 (1958): 192-201.

Warner, N. O. "The Texture of Time in *War and Peace*." *Slavic and East European J* 28 (Summer 1984): 192-204.

Warren, Robert Penn. "Faulkner: The South, the Negro, and Time." *Faulkner*. Ed. R. P. Warren. Englewood Cliffs: Prentice-Hall, 1966. 251-71.

Wasserman, L. "The Music of Time: Henri Bergson and Willa Cather." *American Literature* 57 (May 1985): 226-39.

Whitehead, Lee M. "The Moment of Time: Golding's *Pincher Martin*." *Contemporary Literature* 12 (1971): 18-41.

Widmer, Kingsley. "Timeless Prose." *Twentieth-Century Literature* 4 (1958): 3-8. About e.e. cummings's *The Enormous Room*.

Williams, Mary Ellen. *A Vast Landscape: Time in the Novels of Thornton Wilder*. Pocatello: Idaho State UP, 1979.

Wilson, Jame S. "Time and Virginia Woolf." *Virginia Quarterly Review* 18 (1942): 267-76.

Winter, J. L. "Chronology of James's *Washington Square*." *Notes and Queries* n.s., 28 (October 1981): 426-28.

Zink, Karl E. "Flux and the Frozen Moment: The Imagery of Stasis in Faulkner's Prose." *Publications of the Modern Language Association* 71 (1956): 285-301.

Ziolkowski, Theodore. "The Discordant Clocks." In his *Dimensions of the Modern Novel: German Texts and European Contexts*. Princeton: Princeton UP, 1969. Ch. 6.

——. "Hermann Broch and Relativity in Fiction." *Wisconsin Studies in Contemporary Literature* 8 (1967): 365-76.

E. LANGUAGE

Baldo, J. "Solitude as an Effect of Language in Garcia Márquez's *Cien años de soledad.*" *Criticism* 30 (Fall 1988): 467-96.

Barton, W. B., Jr. "Time and Language: Another Look at St. Augustine's Question." *Southern J of Philosophy* 5 (1967): 200-05.

Black, Max. *Models and Metaphors: Studies in Language and Philosophy.* Ithaca: Cornell UP, 1962.

Braude, S. E. "Tensed Sentences and Free Repeatability." *Philosophical Review* 82.4 (1973): 188-214.

——. "Are Verbs Tensed or Tenseless?" *Philosophical Studies* 25.8 (1974): 373-90.

Bronckart, J. P., and H. Sinclair. "Time, Tense and Aspect." *Cognition* 2.1 (1973): 107-30.

Bruneau, Tom. "Chronemics and the Verbal-Nonverbal Interface." *The Relationship of Verbal and Nonverbal Communication.* Ed. M. Key. The Hague: Mouton, 1980. 101-17.

——. "Communicative Silences: Forms and Functions." *J of Communication* 23 (1973): 17-46.

——. "Time and Nonverbal Communication." *J of Popular Culture* 8 (1974): 658-66.

Capra, S. "Il problema del linguaggio in Maurice Merleau-Ponty." *Rivista di Filosofia Neo-Scolastica* 64.3 (1972): 446-70.

Chomsky, Noam. *Language and Mind.* New York: Harcourt, Brace, 1972.

——. "Verbal Behaviour." *Language* 35 (1959): 26-58.

Edgerton, W. D. "Words and the Word." *Theology Today* 44 (January 1988): 462-78.

Ferreiro, E., and H. Sinclair. "Temporal Relationships in Language." *International J of Psychology* 6.1 (1971): 39-47.

Fillenbaum, S. "Syntactic Locus as a Determinant of Judged Pause Duration." *Perception and Psychophysics* 9.2B (1971): 219-21.

Gibbons, K. "Planes of Language and Time: The Surfaces of the Miranda Stories." *Kenyon Review* n.s., 10 (Winter 1988): 74-79.

Goldblatt, Robert. *Logics of Time and Computation.* Chicago: U of Chicago P, 1987.

Gonseth, F. "Temps et syntaxe." *Dialectica* 22 (1968): 206-13.

Huggins, A. W. "On the Perception of Temporal Phenomena In Speech." *J. Acoustical Society of America* 51.4 Part 2 (1972): 1279-90.

Jennings, R. E. "Subjunctive in Conditionals and Elsewhere." *Pacific*

Philosophical Quarterly 63 (April 1982): 146-56.

Koenig, E. "The Semantic Structure of Time Propositions in English." *International J of Language and Philosophy. Foundations of Language* 11.7 (1974): 551-63.

Krauss, H. H., and R. W. Persons. "Relationship of Time Focus to Distribution of Verb Tenses." *Psychological Reports* 25.1 (1969): 325-26.

Lawrence, N. "Levels of Language in Discourse about Time." *The Study of Time III*. Ed. J. T. Fraser, N. Lawrence, and D. Park. New York: Springer, 1978. 22-49.

Lenneberg, E. H. *Biological Foundations of Language*. New York: Wiley, 1967.

——. "A Biological Perspective of Language." *New Directions in the Study of Language*. Ed. E. H. Lenneberg. Cambridge: MIT P, 1964.

——. "Language in the Light of Evolution." *Animal Communication*. Ed. T. Sebok. Bloomington: Indiana UP, 1968.

Lunney, H. W. M. "Time as Heard in Speech and Music." *Nature* 249 (1974): 592.

Mackay, C. K., and W. P. Brown. "Metaphor Preference Versus Semantic Ratings as Measures of Attitude Toward Time." *J of General Psychology* 83.2 (1970): 207-12.

Mouton, D. L. "Thinking and Time." *Mind* 78.1 (1969): 60-76.

Prior, A. N. "Recent Advances in Tense Logic." *Monist* 53.7 (1969): 325-39.

Pyles, Thomas. *The Origins and Development of the English Language*. 2nd ed. New York: Harcourt, Brace, 1971.

Rebhorn, W. A. "Enduring Word: Language, Time and History in *Il libro del cortegiano*." *Modern Language Notes* 96 (January 1981): 23-40.

Rigotto, F. "Metaphors of Time." *Etc* 43 (Summer 1986): 157-68.

Seigman, Aron W., and Stanley Feldstein. *Of Speech and Time: Temporal Speech Patterns in Interpersonal Contexts*. Hillsdale: L. Erlbaum, 1979.

Simon, Josef. *Sprache und Raum: Philosophische Untersuchungen zum Verhältnis zwischen Wahrheit und Bestimmtheit von Sätzen*. Berlin: De Gruyter, 1969.

Stutterheim, C. F. P. "Time in Language and Literature." *The Voices of Time*. Ed. J. T. Fraser. 2nd ed. Amherst: U of Massachusetts P, 1981. 163-79.

Thass-Thienemann, Theodore. *The Subconscious Language*. New York: Washington Square, 1967.

Thornton, Harry, and Agathe Thornton. *Time and Style: A Psycho-*

Linguistic Essay in Classical Literature. London: Methuen, 1962.
Tiedge, J. T., and K. J. Ksobiech. "Timeliness Cues and Perceived Immediacy in Radio News." *Journalism Quarterly* 59 (Summer 1982): 223-29.
Tschumi, R. "Langage, temps et liberté." *Dialectica* 24 (1970): 87-110.
Urey, D. S. "Language and Temporal Texture." *Research in Phenomenology* 13 (1983): 221-30.

F. POETRY

Adelbert, Claud. "'That Two-Handed Engine' Will Smite: Time Will Have a Stop." *Studies in Philology* 59 (April 1962): 184-200.
Alexander, Ian W. "Valéry and Yeats: The Rehabilitation of Time." *Scottish Periodical* 1.1 (Summer 1947): 77-106.
Baker, Jeffrey. "Time and Judgment in 'The Ruined Cottage': A Reading of Wordsworth's *Excursion*, Book I." *Antigonish Review* 1.3 (1970): 3-23.
——. *Time and Mind in Wordsworth's Poetry*. Detroit: Wayne State UP, 1980.
Barolini, T. "The Making of a Lyric Sequence: Time and Narrative in Petrarch's *Rerum vulgarium fragmenta*." *Modern Language Notes* 104 (January 1989): 1-38.
Bate, W. Jackson. *The Burden of the Past and the English Poet*. New York: Norton, 1970.
Bergsten, Staffan. *Time and Eternity: A Study of the Structure and Symbolism of T. S. Eliot's* Four Quartets. Stockholm: Svenska bokförlaget, 1960.
Blackstone, Bernard. "'The Loops of Time': Spatio-Temporal Patterns in [Byron's] *Childe Harold*." *Ariel* 2.4 (1971): 5-17.
Boenig, R. "Time Markers and Treachery: The Crux at *Beowulf* 1130." *English Language Notes* 24 (June 1987): 1-9.
Brisman, Leslie. *Milton's Poetry of Choice and Its Romantic Heirs*. Ithaca: Cornell UP, 1973.
Chavkin, A. "Wordsworth's Secular Imagination and Spots of Time." *College Language Association J* 26 (June 1983): 452-64.
Chell, Samuel L. *The Dynamic Self: Browning's Poetry of Duration*. Victoria: ELS Monograph Series, 1984.
Cirillo, Albert R. "'Hail Holy Light' and Divine Time in *Paradise Lost*." *J of English and Germanic Philology* 68 (1969): 45-56.
——. "Noon-Midnight and the Temporal Structure of *Paradise Lost*."

English Literary History 29 (1962): 372-95.

Clark, K. "Synchronous Isolations: Norman Dubie's 'Elegies for the Ochre Deer on the Walls at Lascaux.'" *American Poetry* 5 (Winter 1988): 12-32.

Clendenning, John. "Time, Doubt and Vision: Notes on Emerson and T. S. Eliot." *American Scholar* 36 (1967): 125-32.

Colie, Rosalie L. "Time and Eternity: Paradox and Structure in *Paradise Lost*." *J of the Warburg and Courtauld Institutes* 23 (1960): 127-38.

Cope, Jackson I. "Time and Space in Miltonic Symbol." *English Literary History* 26 (1959): 497-513.

Curtis, F. B. "Blake and the 'Moment of Time': An 18th Century Controversy in Mathematics." *Philological Quarterly* 51 (1972): 460-70.

Daiches, David. *Time and the Poet*. Cardiff: U of Wales P, 1965.

Deane, P. "The Fate of Narrative in David Jones's *Anathemata*." *U of Toronto Quarterly* 57 (Winter 1987-88): 306-20. Includes bibliography.

Decavelles, A. "Time Versus Eternity: Odysseus Elytis in the 1980s." *World Literature Today* 62 (Winter 1988): 22-32.

De Luca, V. A. "Lost Traveller's Dream: Blake and the Seductions of Continuity." *Ariel* 11 (January 1980): 49-69.

Dorenkamp, Angela G. "Time and Sacrament in [David Jones's] *The Anathemata*." *Renascence* 23 (1971): 183-91.

Durr, Robert A. "Spenser's Calendar of Christian Time." *J of English Literary History* 24 (1957): 269-93.

Farrelly, James. "*Gerontion*: Time's Eunuch." *U of Dayton Review* 6.2 (1969): 27-34.

Findlay, L. M. "Culler and Byron on Apostrophe and Lyric Time." *Studies in Romanticism* 24 (Fall 1985): 335-53.

Fisher, P. "A Museum with One Work Inside: Keats and the Finality of Art." *Keats-Shelley J* 33 (1984): 85-102.

Foster, Steven. "Relativity and *The Waste Land*: A Postulate." *Texas Studies in Literature and Language* 7 (1965): 77-97.

Gish, Nancy K. *Time in the Poetry of T. S. Eliot: A Study in Structure and Theme*. Totowa: Barnes and Noble, 1981.

Goodkin, R. E. "Zeno's Paradox: Mallarmé, Valéry, and the Symbolist 'Movement.'" *Yale French Studies* 74 (1988): 133-56.

Grafton, A., and N. M. Swerdlow. "Greek Chronography in Roman Epic: The Calendrical Date of the Fall of Troy in the *Aeneid*." *Classical Quarterly* n.s., 36.1 (1986): 212-18.

Grant, Patrick. "Time and Temptation in *Paradise Regained*." *U of*

Toronto Quarterly 43 (1973): 32-47.

Grimes, Ronald L. "Time and Space in Blake's Major Prophecies." *Blake's Sublime Allegory*. Ed. Stuart Curran, and Joseph A. Wittreich. Madison: U of Wisconsin P, 1973. 59-81.

Gruber, W. "Temporal Perspectives in Robert Browning's *A Death in the Desert*." *Victorian Poetry* 17 (Winter 1979): 329-42.

Halli, R. W., Jr. "The Double Hieroglyph in George Herbert's 'Easter-Wings.'" *Philological Quarterly* 63 (Spring 1984): 265-72.

Harmon, William. *Time in Ezra Pound's Work*. Chapel Hill: U of North Carolina P, 1977.

Hatto, E. T., ed. *Eos: An Inquiry Into the Theme of Lovers' Meetings and Partings at Dawn in Poetry*. The Hague: Mouton, 1965.

Heninger, S. K., Jr. *Touches of Sweet Harmony: Pythagorean Cosmology and Renaissance Poetics*. San Marino: Huntington Library, 1974.

Hirsch, David H. "T. S. Eliot and the Vexation of Time." *Southern Review* n.s., 3 (1967): 608-24.

Hoeweler, C. *Het Rhythme in Vers en Muziek*. The Hague: Mouton, 1952.

Jackson, Richard. "Deconstructed Moment in Modern Poetry." *Contemporary Literature* 23 (Summer 1982): 306-22.

——. *The Dismantling of Time in Contemporary Poetry*. Tuscaloosa: U of Alabama P, 1988.

Kamholtz, J. Z. "Imminence and Eminence in Donne." *J of English and German Philology* 81 (October 1982): 480-91.

Korg, Jacob. "Modern Art Techniques in *The Waste Land*." *J of Aesthetics and Art Criticism* 18 (1959-60): 456-63.

Lewalski, Barbara K. "Time and History in *Paradise Regained*." *The Prison and the Pinnacle*. Ed. Balachandra Rajan. Toronto: U of Toronto P, 1973. 49-81.

Lyons, Bridget G. "Milton's 'Il Penseroso' and the Idea of Time." In her *Voices of Melancholy*. London: Routledge and Kegan Paul, 1971. 149-61.

MacCallum, Reid. "Time Lost and Regained." In his *Imitation and Design*. Toronto: U of Toronto P, 1953. 132-61. About T. S. Eliot.

Marken, Ronald. "'Eternity in an Hour'—Blake and Time." *Discourse* 9 (1966): 167-83.

Martino, Mario. *Il problema del tempo nei sonetti di Shakespeare*. Rome: Bulzoni, 1985.

Martz, Louis L. "The Wheel and the Point: Aspects of Imagery and Theme in Eliot's Later Poetry." *Sewanee Review* 55 (1947): 126-47.

Miner, Earl. "Forms of Perception: Time and Place." In his *The Metaphysical Mode from Donne to Cowley*. Princeton: Princeton

UP, 1969. Ch. 2.

———. "The Ruins and Remedies of Time." In his *The Cavalier Mode from Jonson to Cowley*. Princeton: Princeton UP, 1971. Ch. 3.

Morris, R. "Aspects of Time and Place in the French Arthurian Verse Romances." *French Studies* 42 (July 1988): 257-77.

Morse, Donald E. "Meaning of Time in Auden's *For the Time Being*." *Renascence* 22 (1970): 162-68.

Newbolt, Sir Henry. "Poetry and Time." *Proceedings of the British Academy* (1917-18): 487-503.

Olson, Paul R. *Circle of Paradox: Time and Essence in the Poetry of Juan Ramón Jiménez*. Baltimore: Johns Hopkins P, 1967.

Parry, Idris. "Space and Time in Rilke's *Orpheus Sonnets*." *Modern Language Review* 58 (1963): 524-31.

Pearlman, Daniel D. *The Barb of Time: On the Unity of Ezra Pound's Cantos*. New York: Oxford UP, 1969.

Radin, A. P. "Sunrise, Sunset: *èmos* in Homeric Epic." *American J of Philology* 109 (Fall 1988): 293-307.

Rosenblatt, Jason P. "Structural Unity and Temporal Concordance: The War in Heaven in *Paradise Lost*." *Publications of the Modern Language Association* 87 (1972): 31-41.

Ross, Malcolm. "Time, the Timeless, and the Timely: Notes on a Victorian Poem [*In Memoriam*]." *Proceedings and Transactions of the Royal Society of Canada* 4th series, 9 (1971): 219-34.

Russ, Jon R. "Time's Attributes in Shakespeare's Sonnet 126." *English Studies* 52 (1971): 318-23.

Salinger, Herman. "Time in the Lyric." *Studies in German Literature of the 19th and 20th Centuries*. Ed. Siegfried Mews. Chapel Hill: U of North Carolina P, 1970. 157-73.

Sheppard, R. W. "Rilke's *Duiniser Elegien*: A Critical Appreciation in the Light of Eliot's *Four Quartets*." *German Life and Letters* 20 (1966-67): 205-18.

Shichtman, M. B. "*Sir Gawain and the Green Knight*: A Lesson in the Terror of History." *Papers on Language and Literature* 22 (Winter 1986): 3-15.

Smith, Barbara Herrnstein. *Poetic Closure*. Chicago: U of Chicago P, 1968.

Smoot, Jean J. *The Poets and Time: A Comparison of Plays by John Millington Synge and Frederico Garcia Lorca*. Madrid: J. Porrua Turanzas, 1978.

Snyder, S. "Nature, History, and the Waters of *Lycidas*." *Huntington Library Quarterly* 50 (Summer 1987): 323-35.

Spears, Monroe K. "Auden and the Music of Time." *Southern Review*

18 (January 1982): 25-43.

———. "For the Time Being." *Auden*. Ed. M. K. Spears. Englewood Cliffs: Prentice-Hall, 1964. 160-71.

Staiger, Emil. *Die Zeit als Einbildungskraft des Dichters: Untersuchungen zu Gedichten von Brentano, Goethe und Keller*. Zürich: M. Niehans, 1939.

Stapleton, Laurence. "Perspectives of Time in *Paradise Lost*." *Philological Quarterly* 45 (1966): 734-48.

Stelzig, Eugene L. *All Shades of Consciousness: Wordsworth's Poetry and the Self in Time*. The Hague: Mouton, 1975.

Stewart, Stanley. "Time and [Herbert's] *The Temple*." *Studies in English Literature* 6 (1966): 97-110.

Tayler, Edward William. *Milton's Poetry: Its Development in Time*. Pittsburgh: Dusquesne UP, 1979.

Tennenhouse, Leonard. "*Beowulf* and the Sense of History." *Bucknell Review* 19.3 (1971): 137-46.

Turner, Frederick. "Poiesis: Time and Artistic Discourse." *The Study of Time III*. Ed. J. T. Fraser, N. Lawrence, and D. Park. New York: Springer, 1978. 614-33.

———. "Space and Time in Chinese Verse." *Time, Science, and Society in China and the West: The Study of Time V*. Ed. J. T. Fraser, N. Lawrence, and F. C. Haber. Amherst: U of Massachusetts P, 1986. 241-51.

Underwood, Horace H. "Time and Space in the Poetry of Vaughan." *Studies in Philology* 69 (1972): 231-41.

Verheul, Kees. *The Theme of Time in the Poetry of Anna Axmatova*. The Hague: Mouton, 1971.

Verma, Rajendra. *Time and Poetry in Eliot's* Four Quartets. Atlantic Highlands: Humanities P, 1979.

Whitaker, Thomas R. *Swan and Shadow: Yeats' Dialogue with History*. Chapel Hill: U of North Carolina P, 1964.

Wilson, Robert R. "The Deformation of Narrative Time in *The Faerie Queene*." *U of Toronto Quarterly* 41 (1971): 48-62.

———. "Images and 'Allegoremes' of Time in the Poetry of Spenser." *English Literary Renaissance* 4 (1974): 56-82.

Wittreich, Joseph Anthony, Jr. *Visionary Poetics: Milton's Tradition and His Legacy*. San Marino: Huntington Library, 1979.

Yacobi, T. "Time Denatured Into Meaning: New Worlds and Renewed Themes in the Poetry of Dan Pagis." *Style* 22 (Spring 1988): 93-115. Includes bibliography.

G. PROPHECY AND ALTERNATE HISTORY (UCHRONIA)

Churchill, Winston S. "If Lee Had Not Won the Battle of Gettysburg." *If It Had Happened Otherwise: Lapses into Imaginary History*. Ed. John Collings Squire. London: Longmans, 1931.

Collins, W. J. "Hank Morgan in the Garden of Forking Paths: *A Connecticut Yankee in King Arthur's Court* as Alternative History." *Modern Fiction Studies* 32 (Spring 1986): 109-14.

Davidson, James West. *The Logic of Millennial Thought: Eighteenth-Century New England*. New Haven: Yale UP, 1977.

Decouflé, André-Clément, and Alain-Michel Villemur. *Les Millésimes du futur: Contributions à une bibliographie des anticipations datées*. Paris: Laboratoire de Prospective Appliquée et Temps Futurs, 1978.

Dick, Philip K. *The Man in the High Castle*. New York: Putnam's, 1962.

Galbreath, Robert. "Ambiguous Apocalypse: Transcendental Versions of the End." *The End of the World*. Ed. Eric S. Rabkin, et al. Carbondale: Southern Illinois UP, 1983.

Geoffroy, Louis. *Napoléon et la conquête du monde—1812-1832—Histoire de la monarchie universelle (Napoléon apocryphe)*. 1836: reprint, Paris: Éditions Tallandier, 1983.

Hacker, Barton C., and Gordon B. Chamberlain. "Past that Might Have Been: An Annotated Bibliography of Alternate History." *Extrapolation* 22 (Winter 1981): 334-78.

Ketterer, David. *New Worlds for Old: The Apocalyptic Imagination, Science Fiction, and American Literature*. Bloomington: Indiana UP, 1974.

Le Brun, Claire. "Les chansons de geste: La tentation de l'uchronie au moyen âge." *Imagine: Revue de Science-Fiction Québécoise* 4 (Autumn 1982): 44-49.

McKeon, Michael. "Eschatological Prophecy." In his *Politics and Poetry in Restoration England: The Case of Dryden's* Annus Mirabilis. Cambridge: Harvard UP, 1975. Part 2.

Manuel, Frank E. "Prophecy and History." In his *A Portrait of Isaac Newton*. Cambridge: Harvard UP, 1968. 361-80.

Newton, Isaac. *Observations Upon the Prophecies of Daniel, and the Apocalypse of St. John*. London, 1733.

Newton, Thomas. *Dissertations on the Prophecies, Which Have Remarkably Been Fulfilled, and at This Time Are Fulfilling in the World*. 3rd ed. 3 vols. London, 1766.

Nicolson, Marjorie Hope. *Mountain Gloom and Mountain Glory: The*

Development of the Aesthetics of the Infinite. Ithaca: Cornell UP, 1959.

Patrides, C. A., and Joseph Wittreich, eds. *The Apocalypse in English Renaissance Thought and Literature: Patterns, Antecedents and Repercussions*. Ithaca: Cornell UP, 1984.

Scouten, Arthur H., and Robert D. Hume. "Pope and Swift: Text and Interpretation of Swift's Verses on His Death." *Philological Quarterly* 52 (1973): 205-31.

Tuveson, Ernest L. *Millennium and Utopia: A Study in the Background of the Idea of Progress*. Berkeley: U of California P, 1949.

Waugh, Charles G., and Martin H. Greenberg, eds. *Alternative Histories: Eleven Stories of the World as It Might Have Been*. New York: Garland, 1986. This also includes the bibliography by Hacker and Chamberlain listed above.

H. UTOPIAN, FUTURISTIC, AND SCIENCE FICTION

Aldiss, Brian W. *Billion Year Spree: The True History of Science Fiction*. 1973; reprint, New York: Schocken Books, 1975.

——. *Trillion Year Spree: The History of Science Fiction*. New York: Atheneum, 1986.

Alkon, Paul K. *Origins of Futuristic Fiction*. Athens: U of Georgia P, 1987.

——. "Origins of Futuristic Fiction: Felix Bodin's *Novel of the Future*." *Storm Warnings: Science Fiction Confronts the Future*. Ed. George E. Slusser, Colin Greenland, and Eric S. Rabkin. Carbondale: Southern Illinois UP, 1987. 21-33.

——. "The Paradox of Technology in Mercier's *L'An 2440*." *Utopian Vision, Technological Innovation and Poetic Imagination*. Ed. Klaus L. Berghahn, and Reinhold Grimm. Heidelberg: Carl Winter Universitätsverlag, 1990. 43-62.

——. "Samuel Madden's *Memoirs of the Twentieth Century*." *Science-Fiction Studies* 12 Part 2 (July 1985): 184-201.

Angenot, Marc. "The Emergence of the Anti-Utopian Genre in France: Souvestre, Giraudeau, Robida, et al." *Science-Fiction Studies* 12 (July 1985): 129-35.

[Banim, John]. *Revelations of the Dead-Alive*. London, 1824.

Bodin, Félix. *Le roman de l'avenir*. Paris, 1834.

A Brief Description of the Future History of Europe, from Anno 1650, to An. 1710. [London], 1650.

Burnet, Thomas. *The Sacred History of the Earth: Containing an Account of the Original of the Earth and of All the General Changes Which It Hath Already Undergone or Is to Undergo, till the Consumation of All Things*. London, 1684.

Cioranescu, Alexandre. "*Epigone*, le premier roman de l'avenir." *Revue des Sciences Humaines* 39 (July-September 1974): 441-48.

Clarke, I. F. *The Pattern of Expectation, 1644-2001*. New York: Basic Books, 1979.

——. *Tale of the Future from the Beginning to the Present Day*. 3rd ed. London: Library Association, 1978. Includes bibliography for 1644-1976.

Craig, W. L. "Tachyons, Time Travel, and Divine Omniscience." *J of Philosophy* 85 (March 1988): 135-50.

Crick, Bernard. *George Orwell: A Life*. 1980; reprint, Harmondsworth: Penguin, 1982. See pp. 546 and 582-85 for Orwell "hesitating between [the titles] NINETEEN EIGHTY-FOUR and THE LAST MAN IN EUROPE."

De Lesser, A. Creuzé. *Le dernier homme, Poëme imité de Grainville*. Paris, 1831.

De Palacio, Jean. "Mary Shelley and the 'Last Man': A Minor Romantic Theme." *Revue de Littérature Comparée* 42 (January-March 1968): 37-49.

Duyfhulzen, B. "On the Writing of Future-History: Beginning the Ending in Doris Lessing's *The Memoirs of a Survivor*." *Modern Fiction Studies* 26 (Spring 1980): 147-56.

Eizykman, B. "Temporality in Science-Fiction Narrative." Ed. R. M. Philmus. Trans. R. Rosenthall, and B. Eizykman. *Science-Fiction Studies* 12 (March 1985): 66-87.

Flammarion, Camille. *La fin du monde*. Paris, 1894.

Fleischman, Suzanne. *The Future in Thought and Language: Diachronic Evidence from Romance*. Cambridge: Cambridge UP, 1982.

Fredericks, Casey. *The Future of Eternity: Mythologies of Science Fiction and Fantasy*. Bloomington: Indiana UP, 1982.

Gagne, Elise. *Omégar ou le dernier homme: Proso-poésie dramatique de la fin des temps en douze chants*. [Paris, 1859].

Gordon, A. "Back to the Future: Oedipus as Time Traveller." *Science-Fiction Studies* 14 (November 1987): 372-85. Includes bibliography.

Gove, Philip Babcock. *The Imaginary Voyage in Prose Fiction: A History of Its Criticism and a Guide for Its Study, with an Annotated Check List of 215 Imaginary Voyages from 1700 to 1800*. London: Holland P, 1961. Bibliography 403-20.

Grainville, Jean-Baptiste Cousin de. *The Last Man, or Omegarus and*

Syderia, A Romance in Futurity. Anonymous translation. 2 vols. London: R. Dutton, 1806.

[Guttin, Jacques]. *Epigone, histoire du siècle futur*. Paris: Pierre Lamy, 1659.

Hansen, Arlen J. "The Meeting of Parallel Lines: Science, Fiction, and Science Fiction." *Bridges to Fantasy*. Ed. George E. Slusser, et al. Carbondale: Southern Illinois UP, 1982. 51-58.

Hollinger, V. "Deconstructing the Time Machine." *Science-Fiction Studies* 14 (July 1987): 201-21. Includes bibliography.

Hume, Kathryn. *Fantasy and Mimesis*. London: Methuen, 1984.

Huntington, John. "Olaf Stapledon and the Novel About the Future." *Contemporary Literature* 22.3 (1981): 349-65.

——. *The Logic of Fantasy: H. G. Wells and Science Fiction*. New York: Columbia UP, 1982.

——. "Remembrance of Things to Come: Narrative Technique in *Last and First Men*." *Science-Fiction Studies* 9 (1982): 257-64.

Jackson, Rosemary. *Fantasy: The Literature of Subversion*. London: Methuen, 1981. Includes bibliographical references.

Jameson, Fredric. "Progress Versus Utopia; or, Can We Imagine the Future?" *Science-Fiction Studies* 9 (July 1982): 147-58.

Lake, D. J. "Wells's Time Traveller: An Unreliable Narrator?" *Extrapolation* 22 (Summer 1981): 117-26.

The Last Peer. 3 vols. London, 1851.

McConnell, Frank. *The Science Fiction of H. G. Wells*. Oxford: Oxford UP, 1981.

Macey, Samuel L. "Clocks and Time in the Dystopias: Zamyatin's *We* and Huxley's *Brave New World*." *Explorations: Essays in Comparative Literature*. Ed. Makoto Ueda. New York: UP of America, 1986. 24-43.

——. "George Orwell's *1984*: The Future that Becomes the Past." *English Studies in Canada* 11.4 (December 1985): 450-58. Reprinted in *George Orwell: A Reassessment*. Ed. Peter Buitenhuis, and Ira B. Nadel. London: Macmillan, 1988. 23-31, and 189.

[Madden, Samuel]. *Memoirs of the Twentieth Century*. London: Osborn, et al., 1733.

Majewski, Henry F. "Grainville's *Le dernier homme*." *Symposium* 17 (Summer 1963): 114-22.

Manuel, Frank E., and Fritzie P. Manuel. *Utopian Thought in the Western World*. Cambridge: Belknap P of Harvard UP, 1979.

Mercier, Louis-Sébastien. *L'an deux mille quatre cent quarante: Rêve s'il en fût jamais*. 3 vols. Paris: Brosson and Carteret, An VII [actually 1798].

——. *L'an deux mille quatre cent quarante: Rêve s'il en fût jamais*. Ed. Raymond Trousson. Bordeaux: Éditions Ducros, 1971.

Morgan, Chris. *The Shapes of Futures Past: The Story of Prediction*. Exeter, England: Webb and Bower, 1980.

Mrs. Maberly; or, The World as It Will Be. 3 vols. London, 1836.

Nicolson, Marjorie Hope. *Voyages to the Moon*. 1948; reprint, New York: Macmillan, 1960.

Philmus, Robert M. *Into the Unknown: The Evolution of Science Fiction from Francis Godwin to H. G. Wells*. Berkeley: U of California P, 1970.

Porter, Charles A. *Restif's Novels, or An Autobiography in Search of an Author*. New Haven: Yale UP, 1967.

Poster, Mark. *The Utopian Thought of Restif de la Bretonne*. New York: New York UP, 1971.

Rabkin, Eric S., Martin H. Greenberg, and Joseph D. Olander, eds. *The End of the World*. Carbondale: Southern Illinois UP, 1983.

The Reign of George VI, 1900-1925. London, 1763.

[Restif de la Bretonne, Nicolas-Edme]. *Les posthumes; Lettres recues après la mort du mari, par sa femme, qui le croit à Florence*. Par feù Cazotte. 4 vols. Paris, 1802.

Rose, Mark. *Alien Encounters: Anatomy of Science Fiction*. Cambridge: Harvard UP, 1981.

Russell, M. S. "Time in Folklore and Science Fiction." *Foundation: The Review of Science Fiction* 43 (Summer 1988): 5-24.

Russen, David. *Iter Lunare: Or, A Voyage to the Moon....* London, 1703.

Salewski, Michael. *Zeitgeist und Zeitmaschine: Science Fiction und Geschichte*. Munich: Deutscher Taschenbuch Verlag, 1986.

Sambrook, A. J. "A Romantic Theme: The Last Man." *Forum For Modern Language Studies* 2 (January 1966): 25-33.

Scholes, Robert. *Structural Fabulation: An Essay on Fiction of the Future*. Notre Dame: U of Notre Dame P, 1975.

Scholes, Robert, and Eric S. Rabkin. *Science Fiction: History, Science, Vision*. New York: Oxford UP, 1977.

Shelley, Mary. *The Last Man*. Ed. Hugh J. Luke, Jr. Lincoln: U of Nebraska P, 1965.

Slusser, George E. "History, Historicity, Story." *Science-Fiction Studies* 15 (July 1988): 187-213.

Slusser, George E., Colin Greenland, and Eric S. Rabkin, eds. *Storm Warnings: Science Fiction Confronts the Future*. Carbondale: Southern Illinois UP, 1987.

Souvestre, Emile. *Le monde tel qu'il sera*. Paris: W. Coquebert, [1846].

Stone-Blackburn, S. "Science and Humanism in Gregory Benford's *Timescape.*" *Science-Fiction Studies* 15 (November 1988): 295-311.

Suvin, Darko. *Metamorphoses of Science Fiction: On the Poetics and History of a Literary Genre.* New Haven: Yale UP, 1979.

Todorov, Tzvetan. *Introduction à la littérature fantastique.* Paris: Éditions du Seuil, 1970.

Trousson, Raymond. *Voyages aux pays de nulle part: Histoire littéraire de la pensée utopique.* 2nd ed. Brussels: Éditions de L'Université de Bruxelles, 1979.

Versins, Pierre. *Encyclopédie de l'utopie, des voyages extraordinaires et de la science fiction.* Lausanne: Éditions L'Âge d'Homme, 1972.

Voss, Julius von. *Ini: Ein Roman aus dem einundzwanzigsten Jahrhundert.* Berlin, 1810.

Wagar, W. Warren. *Terminal Visions: The Literature of Last Things.* Bloomington: Indiana UP, 1982.

Mrs. Loudon [Jane Webb]. *The Mummy! A Tale of the Twenty-Second Century.* London: Frederick Warne, [1872].

Wells, H. G. *The Time Machine.* 1895; reprint, New York: Bantam Books, 1982.

Wilkie, Everett C., Jr. "Mercier's *L'An 2440*: Its Publishing History During the Author's Lifetime, Part I." *Harvard Library Bulletin* 32 (Winter 1984): 5-35.

——. "Mercier's *L'An 2440*: Its Publishing History During the Author's Lifetime, Part II: Bibliography." *Harvard Library Bulletin* 32 (Fall 1984): 348-400.

[Williams, R. F.] *Eureka: A Prophecy of the Future.* 3 vols. London, 1837.

14. MATHEMATICS

A. GENERAL (INCLUDING HISTORY OF)

Bede, the Venerable. *Bedae opera de temporibus*. Ed. C. W. Jones. Cambridge: The Mediaeval Academy of America, 1943.

Billingsley, P. *Probability and Measure*. 2nd ed. New York: Wiley, 1986.

Carnahan, B., H. A. Luther, and J. O. Wilkes. *Applied Numerical Methods*. New York: Wiley, 1969.

Clark, A. B., and R. L. Disney. *Probability and Random Processes for Engineers and Scientists*. New York: Wiley, 1970.

Coyle, R. G. *Mathematics for Business Decisions*. London: Nelson, 1971.

Davis, Philip J., and Reuben Hirsh. *Descartes' Dream: The World According to Mathematics*. San Diego: Harcourt, Brace, 1986.

Eddington, A. S. *The Mathematical Theory of Relativity*. 2nd ed. Cambridge: Cambridge UP, 1924.

Ekeland, Ivar. *Mathematics and the Unexpected*. Chicago: U of Chicago P, 1988.

Feller, W. *An Introduction to the Theory of Probability and Its Applications*. Vol. 2. 2nd ed. New York: Wiley, 1971.

Greenberg, M. D. *Foundations of Applied Mathematics*. Englewood Cliffs: Prentice-Hall, 1978.

Grize, Jean Blaise. *Essai sur le rôle du temps en analyse mathématique classique*. Neuchâtel, 1954.

Hendry, J. "The Evolution of William Rowan Hamilton's View of Algebra as the Science of Pure Time." *Studies in History and Philosophy of Science* 15 (March 1984): 63-81.

Hildebrand, F. B. *Advanced Calculus for Applications*. 2nd ed. Englewood Cliffs: Prentice-Hall, 1976.

Huntington, Edward V. *The Continuum and Other Types of Serial Order*. 2nd ed. Cambridge: Harvard UP, 1917.

Ifrah, Georges. *From One to Zero: A Universal History of Numbers*. Trans. Lowell Blair. New York: Viking Penguin, 1985.

Kline, M. *Mathematical Thought from Ancient to Modern Times*. New York: Oxford UP, 1972.

Lautman, Albert. *Symétrie et dissymétrie en mathématiques et en physique: Le problème du temps*. Paris: Hermann, 1946.

Menninger, Karl. *Number Words and Number Symbols: A Cultural History of Numbers*. Trans. Paul Boneer. Boston: MIT P, 1969.

Needham, Joseph. *Mathematics and the Sciences of Heaven and Earth.*

Vol. 3 of his *Science and Civilization in China*. Cambridge: Cambridge UP, 1959.
Newton, Isaac. *Sir Isaac Newton's The Mathematical Principles of Natural Philosophy and His System of the World*. Trans. A. Motte. Ed. F. Cajori. Berkeley: U of California P, 1960.
Nicole Oresme and the Medieval Geometry of Qualities and Motions. Ed. and trans. M. Clagett. Madison: U of Wisconsin P, 1968.
Owen, G. E. L. "Zeno and the Mathematicians." *Proceedings of the Aristotelian Society* n.s., 58 (1957-58): 199-222.
Price, Derek J. de Solla. "An Ancient Greek Computer." *Scientific American* 200 (1959): 60-67.
Schoenberg, Isaac J. *Mathematical Time Exposures*. Washington: Mathematical Association of America, 1983.
Strogatz, S. H. *The Mathematical Structure of the Human Sleep-Wake Cycle*. New York: Springer, 1986.
Tennant-Smith, J. *Mathematics for the Manager*. London: Nelson, 1971.
Weyl, Hermann. "Space and Time, the Transcendental External World." *Philosophy of Mathematics and Natural Science*. New York: Atheneum, 1963. 95-137.
Whittaker, E. T. *From Euclid to Eddington*. Cambridge: Cambridge UP, 1949.
Zongju, Jiang, et al. "Why Did Modern Mathematics Fail to Emerge in China?" *J of Dialectics of Nature* 5.3 (1983): 49-61.

B. DYNAMICS

Agrawal, O. P., and A. A. Shabana. "Application of Deformable Body Mean-Axes to Flexible Multibody System Dynamics." *Computer Methods in Applied Mechanics and Engineering* 56 (1986): 217-45.
——. "Dynamic Analysis of Multibody Systems Using Component Modes." *Computers and Structures* 21.6 (1985): 1301-12.
Andrews, G. C., and H. K. Kesaven. "The Vector Network Model: A New Approach to Vector Dynamics." *Mechanism and Machine Theory* 10 (1975): 57-75.
Ashley, H. "Observation on the Dynamic Behavior of Large Flexible Bodies in Orbit." *AIAA J* 5.3 (1967): 460-69.
Bakr, E. M., and A. A. Shabana. "Geometrically Nonlinear Analysis of Multibody Systems." *Computers and Structures* 23.6 (1986): 739-51.
——. "Timoshenko Beams and Flexible Multibody System Dynamics." *J of Sound and Vibration* 116.1 (1987): 89-107.

Bathe, K. J., E. Ramm, and E. L. Wilson. "Finite Element Formulations for Large Deformation Dynamic Analysis." *International J for Numerical Methods in Engineering* 9 (1975): 353-86.

Baumgarte, J. "Stabilization of Constraints and Integrals of Motion in Dynamical Systems." *Computer Methods in Applied Mechanics and Engineering* 1 (1972): 1-16.

Bayazitoglu, Y. O., and M. A. Chace. "Methods for Automated Dynamic Analysis of Discrete Mechanical Systems." *ASME J of Applied Mechanics* 40 (1973): 809-11.

Benfield, W. A., and R. F. Hruda. "Vibration Analysis of Structures by Component Mode Substitution." *AIAA J* 9.7 (1971): 1255-61.

Benson, D. J., and J. D. Hallquist. "A Simple Rigid Body Algorithm for Structural Dynamics Programs." *International J for Numerical Methods in Engineering* 22 (1986): 723-49.

Boltzmann, L. "Über die Form der Lagrangeschen Gleichungen für Nichtolonome, generalisierte Koordinaten." *Sitzungsberichte der Mathematisch-Naturwissenschaftlichen Klasse der Kaiserlichen Akademie der Wissenschaften* 61 (1902): 1603-14.

Book, W. J. "Recursive Lagrangian Dynamics of Flexible Manipulator Arms." *The International J of Robotic Research* 3 (Fall 1984): 87-101.

Book, W. J., O. Malzza-Neto, and D. E. Whitney. "Feedback Control of Two Beams, Two Joint Systems with Distributed Flexibility." *ASME J of Dynamic Systems, Measurement and Control* 97 (December 1975): 424-31.

Boresi, A. P., and P. P. Lynn. *Elasticity in Engineering Mechanics.* Englewood Cliffs: Prentice-Hall, 1974.

Bottema, O., and B. Roth. *Theoretical Kinematics.* Amsterdam: North-Holland, 1979.

Canavin, J. R., and P. W. Likins. "Floating Reference Frames for Flexible Spacecrafts." *J of Spacecraft* 14.12 (1977): 724-32.

Cavin, R. K., and A. R. Dusto. "Hamilton's Principle: Finite Element Methods and Flexible Body Dynamics." *AIAA J* 15.2 (1977): 1684-90.

Chace, M. A. "Analysis of the Time-Dependence of Multi-Freedom Mechanical Systems in Relative Coordinates." *ASME J of Engineering for Industry* 89 (1967): 119-25.

Chang, C. O., and P. E. Nikravesh. "An Adaptive Constraint Violation Stabilization Method for Dynamic Analysis of Mechanical Systems." *ASME J of Mechanisms, Transmission and Automation in Design* 107.4 (1985): 488-94.

Chang, C. W., and A. A. Shabana. "Hybrid Control of Flexible Multibody Systems." *Computers and Structures* 25.6 (1987): 831-44.

Changizi, K., and A. A. Shabana. "Pulse Control of Flexible Multibody Systems." *Computers and Structures* 24.6 (1986): 875-84.

Christensen, E. R., and S. W. Lee. "Nonlinear Finite Element Modeling of the Dynamics of Unrestrained Flexible Structures." *Computers and Structure* 23 (1986): 819-29.

Christensen, R. M. *Theory of Viscoelasticity: An Introduction*. New York: Academic P, 1982.

Chu, S. C., and K. C. Pan. "Dynamic Response of a High-Speed Slider Crank Mechanism with an Elastic Connecting Rod." *ASME J of Engineering for Industry* (May 1975): 542-50.

Clough, R. W., and J. Penzien. *Dynamics of Structures*. New York: McGraw-Hill, 1975.

Craig, J. J. *Introduction to Robotics: Mechanics and Control*. Reading: Addison-Wesley, 1986.

De Jalon, J. G., et al. "A Simple Numerical Method for the Kinematic Analysis of Spatial Mechanisms." *ASME J of Mechanical Design* 104 (January 1982): 78-82.

Denavit, J., and R. S. Hartenberg. "A Kinematic Notation for Low-Pair Mechanisms Based on Matrices." *ASME J of Applied Mechanics* (June 1955): 215-21.

De Veubeke, B. F. "The Dynamics of Flexible Bodies." *International J of Engineering Science* 14 (1976): 895-913.

D'Souza, A. F., and V. K. Garg. *Advanced Dynamics: Modeling and Analysis*. Englewood Cliffs: Prentice-Hall, 1984.

Dugundji, J. "On the Calculation of Natural Modes of Free-Free Structures." *J of the Aerospace Sciences* (February 1964): 164-66.

Erdman, A. G., and G. N. Sandor. "Kineto-Elastodynamics—A Review of the State of the Art and Trends." *Mechanism and Machine Theory* 7 (1972): 19-33.

Featherstone, R. "The Calculation of Robot Dynamics Using Articulated-Body Inertias." *International J of Robotics Research* 2.1 (Spring 1983): 13-30.

Flügge, W. *Viscoelasticity*. Waltham: Blaisdell, 1967.

Gear, C. W., and L. R. Petzold. "ODE Methods for the Solution of Differential/Algebraic Systems." *ASME J of Numerical Analysis* 21.4 (August 1984): 716-28.

Gelfand, I. M., and S. V. Fomin. *Calculus of Variations*. Englewood Cliffs: Prentice-Hall, 1963.

Goldman, R. L. "Vibration Analysis by Dynamic Partitioning." *AIAA J* 7.6 (1969): 1152-54.

Goldstein, H. *Classical Mechanics*. Reading: Addison-Wesley, 1950.

Graig, R. R., and M. C. Bampton. "Coupling of Substructures for Dynamic Analyses." *AIAA J* 6.7 (1968): 1313-19.

Greenwood, D. T. *Principles of Dynamics*. Englewood Cliffs: Prentice-Hall, 1965.

Gupta, G. K. "Dynamic Analysis of Multi-Rigid-Body Systems." *ASME J of Engineering for Industry* 9 (1974): 809-11.

Hemami, H. "Some Aspects of Euler-Newton Equation of Motion." *Ingenieur-Archiv* 52 (1982): 167-76.

Ho, J. Y. L. "Direct Path Method for Flexible Multibody Spacecraft Dynamics." *J of Spacecraft and Rockets* 14 (February 1977): 102-10.

Ho, J. Y. L., and D. R. Herber. "Development of Dynamics and Control Simulation of Large Flexible Space Systems." *J of Guidance, Control and Dynamics* 8 (May-June 1985): 374-83.

Hooker, W. W. "Equations of Motion for Interconnected Rigid and Elastic Bodies." *Celestial Mechanics* 11.3 (May 1975): 337-59.

Hughes, P. C. "Dynamics of Chain of Flexible Bodies." *J of Astronautical Science* 27.4 (1979): 359-80.

Hurty, W. C. "Dynamic Analysis of Structural Systems Using Component Modes." *AIAA J* 3.4 (1965): 678-85.

Huston, R. L. "Multi-Body Dynamics Including the Effects of Flexibility and Compliance." *Computers and Structures* 14.5-6 (1981): 443-51.

Huston, R. L., and J. W. Kamman. "A Discussion on Constraint Equations in Multibody Dynamics." *Mechanics Research Communications* 9.4 (1982): 251-56.

Huston, R. L., and C. E. Passerello. "Multibody Structural Dynamics Including Translation Between the Bodies." *Computers and Structures* 12.5 (1980): 713-20.

———. "On the Dynamics of a Human Body Model." *J of Biomechanics* 4 (1971): 369-78.

———. "On Multi-Rigid Body System Dynamics." *Computers and Structures* 19 (1979): 439-46.

Huston, R. L., C. E. Passerello, and M. W. Harlow. "Dynamics of Multirigid-Body Systems." *ASME J of Applied Mechanics* 45.4 (1978): 889-94.

Huston, R. L., et al. "On Human Body Dynamics." *Annals of Biomedical Engineering* 4 (1976): 25-43.

Irons, B. "Structural Eigenvalue Problems: Elimination of Unwanted Variables." *AIAA J* 3.5 (1965): 961-62.

Jerkovsky, W. "The Structure of Multibody Dynamics Equations." *J of Guidance and Control* 1.3 (May-June 1978): 173-82.

Kamman, J. W., and R. L. Huston. "Constrained Multibody System

Dynamics." *Computers and Structures* 18 (1984): 999-1003.

——. "Dynamics of Constrained Multibody Systems." *ASME J of Applied Mechanics* 51.4 (1984): 387-404.

Kane, T. R. "Solution of Kinematical Differential Equations for a Rigid Body." *ASME J of Applied Mechanics* 40 (March 1973): 109-13.

Kane, T. R., P. W. Likins, and D. A. Levinson. *Spacecraft Dynamics*. New York: McGraw-Hill, 1983.

Kane, T. R., R. R. Ryan, and A. K. Banerjee. "Dynamics of a Cantilever Beam Attached to a Moving Base." *AIAA J of Guidance, Control, and Dynamics* 10.2 (March-April 1987): 139-51.

Khulief, Y. A., and A. A. Shabana. "Dynamics of Multibody Systems with Variable Kinematic Structure." *ASME J of Mechanisms, Transmissions, and Automation in Design* 108.2 (June 1986): 167-75.

Kim, S. S., and M. J. Vanderploeg. "A General and Efficient Method for Dynamic Analysis of Mechanical Systems Using Velocity Transformation." *ASME J of Mechanisms, Transmissions, and Automation in Design* 108.2 (June 1986): 176-82.

Lanczos, C. *The Variational Principles of Mechanics*. 4th ed. Toronto: U of Toronto P, 1970.

Langrana, N. A., and D. L. Bartel. "An Automated Method for Dynamic Analysis of Spatial Linkages for Biomedical Applications." *ASME J of Engineering for Industry* 97 (1975): 566-74.

Likins, P. W. "Modal Method for Analysis of Free Rotations of Spacecraft." *AIAA J* 5.7 (1967): 1304-08.

Magnus, K. *Dynamics of Multibody Systems*. Berlin: Springer, 1978.

Midha, A., A. G. Erdman, and D. A. Frohrib. "An Approximate Method for the Dynamic Analysis of Elastic Linkages." *ASME J of Engineering for Industry* (May 1977): 449-55.

Milne, R. D. "Some Remarks on the Dynamics of Reformable Bodies." *AIAA J* 6.3 (1968): 556-58.

Mooring, B. W., and J. Genin. "A Kinematic Constraint Method for Stability Analysis of Articulated Vehicles." *International J of Vehicle Design* 3.2 (1982): 190-201.

Morris, N. F. "The Use of Modal Superposition in Nonlinear Dynamics." *Computers and Structures* 7 (1977): 65-72.

Nikravesh, P. E. *Computer Aided Analysis of Mechanical Systems*. Englewood Cliffs: Prentice-Hall, 1988.

Nikravesh, P. E., and I. S. Chung. "Application of Euler Parameters to the Dynamic Analysis of Three-Dimensional Constrained Mechanical Systems." *ASME J of Mechanical Design* 104 (1982): 785-91.

Nikravesh, P. E., R. A. Wehage, and O. K. Kwon. "Euler Parameters in Computational Kinematics and Dynamics, Parts I and II." *ASME J of Mechanisms, Transmissions, and Automation in Design* 107.3 (September 1985): 358-70.

Paul, B. "Analytical Dynamics of Mechanisms—A Computer Oriented Overview." *Mechanism and Machine Theory* 10.6 (1975): 481-507.

Paul, R. P. *Robot Manipulators, Mathematics, Programming and Control*. Cambridge: MIT P, 1981.

Rosenberg, R. M. *Analytical Dynamics of Discrete Systems*. New York: Plenum, 1977.

Shabana, Ahmed A. "Automated Analysis of Constrained Inertia-Variant Flexible Systems." *ASME J of Vibration, Acoustic, Stress, and Reliability in Design* 107.4 (October 1985): 431-40.

——. "Dynamics of Inertia-Variant Flexible Systems Using Experimentally Identified Parameters." *ASME J of Mechanisms, Transmissions, and Automation in Design* 108 (September 1986): 358-66.

——. *Dynamics of Multibody Systems*. New York: Wiley, 1989. Includes bibliography, 449-62.

——. "Transient Analysis of Flexible Multibody Systems, Part I: Dynamics of Flexible Bodies." *Computer Methods in Applied Mechanics and Engineering* 54 (1986): 75-91.

Shabana, Ahmed A., and R. A. Wehage. "A Coordinate Reduction Technique for Transient Analysis of Spatial Substructures with Large Angular Rotations." *J of Structural Mechanics* 11.3 (1983): 401-31.

Silver, W. M. "On the Equivalence of Lagrangian and Newton-Euler Dynamics for Manipulators." *The International J of Robotics Research* 1.2 (Summer 1982): 60-70.

Simo, J. C., and L. Vu-Quoc. "On the Dynamics of Flexible Beams Under Large Overall Motions—The Plane Case: Parts I and II." *ASME J of Applied Mechanics* 53 (December 1986): 849-63.

Singh, R. P., and P. W. Likins. "Singular Value Decomposition for Constrained Dynamic Systems." *ASME J of Applied Mechanics* 52.4 (December 1985): 943-48.

Sokolnikoff, I. S. *Mathematical Theory of Elasticity*. 2nd ed. New York: McGraw-Hill, 1956.

Song, J. O., and E. J. Haug. "Dynamic Analysis of Planar Flexible Mechanisms." *Computer Methods in Applied Mechanics and Engineering* 24 (1980): 359-81.

Spring, K. W. "Euler Parameters and the Use of Quaternion Algebra in the Manipulation of Finite Rotations: A Review." *Mechanism and Machine Theory* 21.5 (1986): 365-73.

Sunada, W., and S. Dubowsky. "The Application of the Finite Element Methods to the Dynamic Analysis of Flexible Spatial and Co-Planar Linkage Systems." *ASME J of Mechanical Design* 103.3 (July 1981): 643-51.

——. "On the Dynamic Analysis and Behavior of Industrial Robotic Manipulator with Elastic Members." *ASME J of Mechanical Design* 105.1 (March 1983): 42-51.

Timoshenko, S., D. H. Young, and W. Weaver. *Vibration Problems in Engineering*. 4th ed. New York: Wiley, 1974.

Turcic, D. A., and A. Midha. "Dynamic Analysis of Elastic Mechanism Systems, Part I: Applications." *ASME J of Dynamic Systems, Measurement, and Control* 106 (December 1984): 243-48.

Vukabratovic, M. "Dynamics of Active Articulated Mechanisms and Synthesis of Artificial Motion." *Mechanism and Machine Theory* 13.1 (1978): 1-56.

Walker, M. W., and D. E. Orin. "Efficient Dynamic Computer Simulation of Robotic Mechanisms." *ASME J of Dynamic Systems, Measurement, and Control* 104 (September 1982): 205-11.

Wampler, C., K. Buffinton, and J. Shu-hui. "Formulation of Equations of Motion for Systems Subject to Constraints." *ASME J of Applied Mechanics* 52 (June 1985): 465-70.

Whittaker, E. J. *A Treatise on the Analytical Dynamics of Particles and Rigid Bodies*. 4th ed. New York: Dover, 1937.

Winfrey, R. C. "Dynamic Analysis of Elastic Link Mechanisms by Reduction of Coordinates." *ASME J of Engineering for Industry* (May 1972): 577-82.

Winfrey, R. C., R. V. Anderson, and C. W. Gnika. "Analysis of Elastic Machinery with Clearance." *ASME J of Engineering for Industry* (August 1973): 695-703.

C. STATISTICS

Bowker, A. H., and G. J. Lieberman. *Engineering Statistics*. Englewood Cliffs: Prentice-Hall, 1959.

De Prins, J., and G. Cornelissen. "Complex Demodulation and Double Demodulation." *Bulletin. Académie Royale de Belgique. Classe des Sciences* 5e serie, 65 (1979): 445-55.

——. "Numerical Signal Averaging for Cycle Determination." *J of Interdisciplinary Cycle Research* 6 (1975): 95-102.

De Prins, J., and W. Malbecq. "Signal Averaging for the Analysis of Irregularly Spaced Data." *J of Interdisciplinary Cycle Research* 15

(1984): 179-88.
Diaconis, P., and B. Efron. "Computer-Intensive Methods in Statistics." *Scientific American* 248.5 (1983): 116-30.
Draper, N. R., and H. Smith. *Applied Regression Analysis*. 2nd ed. New York: Wiley, 1981.
Fletcher, R. *Practical Methods of Optimization*. 2 vols. New York: Wiley, 1980.
Gilchrist, W. *Statistical Forecasting*. New York: Wiley, 1976.
Hamming, R. W. *Numerical Methods for Scientists and Engineers*. 2nd ed. New York: McGraw-Hill, 1973.
Hobson, A. *Concepts in Statistical Mechanics*. New York: Gordon and Breach, 1971.
Kendall, M. G., and A. Stuart. *The Advanced Theory of Statistics*. Vols 1-3. London: Griffin, 1966.
Landau, L. D., and E. M. Lifshitz. *Statistische Physik*. Berlin: Akademie Verlag, 1966.
Melard, G. "A Fast Algorithm for the Exact Likelihood of Moving Average Models." *Applied Statistics* 33 (1984): 104-14.
Mood, A. M., and F. A. Graybill. *Introduction to the Theory of Statistics*. New York: McGraw-Hill, 1963.
Sollberger, A. "Problems in the Statistical Analysis of Short Periodic Time Series." *J of Interdisciplinary Cycle Research* 1.1 (1970): 49-88.
——. "Statistical Aspects of Diurnal Biorhythm." *Acta Anatomia* 23 (1955): 7-127.
Spanos, Aris. *Statistical Foundations of Econometric Modelling*. New York: Cambridge UP, 1986.
Tune, G. S. "A Brief Survey of Variables That Influence Random-Generation." *Perceptual and Motor Skills* 18 (1964): 705-10.

D. TIME SERIES ANALYSIS

Abraham, Bovas, and G. E. P. Box. "Deterministic and Forecast-Adaptive Time-Dependent Models." *Applied Statistics* 27 (1978): 120-30.
Abraham, Bovas, and Johannes Ledolter. *Statistical Methods for Forecasting*. New York: Wiley, 1983.
Akaike, H. "Effect of Timing-Error on the Power Spectrum of Sampled Data." *Annals of the Institute of Statistical Mathematics* 11 (1960): 145-65.
——. "Time Series Analysis and Control Through Parametric Models." *Applied Time Series Analysis*. Ed. D. F. Findley. New York: Aca-

demic P, 1978.

Ali, M. M. "Analysis of Autoregressive Moving Average Models: Estimation and Prediction." *Biometrika* 64 (1977): 535-45.

Anderson, O. D., ed. *Analysing Time Series*. Amsterdam: North-Holland, 1980.

Anderson, O. D., and M. R. Perryman, eds. *Applied Time Series Analysis*. Amsterdam: North-Holland, 1982.

Anderson, O. D., ed. *Time Series Analysis: Theory and Practice 2*. Amsterdam: North-Holland, 1982.

——, ed. *Time Series Analysis: Theory and Practice 3*. Amsterdam: North-Holland, 1983.

——, ed. *Time Series Analysis: Theory and Practice 4*. Amsterdam: North-Holland, 1983.

——, ed. *Time Series Analysis: Theory and Practice 5*. Amsterdam: North-Holland, 1984.

——. *Time Series Analysis and Forecasting: The Box-Jenkins Approach*. London: Butterworths, 1976.

Anderson, T. W. "Estimation for Autoregressive Moving Average Models in the Time and Frequency Domains." *Annals of Statistics* 5 (1977): 842-65.

——. *The Statistical Analysis of Time Series*. New York: Wiley, 1971.

Aoki, Masanao. *State Space Modeling of Time Series*. New York: Springer, 1987.

Astrom, K. J., and C. G. Kallstrom. "Application of System Identification Techniques to the Determination of Ship Dynamics." *Identification and System Parameter Estimation*. Ed. P. Eykhoff. Amsterdam: North-Holland, 1973. 415-24.

Balke, N. S., and R. J. Gordon. "The Estimation of Prewar Gross National Product: Methodology and New Evidence." *J of Political Economy* 97 (February 1989): 38-92. Includes bibliography.

Bartlett, M. S. "On the Theoretical Specification of Sampling Properties of Autocorrelated Time Series." *J of the Royal Statistical Society. Supplement* 8 (1946): 27-41.

——. "Periodogram Analysis and Continuous Spectra." *Biometrika* 37 (1950): 1-16.

——. "Problèmes de l'analyse spectrale des séries temporelles stationnaires." *Publications. Institut de Statistique. Université de Paris* 3.3 (1954): 119-34.

Bartlett, M. S., et al. "Discussion on 'Symposium on Spectral Approach to Time Series.'" *J of the Royal Statistical Society. Series B* 19 (1957): 1-63.

Bee de Dagum, Estela Maria. *The X-II-ARIMA Seasonal Adjustment*

Method. Ottawa: Statistics Canada, 1980.

Bittanti, S., ed. *Time Series and Linear Systems*. Berlin: Springer, 1986. Includes bibliography.

Bloomfield, P. *Fourier Analysis of Time Series: An Introduction*. New York: Wiley, 1976.

Bowerman, Bruce L. *Time Series and Forecasting: An Applied Approach*. Scituate: Duxbury P, 1979.

Box, George E. P., and Gwilym M. Jenkins. *Time Series Analysis: Forecasting and Control*. San Francisco: Holden-Day, 1976.

———. "Some Statistical Aspects of Adaptive Optimisation and Control." *J of the Royal Statistical Society. Series B* 24 (1962): 297-343.

Box, George E. P., and J. F. MacGregor. "The Analysis of Closed-Loop Dynamic-Stochastic Systems." *Technometrics* 16 (1974): 391-98.

Box, George E. P., and D. A. Pierce. "Distribution of Residual Autocorrelations in Autoregressive-Integrated Moving Average Time Series Models." *J of the American Statistical Association* 65 (1970): 1509-26.

Brillinger, D. R. *Time Series Analysis: Data Analysis and Theory*. New York: Holt, Rinehart, 1981.

Brinkman, H. J., et al. "Height and Income: A New Method for the Estimation of Historical National Income Series." *Explorations in Economic History* 25 (July 1988): 227-64. Includes bibliography.

Brockwell, Peter J., and Richard A. Davis. *Time Series: Theory and Method*. New York: Springer, 1987. Includes bibliography.

Brown, R. L., J. Durbin, and J. M. Evans. "Techniques for Testing the Constancy of Regression Relationships Over Time." *J of the Royal Statistical Society. Series B* 37 (1975): 149-92.

Cadzow, James A. *Foundations of Digital Signal Processing and Data Analysis*. New York: Macmillan, 1987.

Campbell, M. J., and A. M. Walker. "A Survey of Statistical Work on the MacKenzie River Series of Annual Canadian Lynx Trappings for the Years 1821-1934, and a New Analysis." *J of the Royal Statistical Society. Series A* 140 (1977): 411-31.

Cartwright, D. E. "A Unified Analysis of Tides and Surges Round North and East Britain." *Philosophical Transactions of the Royal Society of London. Series A* 263 (1968): 1-55.

Catalano, R. A., et al. "Selecting a Time Series Strategy." *Psychological Bulletin* 94 (November 1983): 506-23.

Chaghaghi, François S. *Time Series Package (TSPACK)*. New York: Springer, 1985.

Chan, W.-Y. T., and H. Tong. "A Simulation Study of the Estimation of Evolutionary Spectral Functions." *Applied Statistics* 24 (1975): 333-41.

Chatfield, C. *The Analysis of Time Series: An Introduction*. 3rd ed. London: Chapman and Hall, 1984.

Clinger, W., and J. W. Van Ness. "On Unequally Spaced Time Points in Time Series." *Annals of Statistics* 4 (1976): 736-45.

Comes, P., and G. Michel. "High Resolution Fourier Spectra of Stars and Planets." *Astrophysical J* 190 (1974): L29-L32.

Cooley, J. W., P. A. W. Lewis, and P. D. Welch. "The Application of the Fast Fourier Transform Algorithm to the Estimation of Spectra and Cross-Spectra." *J of Sound and Vibration* 12 (1970): 339-52.

Cooley, J. W., and J. W. Tukey. "An Algorithm for the Machine Calculation of Complex Fourier Series." *Mathematics of Computation* 19 (1965): 297-301.

Cooper, D. M., and R. Thompson. "A Note on the Estimation of the Parameters of the Autoregressive Moving Average Process." *Biometrika* 64 (1977): 625-28.

Cover, J. P., and P. D. Thistle. "Time Series, Homicide, and the Deterrent Effect of Capital Punishment." *Southern Economic J* 54 (January 1988): 615-22. Includes bibliography.

d'Allesandro, P. "Structural Properties Invariance and Insensitivity of Discrete Time Bilinear Systems." *Ricerche di Automatica* 3 (1972): 158-69.

Davidson, J. E. H., et al. "Econometric Modelling of the Aggregate Time-Series Relationship Between Consumers' Expenditure and Income in the United Kingdom." *Economic J* 88 (1978): 661-92.

Dzhaparidze, K. O. *Parameter Estimation and Hypothesis Testing in Spectral Analysis of Stationary Time Series*. Trans. Samuel Kotz. New York: Springer, 1986.

Findley, David F. *Applied Time Series Analysis II*. New York: Academic P, 1981.

Fox, R., and M. S. Taqqu. "Large Sample Properties of Parameter Estimates for Strongly Dependent Stationary Gaussian Time Series." *Annals of Statistics* 14 (1986): 517-32.

Franke, J., W. Härdle, and D. Martin, eds. *Robust and Nonlinear Time Series Analysis*. New York: Springer, 1984.

Fuller, Wayne A. *Introduction to Statistical Time Series*. New York: Wiley, 1976.

Gagnon, J. E. "Short-Run Models and Long-Run Forecasts: A Note on the Permanence of Output Fluctuations." *Quarterly J of Economics* 103 (May 1988): 415-24.

Gardner, William A. *Statistical Spectral Analysis: A Nonprobabilistic*

Theory. Englewood Cliffs: Prentice-Hall, 1988.

Geweke, J., and S. Porter-Hudak. "The Estimation and Application of Long-Memory Time Series Models." *J of Time Series Analysis* 4 (1983): 221-38.

Gilbert, C. L. "LSE and the British Approach to Time Series Econometrics." *Oxford Economic Papers* 41 (January 1989): 108-28.

Godfrey, L. G. "Testing the Adequacy of a Time Series Model." *Biometrika* 66 (1979): 67-72.

Godfrey, M. D. "Computational Methods for Time Series." *Bulletin. Institute of Mathematics and Its Applications* 10 (1974): 224-27.

Godolphin, E. J. "A Direct Representation for the Maximum Likelihood Estimator of a Gaussian Moving Average Process." *Biometrika* 64 (1977): 375-84.

Gottman, John Mordechai. *Time-Series Analysis: A Comprehensive Introduction for Social Scientists*. Cambridge: Cambridge UP, 1981.

Granger, Clive William John. *Forecasting Economic Time Series*. Orlando: Academic P, 1986.

———. "Long Memory Relationships and the Aggregation of Dynamic Models." *Econometrics* 14 (1980): 227-38.

Granger, Clive William John, and Allan Paul Andersen. *An Introduction to Bilinear Time Series Models*. Göttingen: Vandenhoeck and Ruprecht, 1978.

———. "Nonlinear Time Series Modelling." *Applied Time Series Analysis*. Ed. D. F. Findley. New York: Academic P, 1978.

Granger, Clive William John, and M. Hatanaka. *Spectral Analysis of Economic Time Series*. Princeton: Princeton UP, 1964.

Granger, Clive William John, and R. Joyeux. "An Introduction to Long-Memory Time Series Models and Fractional Differencing." *J of Time Series Analysis* 1 (1980): 15-29.

Gregson, Robert Anthony Mills. *Time Series in Psychology*. Hillsdale: Erlbaum, 1983.

Grenander, U., and M. Rosenblatt. *Statistical Analysis of Stationary Time Series*. New York: Wiley, 1957.

Groves, G. W., and E. J. Hannan. "Time Series Regression of Sea Level on Weather." *Reviews of Geophysics* 6 (1968): 129-74.

Gudmundsson, G. "Time Series Analysis of Imports, Exports, and Other Economic Variables." *J of the Royal Statistical Society. Series A* 134 (1971): 383.

Hamilton, D. C., and D. G. Watts. "Interpreting Partial Autocorrelation Functions of Seasonal Time Series Models." *Biometrika* 65 (1978): 135-40.

Hamilton, R. T. "Unemployment and Business Formation Rates: Reconciling Time-Series and Cross-Section Evidence." *Environment and Planning A* 21 (February 1989): 249-55.

Hammond, J. K. "Evolutionary Spectra in Random Vibrations." *J of the Royal Statistical Society. Series B* 35 (1973): 167-88.

Hannan, E. J. *Multiple Time Series.* New York: Wiley, 1970.

——. *Time Series Analysis.* London: Methuen, 1960.

Hannan, E. J., et al., eds. *Time Series in the Time Domain.* New York: Elsevier, 1985.

Harvey, A. C. *The Econometric Analysis of Time Series.* Oxford: P. Allan, 1981.

——. *Time Series Models.* Oxford: P. Allan, 1981.

Hasselman, K., W. Munk, and G. Macdonald. "Bispectrum of Ocean Waves." *Time Series Analysis.* Ed. M. Rosenblatt. New York: Wiley, 1963. 125-39.

Hewitt, Christopher. *The Effectiveness of Anti-Terrorist Policies.* Lanham: UP of America, 1984.

Holt, Charles A., Jr., and Robert W. Shore. *Bayesian Analysis of Economic Theory and Time Series Analysis.* Amsterdam: North-Holland, 1980.

Hsiao, Cheng. *A Time Series Analysis of the Impact of Canadian Wage and Price Controls.* Toronto: Institute for Policy Analysis, U of Toronto, 1978.

Jenkins, Gwilym M. *Practical Experience with Modelling and Forecasting.* St. Helier: Gwilym Jenkins and Partners, 1979.

Jenkins, Gwilym M., and M. B. Priestley. "The Spectral Analysis of Time Series." *J of the Royal Statistical Society. Series B* 19 (1957): 1-12.

Jenkins, Gwilym M., and D. G. Watts. *Spectral Analysis and Its Applications.* San Francisco: Holden-Day, 1968.

Kalman, R. E. "A New Approach to Linear Filtering and Prediction Problems." *Transactions. ASME J of Basic Engineering* 83D (1960): 35-45.

Kanasewich, Ernest Raymond. *Time Sequence Analysis in Geophysics.* Edmonton: U of Alberta P, 1981.

Karlin, Samuel, Takeshi Amemiya, and Leo A. Goodman. *Studies in Econometrics, Time Series, and Multivariate Statistics.* New York: Academic P, 1983.

Kashyap, R. L. "A Bayesian Comparison of Different Classes of Dynamic Models Using Empirical Data." *IEEE. Transactions on Automatic Control* AC-22 (1977): 715-27.

Kedem, Benjamin. *Binary Time Series.* New York: M. Dekker, 1980.

Kendall, Maurice George. *Time Series.* 2nd ed. London: C. Griffin,

1976.

Koopmans, L. H. *The Spectral Analysis of Time Series*. New York: Academic P, 1974.

Krueger, A. B. "The Determinants of Queues for Federal Jobs." *Industrial and Labor Relations Review* 41 (July 1988): 567-81. Includes bibliography.

Lawrance, A. J., and N. T. Kottegoda. "Stochastic Modelling of Riverflow Time Series." *J of the Royal Statistical Society. Series A* 140 (1977): 1-47.

Lehmann, E. L. *Theory of Point Estimation*. New York: Wiley, 1983.

Li, W. K., and A. I. McLeod. "Fractional Time Series Modelling." *Biometrika* 73 (1986): 217-21.

Ljung, G. M., and G. E. P. Box. "On a Measure of Lack of Fit in Time Series Models." *Biometrika* 65 (1978): 297-303.

Lomnicki, Z. A., and S. K. Zaremba. "On the Estimation of Autocorrelation in Time Series." *Annals of Mathematical Statistics* 28 (1957): 1.

Lucas, Robert E., Jr., and Thomas J. Sargent. *Rational Expectations and Econometric Practice*. Minneapolis: U of Minnesota P, 1981.

McCleary, Richard, et al. *Applied Time Series Analysis for the Social Sciences*. Beverly Hills: Sage, 1980.

McDowell, David, et al. *Interrupted Time Series Analysis*. Beverly Hills: Sage, 1980.

MacNeill, I. B. "A Test of Whether Several Series Share Common Periodicities." *Biometrika* 64 (1977): 495-508.

——. "Tests for Periodic Components in Multiple Time Series." *Biometrika* 61 (1974): 57-70.

Makridakis, S., et al. *The Forecasting Accuracy of Major Time Series Methods*. New York: Wiley, 1984.

Mandrekar, V. "A Characterisation of Oscillatory Processes and Their Prediction." *Proceedings. American Mathematical Society* 32 (1972): 280-84.

Mehra, Raman K., and Dimitri Lainiotis, eds. *System Identification: Advances and Case Studies*. New York: Academic P, 1976.

Mélard, Guy. *Analyse de données chronologiques*. Montreal: Presses de l'Université de Montréal, 1985.

Mendel, Jerry M. *Optimal Seismic Deconvolution*. New York: Academic P, 1983.

Mohler, R. R. *Nonlinear Time Series and Signal Processing*. New York: Springer, 1988.

Montgomery, Douglas C. *Forecasting and Time Series Analysis*. New York: McGraw-Hill, 1976.

Moran, P. "The Statistical Analysis of the Canadian Lynx Cycle: I—Structure and Prediction." *Australian J of Zoology* 1 (1958): 163-73.

Morris, J. "Forecasting the Sunspot Cycle." *J of the Royal Statistical Society. Series A* 140 (1977): 437-47.

Neftci, S. N. "Are Economic Time Series Assymetric Over the Business Cycle?" *J of Political Economy* 92 (April 1984): 307-28. Includes bibliography.

Nerlove, Marc, David M. Grether, and José L. Carvalho. *Analysis of Economic Time Series*. New York: Academic P, 1979.

Newbold, P., and C. W. J. Granger. "Experience with Forecasting Univariate Time Series and the Combination of Forecasts." *J of the Royal Statistical Society. Series A* 137 (1974): 131-65.

O'Loughlin, J. "Immigration to Western Europe, 1952-82: A Time-Series Analysis of Movement to Sweden, France, and Essen." *Environment and Planning* A 18 (March 1986): 375-99.

Ostrom, Charles W. *Time Series Analysis: Regression Techniques*. Beverly Hills: Sage, 1978.

Otnes, Robert K., and Loren Enochson. *Applied Time Series Analysis: Basic Techniques*. Vol. 1. New York: Wiley, 1978.

Pandit, Sudhakar M., and Shien-Ming Wu. *Time Series and System Analysis with Applications*. New York: Wiley, 1983.

Pankrantz, Alan. *Forecasting with Univariate Box-Jenkins Models: Concepts and Cases*. New York: Wiley, 1983.

Parzen, Emanuel. "Some Recent Advances in Time Series Modelling." *IEEE Transactions on Automatic Control* AC-19 (1974): 723-29.

——, ed. *Time Series Analysis of Irregularly Observed Data*. New York: Springer, 1984.

Pashigian, B. P. "Demand Uncertainty and Sales: A Study of Fashion and Markdown Pricing." *American Economic Review* 78 (December 1988): 936-53.

Pham-Dinh, T. "Estimation of Parameters of a Continuous Time Gaussian Stationary Process with Rational Spectral Density." *Biometrika* 64 (1977): 385-99.

Priestley, M. B. "Current Developments in Time Series Modelling." *J of Econometrics* 37 (January 1988): 67-86. Includes bibliography.

——. *Spectral Analysis and Time Series*. London: Academic P, 1981. Includes bibliography.

——. "Time-Dependent Spectral Analysis and Its Applications in Prediction and Control." *J of Sound and Vibration* 17 (1971): 139-56.

Prothero, D. L., and K. F. Wallis. "Modelling Macroeconomic Time Series." *J of the Royal Statistical Society. Series A* 139 (1976): 468-500.

Ram, R. "Government Size and Economic Growth: A New Framework and Some Evidence from Cross-Section and Time-Series Data." *American Economic Review* 76 (March 1986): 191-203. Includes bibliography.

Reich, M. "Segmented Labour: Time Series Hypothesis and Evidence." *Cambridge J of Economics* 8 (March 1984): 63-81. Includes bibliography.

Rice, S. O. "Mathematical Analysis of Random Noise." *The Bell System Technical J* 23 (1944): 282-332, and 24 (1945): 46-156.

Robinson, Enders A. *Time Series Analysis and Applications*. Houston: Goose Pond P, 1981.

Robinson, Enders A., and Manuel T. Silvia. *Digital Signal Processing and Time Series Analysis*. San Francisco: Holden-Day, 1978.

Romer, C. D. "The Prewar Business Cycle Reconsidered: New Estimates of Gross National Product, 1869-1908." *J of Political Economy* 97 (February 1989): 1-37. Includes bibliography.

Schuster, A. "On the Investigation of Hidden Periodicities with Application to a Supposed 26-Day Period of Meteorological Phenomena." *Terrestrial Magnetism and Atmospheric Electricity* 3 (1898): 13-41.

Shugart, H. H., Jr., ed. *Time Series and Ecological Processes*. Proceedings of a conference sponsored by SIAM Institute for Mathematics and Society and supported by the National Science Foundation. Philadelphia: Siam, 1978.

Sollberger, Arne. "How Are Biological Time Series Related to the Normal Values Concept?" *Annals of the New York Academy of Sciences* 161.2 (1969): 602-25.

Stuck, B. W., and B. Kleiner. "A Statistical Analysis of Telephone Noise." *The Bell System Technical J* 53 (1974): 1263-1320.

Subba, Rao T., and M. M. Gabr. *An Introduction to Bispectral Analysis and Bilinear Time Series Models*. New York: Springer, 1984.

Subba, Rao T., and Howell Tong. "On Some Tests for the Time-Dependence of a Transfer Function." *Biometrika* 60 (1973): 589-97.

——. "A Test for Time-Dependence of Linear Open-Loop Systems." *J of the Royal Statistical Society. Series B* 34 (1972): 235-50.

Tiao, G. C., and S. C. Hillmer. "Some Considerations of Decomposition of a Time Series." *Biometrika* 65 (1978): 497-502.

Tong, Howell. "On Time Dependent Linear Transformations of Non-Stationary Stochastic Processes." *J of Applied Probability* 11 (1974): 53-62.

——. *Threshold Models in Non-Linear Time Series Analysis*. New York: Springer, 1983.

Tunnicliffe-Wilson, G. "The Estimation of Parameters in Multivariate Time Series Models." *J of the Royal Statistical Society. Series B* (1973): 76-85.

Vandaele, Walter. *Applied Time Series and Box-Jenkins Models*. New York: Academic P, 1983.

Von Furstenberg, G. M., and E. Maasoumi. "Macroeconomic Implications of the Information Revolution." *American Economic Review* 78 (May 1988): 178-81.

Wahba, G. "On the Distribution of Some Statistics Useful in the Analysis of Jointly Stationary Time Series." *Annals of Mathematical Statistics* 39 (1968): 1849-62.

Waldheimer, M. *The Sun-Spot Activity in the Years 1610-1960*. Zürich: Schulthess, 1961.

Walker, A. M., and A. Young. "The Analysis of Observations on the Variations of Latitude." *Monthly Notices of the Royal Astronomical Society* 115 (1955): 443.

Wiener, N. *Interpolation, Extrapolation and Smoothing of Stationary Time Series with Engineering Applications*. Cambridge: Technology P of the Massachusetts Institute of Technology, 1964.

Williams, Esther A., and John M. Gottman. *A User's Guide to the Gottman-Williams Time-Series Analysis Computer Programs for Social Scientists*. Cambridge: Cambridge UP, 1982.

Wold, H. O. A. *A Study in the Analysis of Stationary Time Series*. Uppsala: Almquist and Wiksell, 1938.

Yajima, Y. "On Estimation of Long-Memory Time Series Models." *The Australian J of Statistics* 27 (1985): 303-20.

Young, Peter C. "Recursive Approaches to Time Series Analysis." *Bulletin. Institute of Mathematics and Its Applications* 10 (1974): 209-24.

———. *Recursive Estimation and Time-Series Analysis: An Introduction*. New York: Springer, 1984.

Yule, G. U. "On a Method of Investigating Periodicities in Disturbed Series with Special Reference to Wolfer's Sunspot Numbers." *Philosophical Transactions of the Royal Society. Series A* 226 (1927): 267-98.

Zellner, A., and F. Palm. "Time Series Analysis and Simultaneous Equation Econometric Models." *J of Econometrics* 2 (1974): 17-54.

15. MEDICINE

A. GENERAL

Agren, Hans. "Chinese Traditional Medicine." *Time, Science, and Society in China and the West: The Study of Time V*. Ed. J. T. Fraser, N. Lawrence, and F. C. Haber. Amherst: U of Massachusetts P, 1986. 211-18.

Atkinson, P. "Time and Cool Patients." *Medical Work: Realities and Routines*. Ed. P. Atkinson, and C. Heath. Westmead, Hants.: Gower P, 1981. 41-54.

Bellaby, P., and P. Oribabor. "The History of the Present." *Rewriting Nursing History*. Ed. C. Davies. London: Croom Helm, 1980.

Burton, A. C. "The Clinical Importance of the Physiology of Temperature Regulation." *Canadian Medical Association J* 75 (1956): 715-20.

Calnan, M., and J. R. Butler. "The Economy of Time in General Practice: An Assessment of the Influence of List Size." *Social Science and Medicine* 26.4 (1988): 435-41.

Christensen-Szalanski, J. J. J., and G. B. Northcraft. "Patient Compliance Behavior: The Effects of Time on Patients' Values of Treatment Regimens." *Social Science and Medicine* 21.3 (1985): 263-73.

Coffey, R. M. "Effect of Time-Price on the Demand for Medical-Care Services." *J of Human Resources* 18 (Summer 1983): 407-24. Includes bibliography.

Da Silva, A. Martin, et al., eds. *Biorhythms and Epilepsy*. New York: Raven, 1985.

De Vaney, A. S., et al. "Role of Patient Time in the Pricing of Dental Services: The Fee-Provider Density Relation Explained." *Southern Economic J* 49 (January 1983): 669-80. Includes bibliography.

Dreyfuss, R. "Time as Medication in General Practice" (in French). *Revue Médicale de la Suisse Romande* 102.1 (1982): 3-11.

Goldsmith, J. R. "The Importance of Time in Cancer Epidemiology: Testicular Cancer as an Example." *Public Health Review* 15.3 (1987): 195-214.

Huisheng, Lo. "*Zi Wu* Flow Theory and Time." *Time, Science, and Society in China and the West: The Study of Time V*. Ed. J. T. Fraser, N. Lawrence, and F. C. Haber. Amherst: U of Massachusetts P, 1986. 219-24.

Hull, F. M., et al. "Time and the General Practitioner: The Patient's

View." *J. Royal College of General Practitioners* 34 (February 1984): 71-75.

Koller, M. "Health Risks Related to Shift Work." *International Archives of Occupational and Environmental Health* 53 (1983): 59-75.

Kozoll, Charles E. *Plan for Success: Time Management for the Pre-Med Student*. Champaign: National Association of Advisors for the Health Professions, 1984.

La Roche, A. N., and A. Frankel. "Time Perspective and Health." *Health Education Research* 1.2 (1986): 139-42.

Magun, R. "Die Zeitmessung in der Neurologie." *Zeit in Nervenärztlicher Sicht*. Ed. G. Schaltenbrand. Stuttgart: Ferdinand Enke, 1960.

Schmidt, F. L. *Ergometrie bei Herzkranken: Elektrokardiogramm, Herzschlagfrequenz, und Blutdruck bei Belastungsuntersuchungen*. Farmington: S. Karger, 1977.

Silver, A. "Time Accounting in Community Medicine Education." *J of Community Health* 13 (Winter 1988): 241-48.

Starkey, K. "The Lengthening Hour: Time and the Demise of Psychoanalysis as Therapy." *Social Science and Medicine* 20.9 (1985): 939-43.

"Time: The Fourth Dimension in Medicine." *Acta Diabetologica Latina* 19.2 (April-June 1982): 181-82.

Zerubavel, Eviator. *Patterns of Time in Hospital Life: A Sociological Perspective*. Chicago: U of Chicago P, 1979.

B. CHRONOPHARMACOLOGY

Arendt, J. "Melatonin and Rhythmic Functions in Mammals: Therapeutic and Commercial Potential." *Annual Review of Chronopharmacology VI*. Ed. A. Reinberg, M. Smolensky, and G. Labrecque. Oxford: Pergamon, 1990. 137-54.

Arendt, J., M. Aldhous, and V. Marks. "Alleviation of 'Jet Lag' by Melatonin: Preliminary Results of a Controlled Double-Blind Trial." *Annual Review of Chronopharmacology III*. Ed. A. Reinberg, M. Smolensky, and G. Labrecque. Oxford: Pergamon, 1986. 49-52.

Arendt, J., D. S. Minors, and J. M. Waterhouse. *Biological Rhythms in Clinical Practice*. London: Wright, 1989.

Auvil-Novak, S. E., et al. "Temporal Variation in the Self-Administration of Morphine Sulfate via Patient-Controlled Analgesia in Postoperative Gynecologic Cancer Patients." *Annual Review of Chronopharmacology VII*. Ed. A. Reinberg, M. Smolensky, and G. Labrecque. Oxford: Pergamon, 1990. 253-56.

Badran, A. F., and J. M. Echave-Llanos. "Persistence of Mitotic Circadian Rhythm of the Transplantable Mammary Carcinoma after 35 Generations: Its Bearing on the Success of Treatment with Endoxan." *J. National Cancer Institute* 35 (1965): 285-90.

Barnes, P. J., G. A. Fitzgerald, and C. T. Dollery. "Circadian Variation in Adrenergic Responses in Asthmatic Subjects." *Clinical Science* 62 (1982): 349-54.

Barnes, P. J., et al. "Single Dose Slow-Release Aminophylline at Night Prevents Nocturnal Asthma." *Lancet*. February 6, 1982: 299-301.

Bélanger, P.-M. "Chronobiological Variation in the Hepatic Elimination of Drugs and Toxic Chemical Agents." *Annual Review of Chronopharmacology IV*. Ed. A. Reinberg, M. Smolensky, and G. Labrecque. Oxford: Pergamon, 1988. 1-46.

Bélanger, Pierre-M., François Doré, and Gaston Labrecque. "Monthly Variations in the Clearance of Antipyrene in the Rat." *Research Communications in Chemical Pathology and Pharmacology* 46.1 (October 1984): 53-65.

Bélanger, Pierre-M., and Gaston Labrecque. "Temporal Aspects of Drug Metabolism." *Chronopharmacology: Cellular and Biochemical Interactions*. Ed. Björn Lemmer. New York: Dekker, 1989. 15-34.

Bogaty, Peter, and David D. Waters. "Possible Mechanisms Accounting for the Temporal Distribution of Anginal Attacks." *Chronopharmacology: Cellular and Biochemical Interactions*. Ed. Björn Lemmer. New York: Dekker, 1989. 509-24.

Boissier, C., et al. "Timing Optimizes Sustained-Release Ketoprofen Treatment of Osteoarthritis." *Annual Review of Chronopharmacology VII*. Ed. A. Reinberg, M. Smolensky, and G. Labrecque. Oxford: Pergamon, 1990. 289-92.

Bourdalle-Badie, C., et al. "Biological Rhythms in Pain and Anesthesia." *Annual Review of Chronopharmacology VI*. Ed. A. Reinberg, M. Smolensky, and G. Labrecque. Oxford: Pergamon, 1990. 155-82.

Bruguerolle, B. "Chronopharmacokinetics." *Pathologie Biologie* 35.6 (June 1987): 925-34.

———. *Chronopharmacologie*. Paris: Ellipses Éditions Marketing, 1984.

———. "Time Dependence of General and Local Anesthetic Drugs." *Chronopharmacology: Cellular and Biochemical Interactions*. Ed. Björn Lemmer. New York: Pergamon, 1989. 581-96.

Bryson, S. M., J. R. Lawrence, and B. Whiting. "Disopyramide Plasma Kinetics and Pharmacodynamics Applied to the Assessment of Bioavailability." *British J of Clinical Pharmacology* 4 (1977): 633.

Cal, J. C., C. Dorian, and J. Cambar. "Circadian and Circannual Changes in Nephrotoxic Effects of Heavy Metals and Antibiotics." *Annual Review of Chronopharmacology II*. Ed. A. Reinberg, M. Smolensky, and G. Labrecque. Oxford: Pergamon, 1986. 143-76.

Cal, J. C., et al. "Circadian Rhythm in Serum Zinc Concentrations in Healthy Humans." *Annual Review of Chronopharmacology V*. Ed. A. Reinberg, M. Smolensky, and G. Labrecque. Oxford: Pergamon, 1988. 231-34.

Cole, R. B., and A. Al-Khader. "Effect of Slow-Release Oral Aminophylline on Circadian Variation in Airflow Obstruction in Asthmatics." *J of International Medical Research* 7, Supplement 1 (1979): 40-44.

Cornelissen, G., and J. De Prins. "Cycle Detection." *J of Interdisciplinary Cycle Research* 4 (1973): 159-77.

Coulthard, K. P., et al. "Bioavailability and Diurnal Variation and Absorption of Sustained Release Theophylline in Asthmatic Children." *European J of Clinical Pharmacology* 25 (1983): 667-72.

Cui, F., G. Jin, and S. Qiu. "Chronotherapy of Primary Open-Angle Glaucoma." *Annual Review of Chronopharmacology VII*. Ed. A. Reinberg, M. Smolensky, and G. Labrecque. Oxford: Pergamon, 1990. 297-300.

Czeisler, C. A., et al. "Human Sleep: Its Duration and Organization Depend on Its Circadian Phase." *Science* 210 (1980): 1264-67.

Delius, K., et al. "Effects of Lithium Salts on the Behavior and the Circadian System of *Mesocricetus Auratus*." *J of Interdisciplinary Cycle Research* 15 (1984): 289-99.

De Prins, J., G. Cornelissen, and W. Malbecq. "Statistical Procedures in Chronobiology and Chronopharmacology." *Annual Review of Chronopharmacology II*. Ed. A. Reinberg, M. Smolensky, and G. Labrecque. Oxford: Pergamon, 1986. 27-141.

De Silva, M., A. Binder, and B. L. Hazelman. "The Timing of Prednisolone Dosage and Its Effect on Morning Stiffness in Rheumatoid Arthritis." *Annals of the Rheumatic Diseases* 43 (1984): 790-93.

DeVries, K., et al. "Changes during 24 Hours in the Lung Function and Histamine Hyperreactivity of the Bronchial Tree in Asthmatic and Bronchitic Patients." *International Archives of Allergy* 20 (1962): 93-101.

Duncan, W. C., Jr., and T. A. Wehr. "Pharmacological and Non-Pharmacological Chronotherapies of Depression." *Annual Review of Chronopharmacology IV*. Ed. A. Reinberg, M. Smolensky, and G. Labrecque. Oxford: Pergamon, 1988. 137-71.

Ehret, C. F., V. R. Potter, and K. W. Dobra. "Chronotypic Action of

Theophylline and of Phenobarbital as Circadian Zeitgebers in the Rat." *Science* 188 (1975): 1212-15.

Elliott, J. A., and B. D. Goldman. "Seasonal Reproduction: Photoperiodism and Biological Clocks." *Neuroendocrinology of Reproduction*. Ed. N. T. Adler. New York: Plenum, 1981. 377-423.

Engelmann, W. "A Slowing Down of Circadian Rhythms by Lithium Ions." *Zeitschrift für Naturforschung* 28 (1973): 733-36.

Englund, C., R. Loving, and D. Kripke. "Bright Light Amelioration of Shift Work." *Annual Review of Chronopharmacology VII*. Ed. A. Reinberg, M. Smolensky, and G. Labrecque. Oxford: Pergamon, 1990. 33-36.

Evans, W. E., J. J. Schentag, and W. J. Lusko, eds. *Applied Pharmacokinetics*. San Francisco: Applied Pharmacokinetic, 1980.

Farr, L., C. Todero, and L. Boen. "Effects of Caffeine on Circaseptan Rhythmicity Following Surgery in Laboratory Rats." *Annual Review of Chronopharmacology VII*. Ed. A. Reinberg, M. Smolensky, and G. Labrecque. Oxford: Pergamon, 1990. 189-92.

Ferrari, E., et al. "Changes of the Circadian Pattern of Plasma Melatonin Related to Aging." *Annual Review of Chronopharmacology VII*. Ed. A. Reinberg, M. Smolensky, and G. Labrecque. Oxford: Pergamon, 1990. 73-75.

Feuers, R. J., and L. E. Scheving. "Chronobiology of Hepatic Enzymes." *Annual Review of Chronopharmacology IV*. Ed. A. Reinberg, M. Smolensky, and G. Labrecque. Oxford: Pergamon 1988. 209-56.

Focan, C. "How to Accurately Evaluate Growth Fraction Potential Doubling Time and Cell Loss in Animal and Human Tumor Systems with Circadian Kinetics." *Annual Review of Chronopharmacology VII*. Ed. A. Reinberg, M. Smolensky, and G. Labrecque. Oxford: Pergamon, 1990. 215-18.

Fraga, J. M., et al. "Correlation Between Blood Pressure Characteristics and Neonatal Family History of Blood Pressure." *Annual Review of Chronopharmacology VII*. Ed. A. Reinberg, M. Smolensky, and G. Labrecque. Oxford: Pergamon, 1990. 141-44.

French, J., P. Hannon, and G. C. Brainard. "Effects of Bright Illuminance on Body Temperature and Human Performance." *Annual Review of Chronopharmacology VII*. Ed. A. Reinberg, M. Smolensky, and G. Labrecque. Oxford: Pergamon, 1990. 37-40.

Gervais, P., and A. Reinberg. "Chronotherapy of Allergic Diseases." *Pathologie Biologie* 35.6 (June 1987): 969-76.

Gervais, P., et al. "Twenty-Four-Hour Rhythm in the Bronchial Hyperreactivity to House Dust in Asthmatics." *J of Allergy and Clinical*

Immunology 59 (1977): 207-13.

Gillin, J. C., et al. "Age-Related Changes in Sleep in Depressed and Normal Subjects." *Psychiatry Research* 4 (1981): 73-78.

Goldman, B. D. "The Physiology of Melatonin in Mammals." *Pineal Research Reviews*. Ed. R. J. Reiter. New York: Alan R. Liss, 1983. 145-82.

Guissou, P., et al. "Chronopharmokinetic Study of a Prolonged Release Form of Indomethacin." *European J of Clinical Pharmacology* 24 (1983): 667-72.

Halberg, F., and E. Halberg. "Chronopharmacology and Further Steps toward Chronotherapy." *Pharmacokinetic Basis for Drug Treatment*. Ed. L. Z. Benet, N. Massoud, and J. G. Gambertoglio. New York: Raven, 1984. 221-48.

Halberg, F., J. K. Lee, and W. Nelson. "Time-Qualified Reference Intervals—Chronodesms." *Experientia* 34 (1978): 713-16.

Halberg, F., A. Reinberg, and A. Reinberg. "Chronobiologic Serial Sections Gauge Circadian Rhythm Adjustments following Transmeridian Flight and Life in Novel Environment." *Waking and Sleeping* 1 (1977): 259-79.

Halberg, F., et al. "Autorhythmometry—Procedures for Physiologic Self-Measurements and Their Analysis." *Physiology Teacher* 1 (1972): 1-11.

———. "On Methods for Testing and Achieving Cancer Chronotherapy." *Cancer Treatment Reports* 63 (1979): 1428-30.

———. "Toward a Chronotherapy of Neoplasia: Tolerance of Treatment Depends on Host Rhythms." *Experientia* 29 (1973): 909-1044.

Harkness, J. A. L., et al. "Circadian Variation in Disease Activity in Rheumatoid Arthritis." *British Medical J* 284 (1982): 551-54.

Hastings, J. W. "Biochemical Aspects of Rhythms: Phase Shifting by Chemical." *Cold Spring Harbor Symposia on Quantitative Biology* 25 (1960): 131-43.

Haus, E., et al. "Chronotherapy of Cancer—A Critical Evaluation." *International J of Chronobiology* 6 (1979): 67-107.

———. "Circadian Rhythm Parameters of Clinical and Endocrine Functions in Elderly Subjects Under Treatment with Various Commonly Used Drugs." *Annual Review of Chronopharmacology V*. Ed. A. Reinberg, M. Smolensky, and G. Labrecque. Oxford: Pergamon, 1988. 77-80.

———. "Reference Values for Chronopharmacology." *Annual Review of Chronopharmacology IV*. Ed. A. Reinberg, M. Smolensky, and G. Labrecque. Oxford: Pergamon, 1988. 333-424.

Hayes, D. K., and N. O. Morgan. "Insects as Animal Models for

Chronopharmacological Research." *Annual Review of Chronopharmacology V*. Ed. A. Reinberg, M. Smolensky, and G. Labrecque. Oxford: Pergamon, 1988. 243-46.

Hecquet, B. "Chronoavailability: A Concept to Determine Optimal Time for Drug Administration." *Annual Review of Chronopharmacology III*. Ed. A. Reinberg, M. Smolensky, and G. Labrecque. Oxford: Pergamon, 1986. 285-88.

———. "Models for Chronopharmacology." *Pathologie Biologie* 35.6 (June 1987): 937-41.

———. "Theoretical Aspects of Chronopharmacology." *Annual Review of Chronopharmacology VI*. Ed. A. Reinberg, M. Smolensky, and G. Labrecque. Oxford: Pergamon, 1990. 1-24.

Hetzel, M. R. "The Pulmonary Clock." *Thorax* 36 (1981): 481-86.

Hetzel, M. R., and T. J. H. Clark. "Comparison of Normal and Asthmatic Circadian Rhythms in Peak Expiratory Flow Rate." *Thorax* 35 (1980): 732-38.

Hoeschen, R. J., and T. E. Cully. "Dose-Response Relation between Therapeutic Levels of Serum Digoxin and Systolic-Time Intervals." *American J of Cardiology* 35 (1975): 469-72.

Hrushesky, W. J. M. "Circadian Timing of Cancer Chemotherapy." *Science* 228 (1985): 73-75.

———. "The Clinical Application of Chronobiology to Oncology." *American J of Anatomy* 168 (1983): 519-42.

Hrushesky, William J. M., Reinhard von Roemeling, and Robert B. Sothern. "Circadian Chronotherapy: From Animal Experiments to Human Cancer Chemotherapy." *Chronopharmacology: Cellular and Biochemical Interactions*. Ed. Björn Lemmer. New York: Dekker, 1989. 439-76.

Hrushesky, W. J. M., et al. "Clinical Chrono-Oncology." *Chronobiology: Principles and Applications to Shifts in Schedules*. Ed. L. E. Scheving, and F. Halberg. Alphen aan den Rijn, The Netherlands: Sijthoff and Noordhoff, 1980. 513-33.

Hunter, J. D., et al. "Circadian Variations in Insulin and Alloxan Sensitivity Noted in Blood Glucose Alterations in Normal and Diabetic Mice." *Annual Review of Chronopharmacology V*. Ed. A. Reinberg, M. Smolensky, and G. Labrecque. Oxford: Pergamon, 1988. 295-98.

Huskisson, E. C. "Chronopharmacology of Antirheumatic Drugs, with Special Reference to Indomethacin." *Inflammatory Arthropathies*. Ed. E. C. Huskisson, and G. P. Velo. Amsterdam: Excerpta Medica, 1976. 99-105.

Isles, A. F., and P. von Wichert, eds. *Sustained Release Theophylline and Nocturnal Asthma*. Amsterdam: Excerpta Medica, 1985.

Jackson, S. H. D., et al. "Is Circadian Variation in Theophylline Trough Serum Concentrations Determined by Time of Dosing?" *British J of Clinical Pharmacology* 17 (1984): 777-79.

Jarrett, R. J. "Temporal Aspects of Blood Glucose Regulation: Implications for Diabetic Patients." *Chronopharmacology: Cellular and Biochemical Interactions*. Ed. Björn Lemmer. New York: Dekker, 1989. 615-20.

Job-Deslandre, C., A. Reinberg, and F. Delbarre. "Chronoeffectiveness of Indomethacin in Four Patients Suffering from Evolutive Osteoarthritis of Hip or Knee." *Chronobiologia* 10 (1983): 245-54.

Johnsson, A., et al. "Period Lengthening of Human Circadian Rhythms by Lithium Carbonate, a Prophylactic for Depressive Disorders." *International J of Chronobiology* 8 (1983): 129-47.

Jonkman, J. H. G., et al. "Chronopharmacokinetics of Theophylline after Sustained Release and Intravenous Administration to Adults." *European J of Clinical Pharmacology* 26 (1984): 215-25.

Källstrand, G. "The Effect of Food Intake on the Absorption of Different Sustained Release Formulations." *Sustained Release Theophylline*. Ed. A. F. Isles, and P. von Wichert. Amsterdam: Excerpta Medica, 1985. 169-76.

Karanian, J. W., E. R. Ramey, and P. W. Ramwell. "Effect of Gender on Development and Diurnal Rhythm of Prostaglandin Receptor in Rat Aorta." *British J of Pharmacology and Chemotherapy* 73 (1981): 903-07.

Kelley, H. W., and S. Murphy. "Efficacy of a 12-Hour Sustained-Release Preparation in Maintaining Therapeutic Serum Theophylline Levels in Asthmatic Children." *Pediatrics* 66 (1980): 97-102.

Klemfuss, Harry, and Daniel F. Kripke. "Effects of Lithium on Circadian Rhythms." *Chronopharmacology: Cellular and Biochemical Interactions*. Ed. Björn Lemmer. New York: Dekker, 1989. 281-98.

Klevecz, R. R., and P. S. Braly. "Synchronous Waves of Proliferation in Human Ovarian Cancers." *Annual Review of Chronopharmacology III*. Ed. A. Reinberg, M. Smolensky, and G. Labrecque. Oxford: Pergamon, 1986. 175-78.

Kollop, M. A., et al. "Ultradian, Circadian and Circannual Rhythms of Blood Glucose and Injected Insulins Documented in Six Self-Controlled Diabetics." *Annual Review of Chronopharmacology III*. Ed. A. Reinberg, M. Smolensky, and G. Labrecque. Oxford: Pergamon, 1986. 389-90.

Kost, J. "Implantable Therapeutic Systems." *Annual Review of Chronopharmacology V*. Ed. A. Reinberg, M. Smolensky, and G. Labrecque. Oxford: Pergamon, 1988. 415-18.

Kowanko, I. C., et al. "Circadian Variations in the Signs and Symptoms of Rheumatoid Arthritis and in the Therapeutic Effectiveness of Flurbiprofen at Different Times of the Day." *British J of Clinical Pharmacology* 11 (1981): 477-84.

——. "Time of the Day of Prednisolone Administration in Rheumatoid Arthritis." *Annals of the Rheumatic Diseases* 41 (1982): 445-52.

Kripke, D. F., and V. G. Wyborney. "Lithium Slows Rat Circadian Activity Rhythms." *Life Sciences* 26 (1980): 1319-21.

Kyle, G. M. "Circadian Rhythm in the Pharmacokinetics of Orally Administered Theophylline." *Recent Advances in the Chronobiology of Allergy and Immunology*. Ed. M. H. Smolensky, A. Reinberg, and J. P. McGovern. Oxford: Pergamon, 1980. 95-111.

Labrecque, G. "Chronopharmacology of Non-Steroid Anti-Inflammatory Drugs." *Clinical Chronopharmacology: Concepts, Kinetics, Application*. Ed. A. Reinberg. New York: Hemisphere Publishing, 1990. 67-84.

Labrecque, G., and P. M. Bélanger. "Basic Mechanisms of Chronopharmacology." *Pathologie Biologie* 35.6 (June 1987): 917-23.

Labrecque, G., and P. M. Bélanger. "The Chronopharmacology of Inflammatory Process." *Annual Review of Chronopharmacology II*. Ed. A. Reinberg, M. Smolensky, and G. Labrecque. Oxford: Pergamon, 1986. 291-325.

Labrecque, G., P.-M. Bélanger, and F. M. Doré. "Chronopharmacological Studies of Phenylbutazone in the Rat." *Canadian J of Physiology and Pharmacology* 61.6 (1983): 649-52.

Labrecque, G., Pierre-M. Bélanger, and François Doré. "Circannual Variations in Carrageenan-Induced Paw Edema and in the Anti-Inflammatory Effect of Phenylbutazone in the Rat." *Pharmacology* 24 (1982): 169-74.

Labrecque, G., François Doré, and Pierre-Maxime Bélanger. "Circadian Variation of Carrageenan-Paw Edema in the Rat." *Life Sciences* 28 (1981): 1337-43.

Labrecque, G., D. Lepage-Savary, and E. Poulin. "Time-Dependent Variations in Morphine-Induced Analgesia." *Annual Review of Chronopharmacology V*. Ed. A. Reinberg, M. Smolensky, and G. Labrecque. Oxford: Pergamon, 1988. 135-38.

Labrecque, G., et al. "Chronobiological Study of Plasma Exudation in Carrageenan-Paw Oedema in the Rat." *Agents and Actions* 14.5/6 (1984): 719-22.

Lagas, M., and J. H. G. Jonkman. "Greatly Enhanced Bioavailability of Theophylline on Post-Prandial Administration of a Sustained

Release Tablet." *European J of Clinical Pharmacology* 24 (1983): 761-67.

Lakatua, D. J. "Diet and Mealtiming as Circadian Synchronizers." *Annual Review of Chronopharmacology V*. Ed. A. Reinberg, M. Smolensky, and G. Labrecque. Oxford: Pergamon, 1988. 303-06.

Lee, E. S., and M. Rawdin. "Seasonal Variation in Oral Contraceptive Failure: An Analysis of Survey Data." *Annual Review of Chronopharmacology III*. Ed. A. Reinberg, M. Smolensky, and G. Labrecque. Oxford: Pergamon, 1986. 273-76.

Lee, R. E. "Circadian Rhythms in the Cutaneous Reactivity to Histamine and Selected Antigens Including Phase Relationship to Urinary Cortisol Excretion." *Annals of Allergy* 38 (1977): 231-36.

Leeds, N. H., et al. "Effect of Food on the Bioavailability and Pattern of Release of a Sustained Release Theophylline Tablet." *J of Clinical Pharmacology* 22 (1982): 196-200.

Lemmer, Björn, ed. *Chronopharmacology: Cellular and Biochemical Interactions*. New York: Dekker, 1989.

———. "The Chronopharmacology of Cardiovascular Medications." *Annual Review of Chronopharmacology II*. Ed. A. Reinberg, M. Smolensky, and G. Labrecque. Oxford: Pergamon, 1986. 199-228.

———. *Chronopharmakologie*. 2nd ed. Stuttgart: WVG, 1984.

———. "Temporal Aspects of the Effects of Cardiovascular Active Drugs in Humans." *Chronopharmacology: Cellular and Biochemical Interactions*. Ed. Björn Lemmer. New York: Dekker, 1989. 525-44.

Lemmer, Björn, and Gaston Labrecque. "Chronopharmacology and Chronotherapeutics: Definitions and Concepts." *Chronobiology International* 4.6 (1987): 319-29.

Lesko, L. J., et al. "Temporal Variations in Trough Serum Theophylline Concentrations at Steady State." *J of Pharmaceutical Sciences* 69 (1980): 357-59.

Lévi, F. "Chronobiology in Oncology." *Pathologie Biologie* 35.6 (June 1987): 960-68.

Lévi, F., N. A. Boughattas, and I. Blazsek. "Comparative Murine Chronotoxicity of Anticancer Agents and Related Mechanisms." *Annual Review of Chronopharmacology IV*. Ed. A. Reinberg, M. Smolensky, and G. Labrecque. Oxford: Pergamon, 1988. 283-332.

Lévi, F., C. Le Louarn, and A. Reinberg. "Timing Optimizes Sustained-Released Indomethacin Treatment of Osteoarthritis." *Clinical Pharmacology and Therapeutics* 37 (1985): 77-84.

Markiewicz, A., and K. Semenowicz. "Time-Dependent Changes in the Pharmacokinetics of Aspirin." *International J of Clinical Pharmacology and Biopharmacology* 17 (1979): 409-11.

Mascia, M. "Evaluation of the Night Coughing in Asthmatic Children." *J of Asthma Disease* 5 (1968): 163-69.

Mayersbach, H. von, ed. *The Cellular Aspects of Biorhythms.* Berlin: Springer, 1972.

Merkus, F. W. H. M., and L. Hendeles, eds. *Sustained Release Theophylline: A Biopharmaceutical Challenge to a Clinical Need.* Amsterdam: Excerpta Medica, 1983.

Minors, D. S., and J. M. Waterhouse. "Analysis of Biological Time Series." *Biological Rhythms in Clinical Practice.* Ed. J. Arendt, D. S. Minors, and J. M. Waterhouse. London: Wright, 1989. 272-94.

Monge-Strauss, M. F., and F. Mikol. "Biorhythmicity and EEG Paroxysmal Abnormalities in Various Forms of Epileptic Seizures." *Annual Review of Chronopharmacology V.* Ed. A. Reinberg, M. Smolensky, and G. Labrecque. Oxford: Pergamon, 1988. 85-88.

Moore, J. G., and M. H. Smolensky. "The Chronobiology of Peptic Ulcer Disease and Implications for Its Chronotherapy with H_2-Receptor Antagonist Medication." *Annual Review of Chronopharmacology VI.* Ed. A. Reinberg, M. Smolensky, and G. Labrecque. Oxford: Pergamon, 1990. 113-36.

Moore-Ede, M. C. "Circadian Rhythms of Drug Effectiveness and Toxicity." *Clinical Pharmacology and Therapeutics* 14 (1973): 925-35.

Motohashi, Y., and T. Takano. "Alteration of Circadian Rhythms of Blood Pressure by Delayed Bed-Time Sleep Schedule." *Annual Review of Chronopharmacology V.* Ed. A. Reinberg, M. Smolensky, and G. Labrecque. Oxford: Pergamon, 1988. 453-56.

Nagashima, R., R. A. O'Reilly, and G. Levy. "Kinetics of Pharmacologic Effect in Man: The Anticoagulant Action of Warfarin." *Clinical Pharmacology and Therapeutics* 10 (1969): 22-35.

Nelson, W., et al. "Methods for Cosinor Rhythmometry." *Chronobiologia* 6 (1979): 305-23.

Nicholson, P. A., and W. Bogie. "Diurnal Variation in the Symptoms of Hay Fever: Implications for Pharmaceutical Development." *Current Medical Research and Opinion* 1 (1973): 395-400.

Okudaira, N., D. F. Kripke, and D. J. Mullaney. "Theophylline Delays Human Sleep Phase." *Life Sciences* 34 (1984): 933-39.

Osman, M. A., et al. "Absorption of Theophylline from Enteric Coated and Sustained Release Formulations in Fasted and Non-Fasted Subjects." *Biopharmaceutics and Drug Disposition* 4 (1983): 63-72.

Perry, E. K., R. H. Perry, and B. E. Tomlinson. "Circadian Variations in Chronergic Enzymes and Muscarinic Receptors Binding in Human Cerebral Cortex." *Neuroscience Letters* 4 (1977): 185-89.

Pfeiffer, A., et al. "Circadian Variation of the Effect of Aspirin on

Gastric Potential Difference." *Annual Review of Chronopharmacology VII*. Ed. A. Reinberg, M. Smolensky, and G. Labrecque. Oxford: Pergamon, 1990. 87-90.

Philippens, K. M. H. "Morphology, Function and Time: Its Relation to Drug Action." *British J of Clinical Practice* 38 (1984): 8-13.

Philippens, K. M. H., H. von Mayersbach, and L. E. Scheving. "Effects of the Scheduling of Meal-Feeding at Different Phases of the Circadian System in Rats." *J of Nutrition* 107 (1977): 176-93.

Place, V. A. "Testosterone Replacement with a Circadian Pattern." *Annual Review of Chronopharmacology V*. Ed. A. Reinberg, M. Smolensky, and G. Labrecque. Oxford: Pergamon, 1988. 93-94.

Pöllmann, L. "Chronobiologic Considerations of Aspirin and Indomethacin." *Annual Review of Chronopharmacology I*. Ed. A. Reinberg, M. Smolensky, and G. Labrecque. Oxford: Pergamon, 1984. 349-52.

Pöllmann, L., and P. H. P. Harris. "Rhythmic Changes in Pain Sensitivity in Teeth." *International J of Chronobiology* 5 (1978): 459-64.

Pöllmann, L., and G. Hildebrand. "Circadian Variations of Potency of Placebos on Pain Threshold in Healthy Teeth." *Chronobiologia* 6 (1979): 145.

Pöllmann, L., and B. Pöllmann. "Circadian Variations of the Efficiency of Hepatitis B Vaccination." *Annual Review of Chronopharmacology V*. Ed. A. Reinberg, M. Smolensky, and G. Labrecque. Oxford: Pergamon, 1988. 45-48.

Pownall, R., and M. S. Knapp. "Immunology Has Rhythms: Are They Important?" *Immunology Today* 1 (1980): 8-10.

Proccacci, P., et al. "Rhythmic Changes of the Cutaneous Pain Threshold in Man: A General Review." *Chronobiologia* 1 (1974): 77-96.

Redfern, P. H., and P. C. Moser. "Factors Affecting Circadian Variation in Responses to Psychotropic Drugs." *Annual Review of Chronopharmacology IV*. Ed. A. Reinberg, M. Smolensky, and G. Labrecque. Oxford: Pergamon, 1988. 107-36.

Redon, P., M. Laval, and J. Cambar. "Chronobiological Aspects of the Renin-Angiotensin System: Physiology and Physiopathology." *Annual Review of Chronopharmacology VI*. Ed. A. Reinberg, M. Smolensky, and G. Labrecque. Oxford: Pergamon, 1990. 183-224.

Reinberg, A. "Advances in Human Chronopharmacology." *Chronobiologia* 3 (1976): 151-66.

——. "Clinical Chronopharmacology, an Experimental Basis for Chronotherapy." *Arzneimittel Forschung* 28 (1978): 136-52.

——. "Concepts of Chronopharmacology." *Clinical Chronopharmacology: Concepts, Kinetics, Application*. Ed. A. Reinberg. New

York: Hemisphere Publishing, 1990. 19-45.
——. "Hours of Changing Responsiveness and Susceptibility." *Perspectives in Biology and Medicine* 11 (1967): 111-28.
——. *Les rhythmes biologiques*. Paris: Presses Universitaires de France, 1989.
——, ed. *Clinical Chronopharmacology: Concepts, Kinetics, Application*. New York: Hemisphere Publishing, 1990.
Reinberg, A., J. Ghata, and E. Sidi. "Circadian Variations in the Activity of the Adrenal Cortex and the Nocturnal Exacerbation of Asthma." *Année Endocrinologique* 24 (1963): 452-59.
——. "Nocturnal Asthma Attacks: Their Relationship to the Circadian Adrenal Cycle." *J of Allergy* 34 (1963): 323-30.
Reinberg, A., and F. Halberg. *Chronopharmacology*. New York: Pergamon, 1979.
——. "Circadian Chronopharmacology." *Annual Review of Pharmacology 11 (1971): 455-92.*
Reinberg, A., F. Halberg, and C. Falliers. "Circadian Timing of Methylprednisolone Effects in Asthmatic Boys." *Chronobiologia* 1 (1974): 333-47.
Reinberg, A., and F. Lévi. "Dose-Response Relationships in Chronopharmacology." *Annual Review of Chronopharmacology VI*. Ed. A. Reinberg, M. Smolensky, and G. Labrecque. Oxford: Pergamon, 1990. 25-46.
Reinberg, A., F. Lévi, and M. Smolensky. "Chronopharmacocinétique clinique." *J de Pharmacologie* 15 (1984): 95-124.
Reinberg, A., and M. Reinberg. "Circadian Changes of the Duration of Action of Local Anaesthetic Agents." *Naunyn-Schmiedebergs Archiv für Pharmakologie* 207 (1977): 149-59.
Reinberg, A., and M. H. Smolensky. *Biological Rhythms and Medicine: Cellular, Metabolic, Physiopathologic and Pharmacologic Aspects*. New York: Springer, 1983.
——. "Circadian Changes of Drug Disposition in Man." *Clinical Pharmacokinetics* 7 (1982): 401-20.
Reinberg, A., M. H. Smolensky, and G. Labrecque, eds. *Annual Review of Chronopharmacology I*. Oxford: Pergamon, 1984.
——, eds. *Annual Review of Chronopharmacology II*. Oxford: Pergamon, 1986.
——, eds. *Annual Review of Chronopharmacology III*. Oxford: Pergamon, 1986.
——, eds. *Annual Review of Chronopharmacology IV*. Oxford: Pergamon, 1988.
——, eds. *Annual Review of Chronopharmacology V*. Oxford: Perg-

amon, 1988.

——, eds. *Annual Review of Chronopharmacology VI*. Oxford: Pergamon, 1990.

——, eds. *Annual Review of Chronopharmacology VII*. Oxford: Pergamon, 1990.

——. "The Hunting of a Wonder Pill for Resetting All Biological Clocks." *Annual Review of Chronopharmacology IV*. Ed. A. Reinberg, M. Smolensky, and G. Labrecque. Oxford: Pergamon, 1988. 171-208.

——. "New Aspects of Chronopharmacology." *Annual Review of Chronopharmacology II*. Ed. A. Reinberg, M. H. Smolensky, and G. Labrecque. Oxford: Pergamon, 1986. 3-26.

Reinberg, Alain, et al. "Chronokinetics." *Comprehensive Medicinal Chemistry: The Rational Design, Mechanistic Study and Therapeutic Application of Chemical Compounds. Vol. 5: Biopharmaceutics*. Ed. John B. Taylor. Oxford: Pergamon, 1990.

——. "Tenoxicam Chronotherapy of Rheumatic Diseases." *Annual Review of Chronopharmacology VII*. Ed. A. Reinberg, M. Smolensky, and G. Labrecque. Oxford: Pergamon, 1990. 293-96.

Reiter, R. J. "Intrinsic Rhythms of the Pineal Gland and Associated Hormone Cycles in Body Fluids." *Annual Review of Chronopharmacology IV*. Ed. A. Reinberg, M. Smolensky, and G. Labrecque. Oxford: Pergamon, 1988. 77-106.

——. "The Mammalian Pineal Gland: Structure and Function." *American J of Anatomy* 162 (1981): 287-313.

Reiter, R. J., B. A. Richardson, and T. S. King. "The Pineal Gland and Its Indole Products: Their Importance in the Control of Reproduction in Mammals." *The Pineal Gland*. Ed. R. Relkin. New York: Elsevier Biomedical, 1983. 151-99.

Rogers, C., S. Szabo, and K. J. Vener. "Drug-Induced Gastric and Duodenal Ulcers: Possible Mechanisms Involving Circadian Variation." *Chronopharmacology: Cellular and Biochemical Interactions*. Ed. Björn Lemmer. New York: Dekker, 1989. 621-30.

St.-Pierre, M. V., et al. "Temporal Variation in the Disposition of Theophylline and Its Metabolites." *Clinical Pharmacology and Therapeutics* 38 (1985): 89-95.

Samaan, S. S., et al. "Comparison of Diurnal and Nocturnal Theophylline Disposition." *Annual Review of Chronopharmacology I*. Ed. A. Reinberg, M. Smolensky, and G. Labrecque. Oxford: Pergamon, 1984. 69-72.

Scheving, L. E. "Chronobiology and How It Might Be Used to Better Understand and Improve Chemotherapy." *Neoplasms—Comparative*

Pathology of Growth in Animals, Plants and Man. Ed. H. E. Kaiser. Baltimore: Williams and Wilkins, 1980. 317-32.

———. "Chronobiology—A New Perspective in Pharmacology and Chemotherapy." *Trends in Pharmacological Sciences* 1 (July 1980): 303-07.

———. "The Dimension of Time and Biology in Medicine: Chronobiology." *Endeavour* 35 (1976): 66-76.

Scheving, L. E., and E. R. Burns. "Some Evidence that a Consideration of Chronobiology of the Cell-Cycle May Improve Chemotherapy." *Die Zeit und das Leben*. Ed. J. H. Scharf and H. von Mayersbach. Nova Acta Leopoldina, Vol. 46, No. 225 (1977): 227-83.

Scheving, L. E., F. Halberg, and J. E. Pauly, eds. *Chronobiology*. Tokyo: Igahu Shoin, 1974.

Scheving, L. E., H. von Mayersbach, and J. E. Pauly. "An Overview of Chronopharmacology (A General Review)." *J of European Toxicology* 7 (1974): 203-27.

Scheving, L. E., J. E. Pauly, and L. A. Scheving. "Chronobiology of Cellular Proliferation: Implication for Cancer Chemotherapy." *Topics in Environmental Physiology and Medicine*. Ed. A. Reinberg, and M. Smolensky. New York: Springer, 1983. 79-130.

Scheving, L. E., T. H. Tsai, and J. E. Pauly. "Chronotoxicology and Chronopharmacology with Emphasis on Carcinostatic Agents." *Annual Review of Chronopharmacology II*. Ed. A. Reinberg, M. Smolensky, and G. Labrecque. Oxford: Pergamon, 1986. 177-97.

Scheving, Lawrence E., et al. "Cellular Mechanisms Involved in the Action of Anticancer Drugs." *Chronopharmacology: Cellular and Biochemical Interactions*. Ed. Björn Lemmer. New York: Dekker, 1989. 317-70.

———. "The Persistence of the Circadian Rhythm in Histamine Response of Guinea Pigs Maintained Under Continuous Illumination." *Anatomical Record* 175 (1973): 1-6.

Schmid, U., et al. "Maturation of Circadian Behavior Rhythms in the First Year of Human Life." *Annual Review of Chronopharmacology V*. Ed. A. Reinberg, M. Smolensky, and G. Labrecque. Oxford: Pergamon, 1988. 273-76.

Schulz, H.-U., V. W. Steinijans, and H. Gabel. "Differences in Steady-State Plasma Levels between Aminophylline and Theophylline Sustained-Release Micropellets after Repeated Circadian Dosing." *International J of Clinical Pharmacology, Therapy and Toxicology* 22 (1984): 621-24.

Simmons, D. J. "Chronotherapeutic Options in the War against Osteoporosis." *Annual Review of Chronopharmacology VI*. Ed. A. Reinberg, M. Smolensky, and G. Labrecque. Oxford: Pergamon,

1990. 47-80.

Smaaland, R., et al. "Glutathione in Human Bone Marrow and Its Circadian Stage Relation to DNA Synthesis." *Annual Review of Chronopharmacology VII*. Ed. A. Reinberg, M. Smolensky, and G. Labrecque. Oxford: Pergamon, 1990. 237-40.

Smolensky, M. H. "Chronobiology and Epidemiology." *Pathologie Biologie* 35.6 (June 1987): 991-1004.

——. "Present Achievements and Future Prospects for Clinical Chronobiology." *Biological Rhythms in Clinical Practice*. Ed. J. Arendt, D. S. Minors, and J. M. Waterhouse. London: Wright, 1989. 253-71.

Smolensky, M. H., G. E. D'Alonzo, and A. Reinberg. "Current Developments in the Chemotherapy of Nocturnal Asthma." *Annual Review of Chronopharmacology VI*. Ed. A. Reinberg, M. Smolensky, and G. Labrecque. Oxford: Pergamon, 1990. 81-112.

Smolensky, M. H., and A. Reinberg. "The Chronotherapy of Corticosteroids: Practical Applications for Chronobiologic Findings to Nursing." *Nursing Clinics of North America* 11 (1976): 609-20.

Smolensky, M. H., A. Reinberg, and J. P. McGovern, eds. *Recent Advances in the Chronobiology of Allergy and Immunology*. Oxford: Pergamon, 1980.

Smolensky, M. H., A. Reinberg, and J. T. Queng. "The Chronobiology and Chronopharmacology of Allergy." *Annals of Allergy* 47 (1981): 234-52.

Smolensky, M. H., et al. "The Chronopharmacology and Chronotherapy of Asthma." *Annual Review of Chronopharmacology II*. Ed. A. Reinberg, M. Smolensky, and G. Labrecque. Oxford: Pergamon, 1986. 229-73.

Soliman, K. F. A., et al. "Diurnal Variation in the Phlogistic Response of Rats to Inflammatory Agents." *J of Pharmacy and Pharmacology* 35 (1983): 388-89.

Soubeyrand, J., et al. "A 'Once-a-Day Administration' Sustained Release Theophylline Formulation: Disposition and Pharmacokinetics." *European J of Clinical Pharmacology* 27 (1984): 325-28.

Soulban, G., and G. Labrecque. "Circadian Rhythms of Blood Clotting Time and Coagulation Factors II, VII, IX, and X in Rats." *Life Sciences* 45.25 (1989): 2485-89.

Soyka, L. F., and G. P. Redmond, eds. *Drug Metabolism in the Immature Human*. New York: Raven, 1981.

Stephan, K., et al. "Analysis of Variance in Human Chronopharmacology. An Old Method in a New Area of Research." *Annual Review of Chronopharmacology V*. Ed. A. Reinberg, M. Smolensky, and G. Labrecque. Oxford: Pergamon, 1988. 281-82.

Straughn, S., et al. "A Chronopharmacokinetic Model for Controlled-Release Formulations." *Annual Review of Chronopharmacology I*. Ed. A. Reinberg, M. Smolensky, and G. Labrecque. Oxford: Pergamon, 1984. 93-96.

Sturtevant, R. P., and S. L. Garber. "Chronopharmacology of Ethanol: Acute and Chronic Administration in the Rat." *Annual Review of Chronopharmacology IV*. Ed. A. Reinberg, M. Smolensky, and G. Labrecque. Oxford: Pergamon, 1988. 47-76.

Swanson, B. N., et al. "Sunlindac Disposition When Given Once and Twice Daily." *Clinical Pharmacology and Therapeutics* 32 (1982): 397-403.

Takahashi, R., F. Halberg, and C. A. Walker, eds. *Toward Chronopharmacology*. Oxford: Pergamon, 1982.

Tamarkin, L., C. J. Baird, and O. F. X. Almeida. "Melatonin: A Coordinating Signal for Mammalian Reproduction?" *Science* 227 (1985): 714-20.

Tamura, K., et al. "Chronobiologic Difference in the Efficacy of Antihypertensive Agents to Essential 'Hypertension.'" *Annual Review of Chronopharmacology VII*. Ed. A. Reinberg, M. Smolensky, and G. Labrecque. Oxford: Pergamon, 1990. 125-28.

Taylor, D. R., et al. "Investigation of Diurnal Changes in the Disposition of Theophylline." *British J of Clinical Pharmacology* 16 (1983): 413-16.

Thuresson, S.-O., et al. "Unequal Divisions of Daily Doses of Theophylline: A Pharmacokinetic Study." *Sustained Release Theophylline and Nocturnal Asthma*. Ed. A. F. Isles, and P. von Wichert. Amsterdam: Excerpta Medica, 1985. 125-34.

Touitou, Y. "Aging of Biological Rhythms in the Man." *Pathologie Biologie* 35.6 (June 1987): 1005-12.

Van Tornout, J. M. A., et al. "Feasibility, Safety and Accuracy of an Automated Drug Delivery System Suitable for Chronochemotherapy in Pediatric Patients." *Annual Review of Chronopharmacology V*. Ed. A. Reinberg, M. Smolensky, and G. Labrecque. Oxford: Pergamon, 1988. 431-32.

Vener, K. J., and J. G. Moore. "Chronobiologic Properties of the Alimentary Canal Affecting Xenobiotic Absorption." *Annual Review of Chronopharmacology IV*. Ed. A. Reinberg, M. Smolensky, and G. Labrecque. Oxford: Pergamon, 1988. 257-82.

Wever, R. A. *The Circadian System of Man: The Results of Experiments Under Temporal Isolation*. New York: Springer, 1979.

Wirz-Justice, A., G. A. Groos, and T. A. Wehr. "The Neuropharmacology of Circadian Timekeeping in Mammals." *Vertebrate*

Circadian Systems. Ed. J. Aschoff, S. Daan, and G. Groos. Berlin: Springer, 1982. 183-93.

Zatz, M., and M. H. Brownstein. "Intraventricular Carbachol Mimics the Effects of Light on the Circadian Rhythm in the Rat Pineal Gland." *Science* 203 (1979): 358-61.

Zulley, J. "Distribution of REM Sleep in Entrained 24 Hour and Free-Running Sleep-Awake Cycles." *Sleep* 2 (1980): 377-89.

C. PSYCHIATRY

See also Psychology: Psychoanalysis and Psychiatry

Ahrens, R. "Störungen und Sukzessivgestalten und Zeiterleben bei Aphasie." *Zeit in nervenärztlicher Sicht*. Ed. G. Schaltenbrand. Stuttgart: Enke, 1963. 100-03.

Anliker, J. "Variations in Alpha Voltage of the Electro-Encephalogram and Time Perception." *Science* 140 (1963): 1307-09.

Arlow, Jacob A. "Disturbances of the Sense of Time With Special Reference to the Experience of Timelessness." *Psychoanalytic Quarterly* 53 (1984): 13-37.

——. "Psychoanalysis and Time." *J of the American Psychoanalytic Association* 34 (1986): 507-28.

——. "The Structure of the Déjà Vu Experience." *J of the American Psychoanalytic Association* 7 (1959): 611-31.

——. "Time as Emotion." *Time and Mind: The Study of Time VI*. Ed. J. T. Fraser. Madison: International Universities P, 1989.

Beck, A. T. *Depression: Clinical, Experimental and Theoretical Aspects*. New York: Harper and Row, 1967.

Boardman, W. K., S. Goldstone, and W. T. Lhamon. "Effects of Lysergic Acid Diethylamide (LSD) on the Time Sense of Normals." *American Medical Association Archives of Neurological Psychiatry* 78 (1957): 321-24.

Dahl, M. "A Singular Distortion of Temporal Orientation." *American J of Psychiatry* 115 (August 1958): 146-49.

Davis, H. "Space and Time in the Central Nervous System." *EEG Clinical Neurophysiology* 8 (1956): 185-91.

De Vane, C. L. "Tricyclic Antidepressants." *Applied Pharmacokinetics*. Ed. W. E. Evans, J. J. Schentag, and W. J. Lusko. San Francisco: Applied Pharmacokinetics, 1980. 549-85.

Dilling, C. A., and A. I. Rabin. "Temporal Experience in Depressive States and Schizophrenia." *J of Consulting Psychology* 31 (1967):

604-08.

Doerr, P., et al. "Relationship Between Mood Changes and Adrenal Cortical Activity in a Patient with 48-Hour Unipolar-Depressive Cycles." *J of Affective Disorders* 1 (1979): 93-104.

Dooley, L. "The Concept of Time in Defense of Ego Integrity." *Psychiatry* 4 (1941): 13-23.

Dubois, F. S. "Rhythms, Cycles and Periods in Health and Disease." *American J of Psychiatry* 116 (July 1959): 114-19.

———. "The Sense of Time and Its Relation to Psychiatric Illness." *American J of Psychiatry* 111 (July 1954): 46-51.

Edlund, Matthew. *Psychological Time and Mental Illness*. New York: Gardner, 1987.

Fischer, R., F. Griffin, and L. Liss. "Biological Aspects of Time in Relation to (Model) Psychoses." *Annals of the New York Academy of Sciences* 96 (1962): 44-65.

Fuse, K. "An Approach to Schizophrenics. III. Time Experiences of Schizophrenics." *Hirosaki Medical J* 18 (1966): 75-86.

Gay, Peter. *Freud—A Life for Our Time*. New York: Norton, 1988.

Gifford, Sanford. "Sleep, Time and the Early Ego. Comments on the Development of the 24-Hour Sleep Wakefulness Pattern as a Precursor of Ego Functioning." *J of the American Psychoanalytic Association* 8 (1960): 5-12.

Gooddy, William. "Disorders of Orientation in Space-Time." *British J of Psychiatry* 112 (1966): 661-70.

———. *Time and the Nervous System*. New York: Praeger, 1988.

Granit, R. "The Time-Memory Complex." *Perspectives in Biology and Medicine* 31.3 (Spring 1988): 313-17.

Hartocollis, P. *Time and Timelessness*. New York: International Universities P, 1983.

———. "Time as a Dimension of Affects." *J of the American Psychoanalytic Association* 20 (1972): 92-108.

Hawkins, W. L., et al. "Depressed Affect and Time Perception." *J of Abnormal Psychology* 97.3 (August 1988): 275-80.

Hoff, H., and O. Pötzl. "Über eine Zeitrafferwirkung bei homonymer linksseitiger Hemianopsie." *Zeitschrift für die Gesamte Neurologie und Psychiatrie* 151 (1934): 599-641.

Holubar, J., and J. Machek. "Time Sense and Epileptic EEG Activity." *Epilepsia* 3 (1962): 323-28.

Hulbert, Ryan J., and Willy Lens. "Time Perspective, Time Attitude, and Time Orientation in Alcoholism: A Review." *International J of the Addictions* 23.3 (1988): 279-98.

Jaques, E. *The Form of Time*. New York: Crane Russak, 1982.

Kafka, John. "On Reality. An Examination of Object Constancy, Ambiguity, Paradox and Time." *Psychiatry and the Humanities, Volume 2*. New Haven: Yale UP, 1977. 133-58.

Kafka, J. S. "A Method for Studying the Organization of Time Experience." *American J of Psychiatry* 114 (1957): 546-53.

Kitamura, Toshinori, and Rajinder Kumar. "Controlled Study on Time Reproduction of Depressive Patients." *Psychopathology* 17.1 (1984): 24-27.

Kitamura, T., et al. "Time Passes Slowly for Patients with Depressive State." *Acta Psychiatrica Scandanavica* 65.6 (June 1982): 415-20.

Korzybski, A. *Time-Binding: The General Theory*. 2nd ed. Lancaster: International Non-Aristotelian Library, 1956.

Kries, J. von. "Über die Bedeutung des Aufmerksamkeitssprunges für den Zeitsinn." *Deutsche Zeitschrift für Nervenheilkunde* 47 and 48 (1913): 352-70.

Kripke, D. F. "Phase Advance Theories for Affective Illnesses." *Circadian Rhythms in Psychiatry*. Ed. T. A. Wehr, and F. K. Goodwin. Pacific Grove: Boxwood P, 1983. 41-69.

——. "Photoperiodic Mechanisms for Depression and Its Treatment." *Biological Psychiatry*. Ed. C. Perris, G. Struwe, and B. Jansson. New York: Elsevier, 1981. 1249-52.

Kripke, D. F., D. J. Mullaney, and S. Gabriel. "The Chronopharmacology of Antidepressant Drugs." *Annual Review of Chronopharmacology II*. Ed. A. Reinberg, M. Smolensky, and G. Labrecque. Oxford: Pergamon, 1986. 275-89.

Kripke, D. F., S. C. Risch, and D. S. Janowsky. "Bright White Light Alleviates Depression." *Psychiatry Research* 10 (1983): 105-12.

Kripke, D. F., et al. "Circadian Rhythm Disorders in Manic-Depressives." *Biological Psychiatry* 13 (1978): 335-51.

——. "Circadian Rhythm Phases in Affective Illnesses." *Chronobiologia* 6 (1979): 365-75.

Linkowski, P., et al. "The 24-Hour Profile of Adrenocorticotropin and Cortisol in Major Depressive Illness." *J of Clinical Endocrinology* 61 (1985): 429-38.

Loewald, Hans. "Superego and Time." *International J of Psychoanalysis* 43 (1962): 264-68.

McEachron, D. L., D. F. Kripke, and V. G. Wyborney. "Lithium Promotes Entrainment of Rats to Long Circadian Light-Dark Cycles." *Psychiatry Research* 5 (1981): 1-9.

McGovern, J. P., M. H. Smolensky, and A. Reinberg. "Circadian and Circamensual Rhythmicity in Cutaneous Reactivity to Histamine and Allergenic Extracts." *Chronobiology in Allergy and Immunology*.

Ed. J. P. McGovern, M. H. Smolensky, and A. Reinberg. Springfield: Charles C. Thomas, 1977. 76-116.

Meerloo, Joost A. M. "The Time Sense in Psychiatry." *The Voices of Time*. Ed. J. T. Fraser. 2nd ed. Amherst: U of Massachusetts P, 1981. 235-52.

Melges, Frederick T. "Disorders of Time and the Brain in Severe Mental Illness." *Time and Mind: The Study of Time VI*. Ed. J. T. Fraser. Madison: International Universities P, 1989.

——. "Identity and Temporal Perspective." *Cognitive Models of Psychological Time*. Ed. Richard A. Block. Hillsdale: Erlbaum, 1990. 255-66.

——. *Time and the Inner Future: A Temporal Approach to Psychiatric Disorders*. New York: Wiley, 1982. Bibliography 325-45.

Melges, Frederick T., and C. E. Fougerousse, Jr. "Time Sense, Emotions, and Acute Mental Illness." *J of Psychiatric Research* 4 (1966): 127-40.

Melges, F. T., et al. "Marihuana and the Temporal Span of Awareness." *Archives of General Psychiatry* 14 (1971): 564-67.

——. "Temporal Disintegration and Depersonalization During Marihuana Intoxication." *Archives of General Psychiatry* 13 (1970): 204-10.

Mo, Suchoon S. "Time Reversal in Human Cognition: Search for a Temporal Theory of Insanity." *Cognitive Models of Psychological Time*. Ed. Richard A. Block. Hillsdale: Erlbaum, 1990. 241-54.

Moore, J. E., H. H. Merritt, and Rollo Masselink, eds. *Neurologic and Psychiatric Aspects of Disorders of Aging*. Baltimore: Williams and Wilkins, 1956.

Murphy, Thomas J., and Alan S. De Wolfe. "Future Time Perspective in Alcoholics, Process and Reactive Schizophrenics, and Normals." *International J of the Addictions* 20.11-12 (1985-86): 1815-22.

Past, E. R., et al. "Time Therapy Technique: The Use of Time as a Catalyst for Treatment." *Hospital and Community Psychiatry* 39.2 (February 1988): 177-80.

Ploeger, A. "Zeiterleben in einer Extremsituation." *Zeitschrift für Psychotherapie und Medizinische Psychologie* 16 (1966): 13-20.

Rabin, Albert I. "Time Estimation of Schizophrenics and Non-Psychotics." *J of Clinical Psychology* 13 (1957): 88-90.

Richter, Curt Paul. *Biological Clocks in Medicine and Psychiatry*. Springfield: Charles C. Thomas, 1965.

Rosenthal, N. E., et al. "Seasonal Affective Disorder: A Description of the Syndrome and Preliminary Findings with Light Therapy." *Archives of General Psychiatry* 41 (1984): 72-80.

Sack, D. A., et al. "Potentiation of Antidepressant Medications by Phase Advance of the Sleep-Wake Cycle." *American J of Psychiatry* 142 (1985): 606-08.

Schaltenbrand, George. "Consciousness and Time." *Annals of the New York Academy of Sciences* 138.2 (1967): 632-45.

——, ed. *Zeit in nervenärztlicher Sicht*. Stuttgart: Enke, 1963.

Scoville, W. B., and B. Milner. "Loss of Recent Memory After Bilateral Hippocampal Lesions." *J of Neurology, Neurosurgery and Psychiatry* 20 (1957): 11-21.

Suddath, R. L., et al. "Temporal Lobe Pathology in Schizophrenia: A Quantitative Magnetic Resonance Imaging Study." *American J of Psychiatry* 146 (April 1989): 464-72.

Terr, L. C. "Time Sense Following Psychic Trauma: A Clinical Study of Ten Adults and Twenty Children." *American J of Orthopsychiatry* 53 (April 1983): 244-61.

Torras i Genis, Carme. *Temporal-Pattern Learning in Neural Models*. New York: Springer, 1985.

Tysk, Lennart. "Estimation of Time and the Subclassification of Schizophrenic Disorders." *Perceptual and Motor Skills* 57.3 (1983): 911-18.

——. "Longitudinal Changes in Time Estimation in Affective Disorders: A Preliminary Study." *Perceptual and Motor Skills* 60.1 (1985): 179-88.

——. "A Longitudinal Study of Time Estimation in Psychotic Disorders." *Perceptual and Motor Skills* 59.3 (1984): 779-89.

——. "Time Estimation by Healthy Subjects and Schizophrenic Patients: A Methodological Study." *Perceptual and Motor Skills* 56 (June 1983): 983-88.

——. "Time Perception and Affective Disorders." *Perceptual and Motor Skills* 58.2 (1984): 455-64.

Wallace, M. "Future Time Perspective in Schizophrenia." *J of Abnormal and Social Psychology* 52 (1956): 240-45.

Webster, F. R., S. Goldstone, and W. W. Webb. "Time Judgment and Schizophrenia: Psychophysical Method as a Relevant Contextual Factor." *J of Psychology* 54 (1962): 159-64.

Wehr, T. A., J. C. Gillin, and F. K. Goodwin. "Sleep and Circadian Rhythms in Depression." *Sleep Disorders: Basic and Clinical Research*. Ed. M. Chase, and E. Weitzman. New York: Spectrum, 1983. 195-225.

Wehr, T. A., et al. "Phase Advance of the Circadian Sleep-Wake Cycle as an Antidepressant." *Science* 206 (1979): 710-13.

——. "48-Hour Sleep-Wake Cycles in Manic-Depressive Illness." *Archives of General Psychiatry* 39 (1982): 559-65.

Werboff, J. "Time Judgment as a Function of Electroencephalographic Activity." *Experimental Neurology* 6 (1962): 152-60.

Wirz-Justice, A. "Antidepressant Drugs: Effects on the Circadian System." *Circadian Rhythms in Psychiatry*. Ed. T. A. Wehr, and F. K. Goodwin. Pacific Grove: Boxwood P, 1983. 235-64.

Wirz-Justice, A., and I. C. Campbell. "Antidepressant Drugs Can Slow or Dissociate Circadian Rhythms." *Experientia* 38 (1982): 1301-09.

16. MUSIC

A. GENERAL

Alperson, Philip. "'Musical Time' and Music as an 'Art of Time.'" *J of Aesthetics and Art Criticism* 38 (1980): 407-17.

Altwein, E. F. "Zeit und Musik." *Schweizerische Musikzeitung* 108 (1968): 305-13.

Ansermet, Ernest. "Le temps musical." In *Écrits sur la musique*. Neuchâtel: Éditions de la Baconnière, 1971. 159-68.

Arnheim, Rudolph. "A Stricture on Space and Time." *Critical Inquiry* 4 (1978): 645-55.

Babbitt, Milton. "The Synthesis, Perception, and Specification of Musical Time." *J of the International Folk Music Council* 16 (1964): 92-95.

Barry, B. R. "Time Levels and Tonal Structures in Schubert's *Die Winterreise*." *Music Review* 46 (August 1985): 170-78.

Bass, E. C. "Musical Time and Space in Berlioz." *Music Review* 30 (1969): 211-24.

Beauvillian, C., and P. Fraisse. "On the Temporal Control of Polyrhythmic Performance." *Music Perception* 1.4 (1984): 485-99.

Bértola, Elena de. "On Space and Time in Music and the Visual Arts." *Leonardo* 5.1 (1972): 27-30.

Bosseur, D. "Introduction à la musique américaine: l'expérience du temps chez Cage." *Musique en jeu* 1 (1970): 16-22.

Brelet, G. *Le temps musical*. Paris: Presses Universitaires de France, 1949.

Briner, A. *Der Wandel der Musik als Zeitkunst*. Vienna: Universal-Edition, 1955.

Budden, J. "Time Stands Still in *Otello*." *Opera* 32 (September 1981): 888-93.

Burrows, David. "Music and the Biology of Time." *Perspectives of New Music* 11.1 (1972): 241-49.

Carpenter, Patricia. "The Musical Object." *Current Musicology* 5 (1967): 56-87.

Carter, Elliott. "Music and the Time Screen." *The Writings of Elliott Carter*. Ed. Kurt Stone, and Else Stone. Bloomington: Indiana UP, 1977. 343-65.

——. "The Time Dimension in Music." *The Writings of Elliott Carter*. Ed. Kurt Stone, and Else Stone. Bloomington: Indiana UP, 1977. 243-47.

Charles, D. "Le son comme image du temps." *Revue d'esthétique* 7 (1984): 103-08.

Chatterjee, Margaret. "Towards a Phenomenology of Time-Consciousness in Music." *Diogenes* 74 (1971): 49-56.

Cherbuliez, A.-E. "L'unité du temps et sa division: problème psychologique fondamental du rhythme musical." *Compte rendu du premier congrès du rhythme*. Geneva: Institut Jacques-Dalcroze, 1926. 33-38.

Childs, Barney. "Time and Music: A Composer's View." *Perspectives of New Music* 15.2 (1977): 194-219.

Clynes, M., and J. Walker. "Music as Time's Measure" (consistent performance lengths related to psychobiological clocks). *Music Perception* 4.1 (1986): 85-119. Includes bibliography.

Colwell, Richard. *The Theory of Expectation Applied to Musical Listening*. Urbana: U of Illinois Cooperative Research Project No. H-106, 1966.

Dahlhaus, Carl. "Zeitstrukturen in der Oper." *Musikforschung* 34 (1981): 2-11.

Davis, Shelley. "Harmonic Rhythm in Mozart's Sonata Form." *Music Review* 27.1 (1966): 25-43.

De Selincourt, Basil. "Music and Duration." Trans. Susanne Langer. *Reflections on Art*. Ed. Susanne Langer. Baltimore: Johns Hopkins P, 1958. 152-60.

Dickinson, George Sherman. "Aesthetic Pace in Music." *J of Aesthetics and Art Criticism* 15 (1956): 311-21.

Drew, James. "Information, Space, and a New Time-Dialectic." *J of Music Theory* 12.1 (Spring 1968): 86-103.

Edwards, Allen. *Flawed Words and Stubborn Sounds: A Conversation with Elliott Carter*. New York: Norton, 1971.

Eisenbrey, Keith E. "Timeshapes." *Perspectives of New Music* 20 (1981-82): 649-50.

Emery, Eric. *Temps et musique*. Lausanne: Éditions l'âge d'homme, 1975.

Epstein, David. "Duration." In his *Beyond Orpheus*. Cambridge: MIT P, 1979. 55-98.

———. "On Musical Continuity." *The Study of Time IV*. Ed. J. T. Fraser, N. Lawrence, and D. Park. New York: Springer, 1981. 180-97.

Erickson, Robert. "Time Relations." *J of Music Theory* 7 (1963): 174-92.

Ernst, Anselm. *Philosophische Untersuchungen zum Zeitbegriff in der Musik*. Aachen: Aachen Technische Hochschule, 1973.

Fischer, Kurt von. "Das Zeitproblem in der Musik." *Das Zeitproblem im 20. Jahrhundert*. Ed. Rudolf W. Meyer. Bern: Franke, 1964. 296-317.

Fraser, J. T. "The Art of the Audible 'Now.'" *Music Theory Spectrum* 7 (1985): 181-84.

Garrant, S. "Le temps musical." *Canadian U Musical Review* 7 (1986): 32-35.

Gass, G. "Elliott Carter's *Second String Quartet*: Aspects of Time and Rhythm." *Indiana Theory Review* 4.3 (1981): 412-23.

Gevaert, F. A. *Histoire et théorie de la musique de l'antiquité*. 2 vols. Gand: Braeckmann, 1875-81.

Gooddy, William. "The Timing and Time of Musicians." *Music and the Brain*. Ed. Macdonald Critchley, and R. A. Henson. London: Heinemann, 1977. 131-40.

Gossett, Philip. "The Mensural System and the Choralis Constantinus." *Studies in Renaissance and Baroque Music in Honour of Arthur Mendel*. Ed. Robert L. Marshall. Kassel: Bärenreiter, 1974. 71-107.

Greene, David B. *Mahler, Consciousness and Temporality*. New York: Gordon and Breach, 1984.

——. *Temporal Processes in Beethoven's Music*. New York: Gordon and Breach, 1982.

Greenhough, M. "The Philosophy and Design of a Real-Time Music-Structure Generating System." *Interface* 17.1 (1988): 25-44. Includes bibliography.

Griffiths, P. *Olivier Messiaen and the Music of Time*. London: Faber, 1985.

Grisley, G. "Tempus ex Machina (Reflexionen über die musikalische Zeit)." *Neuland Ansätze zur Musik der Gegenwart* 3 (1983-84): 190-202. Includes bibliography.

Homick, J. L., L. F. Elfner, and G. G. Bothe. "Auditory Temporal Masking and the Perception of Order." *J of the Acoustical Society of America* 45 (1969): 712-18.

Hsu, Dolores Menstell. "Ernst Kurth and His Concept of Music as Motion." *J of Music Theory* 10 (1966): 2-17.

Hughes, Andrew. "Mensuration and Proportion in Early Fifteenth-Century English Music." *Acta Musicologica* 37 (1965): 48-59.

Hutchinson, William. "Aspects of Musical Time." *Selected Reports of the Institute of Ethnomusicology of UCLA* 1 (1966): 66-76.

Imberty, Michel. *Les écritures du temps*. Paris: Dunod, 1981.

Jackendoff, Ray, and Fred Lerdahl. *A Generative Theory of Tonal Music*. MIT P, 1983.

Jones, Mari Riess. "Musical Events and Models of Musical Time."

Cognitive Models of Psychological Time. Ed. Richard A. Block. Hillsdale : Erlbaum, 1990. 207-40.

——. "Only Time Can Tell: On the Topology of Mental Space and Time." *Critical Inquiry* 7 (1981): 557-76.

Klugmann, Friedhelm. *Die Kategorie der Zeit in der Musik*. Kassel-Wilhelmshöhe: Bärenreiter, 1961.

Koechlin, Charles. "Le temps et la musique." *Revue musicale* 58 (1926): 45-62.

Kohl, Jerome. "The Evolution of Macro- and Micro-Time Relations in Stockhausen's Recent Music." *Perspectives of New Music* 22 (1983-84): 147-86.

Kowzan, Tadeusz. "Music and the Plastic Arts in Conquest of Time and Space." *Diogenes* 73 (1971): 1-20.

Kramer, Jonathan D. "Beginnings and Endings in Western Art Music." *Canadian U Music Review* 3 (1982): 1-14.

——. "The Impact of Technology on Musical Time." *Percussive Notes* 22.3 (1984): 16-25.

——. "Moment Form in Twentieth-Century Music." *Musical Quarterly* 64 (1978): 177-94.

——. "New Temporalities in Music." *Critical Inquiry* 7 (1981): 539-56.

——. "Studies of Time and Music: A Bibliography." *Music Theory Spectrum* 7 (1985): 72-106. Kramer's bibliography could be used in conjunction with this section on music.

——. "Temporal Linearity and Nonlinearity in Music." *Time, Science, and Society in China and the West: The Study of Time V*. Ed. J. T. Fraser, N. Lawrence, and F. C. Haber. Amherst: U of Massachusetts P, 1986. 126-40.

——. *The Time of Music: New Meanings, New Temporalities, New Listening Strategies*. New York: Schirmer, 1988. Includes material from Kramer's articles listed above.

Kümmel, Werner Friedrich. "Puls und Musik (16.-18. Jahrhundert)." *Medizinhistorisches J* 3 (1968): 269-93.

Langer, Susanne. "The Image of Time." *Feeling and Form*. New York: Scribner, 1953. Ch. 7.

Larson, Paul. "The Golden Section in the Earliest Notated Western Music." *Fibonacci Quarterly* 16 (1978): 513-15.

Lewin, D. "On Formal Intervals between Time-Spans." *Music Perception* 1.4 (1984): 414-23.

Lippman, Edward. "Progressive Temporality in Music." *J of Musicology* 3 (1984): 121-41.

Lissa, Zofia. "Aesthetic Functions of Silence and Rests in Music." Trans. Eugenia Tarska. *J of Aesthetics and Art Criticism* 22 (1964):

443-54.

——. "The Temporal Nature of Musical Work." Trans. Eugenia Tarska. *J of Aesthetics and Art Criticism* 26 (1968): 529-38.

Lochhead, Judith I. "The Temporal in Beethoven's *Opus 135*: When Are Ends Beginnings?" *In Theory Only* 4 (January 1979): 3-30.

——. "Temporal Structure in Recent Music." *J of Musicological Research* 6.1-2 (1986): 49-94.

McCreless, Patrick. "Schubert's 'Moment Musical' No. 2: The Interaction of Rhythmic and Tonal Structures." *In Theory Only* 3.4 (1977): 3-11.

Maddrell, John Edward. "Mensura and the Rhythm of Medieval Monodic Song." *Current Musicology* 10 (1970): 64-69.

Maniates, Maria Rika. "Sound, Silence, and Time: Towards a Fundamental Ontology of Music." *Current Musicology* 3 (1966): 59-64.

Marcel, Gabriel. "Bergsonism and Music." Trans. Susanne Langer. In *Reflections on Art*. Ed. Susanne Langer. Baltimore: Johns Hopkins P, 1958. 142-51.

Mayr, Albert. "Creative Time Organization Versus Subsonic Noises." *Diogenes* 122 (1983): 45-63.

——. "Musik, Zeit und Gesundheit." *Music and Health*. Ed. E. Ruud. Oslo: Norsk Musikforlag, 1986. 107-18.

——. "Sketches for a Low-Frequency Solfège." *Music Theory Spectrum* 7 (1985): 107-13.

Mendel, Arthur. "Some Ambiguities of the Mensural System." *Studies in Music History: Essays for Oliver Strunk*. Ed. Harold Powers. Princeton: Princeton UP, 1968. 137-60.

Morgan, Robert P. "Musical Time Musical Space." *Critical Inquiry* 6 (1980): 527-38.

Morgenstein, R. "Rock 'n' Jazz Clinic: In Pursuit of Odd Time." *Modern Drummer* 12 (April, June, and July 1988): 82-83, 78-80, and 42-43.

Newell, Robert. "Four Tiers on the Foundation of Time." *International Review of the Aesthetics and Sociology of Music* 7 (1976): 147-73.

Noorden, Leon P. A. S. van. *Temporal Coherence in the Perception of Tone Sequences*. Eindhoven: Technische Hogeschool, 1975.

Orlov, Henry. "The Temporal Dimensions of Musical Experience." *Musical Quarterly* 65.3 (1979): 368-78.

Pasler, Jann. "Narrative and Narrativity in Music." *Time and Mind: The Study of Time VI*. Ed. J. T. Fraser. Madison: International Universities P, 1989. 233-57.

——. "Debussy, *Jeux*: Playing with Time and Form." *19th Century Music* 6.1 (1982): 60-75.

Perry-Camp, Jane. "Time and Temporal Proportion: The Golden Section Metaphor in Mozart, Music, and History." *J of Musicological Research* 3.1-2 (1979): 133-76.

Povel, Dirk-Jan. "Temporal Structure of Performed Music: Some Preliminary Observations." *Acta Psychologica* 41 (1977): 309-20.

Povel, Dirk-Jan, and P. Essens. "Perception of Temporal Patterns." *Music Perception* 2.4 (1985): 411-40.

Quijano, J. P. "La musique, art du temps." *Vie musicale* 14 (1969): 18-23.

Rahn, Jay. "Ockeghem's Three-Section Motet *Salve Regina*: Problems in Coordinating Pitch and Time Constructs." *Music Theory Spectrum* 3 (1981): 117-31.

Ratner, Leonard. "Eighteenth-Century Theories of Musical Period Structure." *Musical Quarterly* 42 (1956): 439-54.

Reynolds, Roger. "It(')s Time." *Electronic Music Review* 7 (1968): 12-17.

Rochberg, George. "The Concepts of Musical Time and Space." In *The Aesthetics of Survival*. Ann Arbor: U of Michigan P, 1984. 78-136.

———. "Duration in Music." *The Modern Composer and His Works*. Ed. J. Beckwith, and Udo Kasemets. Toronto: U of Toronto P, 1961. 56-64.

———. "The Structure of Time in Music: Traditional and Contemporary Ramifications and Consequences." *The Study of Time II*. Ed. J. T. Fraser, and N. Lawrence. New York: Springer, 1975. 136-49.

Rothärmel, Marion. *Der musikalische Zeitbegriff seit Moritz Hauptmann*. 2nd ed. Regensburg: Bosse, 1968.

Rowell, Lewis. "The Creation of Audible Time." *The Study of Time IV*. Ed. J. T. Fraser, N. Lawrence, and D. Park. New York: Springer, 1981. 198-210.

———. "The Subconscious Language of Musical Time." *Music Theory Spectrum* 1 (1979): 96-106.

———. *Thinking about Music: An Introduction to the Philosophy of Music*. Amherst: U of Massachusetts P, 1983.

———. "Time in the Musical Consciousness of Old High Civilizations—East and West." *The Study of Time III*. Ed. J. T. Fraser, N. Lawrence, and D. Park. New York: Springer, 1978. 578-611.

Rummenhöller, Peter. *Moritz Hauptmann als Theoretiker*. Wiesbaden: Breitkopf and Härtel, 1963.

Schmidt, C. M. "Von Zeitarten tonaler Musik." *Archiv für Musikwissenschaft* 42.4 (1985): 287-99.

Schnebel, Dieter. "Auf der Suche nach der befreiten Zeit: Versuch über Schubert." *Neue Zeitschrift für Musik* 131 (1970): 498-505.

Schoen, M. "Bibliography of Experimental Studies on the Psychology of Music." *Proceedings*. Music Teachers National Association, 1941-43.

Schönberger, Elmer. "On the Conceiving of Time." *Keynotes* 13.1 (1981): 5-11.

Schopenhauer, Arthur. "On the Metaphysics of Music." In his *The World as Will and Representation*. Trans. E. F. J. Payne. Indian Hills, Colorado: Falcon's Wing Press, 1958. 2:447-57.

Schuldt, Agnes Crawford. "The Voices of Time in Music." *American Scholar* 45 (1976): 549-59.

Seidel, Wilhelm. "Descartes Bemerkungen zur musikalischen Zeit." *Archiv für Musikwissenschaft* 27 (1970): 287-303.

Shaffer, H. "Timing in Musical Performance." *Annals of the New York Academy of Sciences* 423 (1984): 420-28.

Smith, Charles J. "(Supra-) Durational Patterns in Chopin's Revolutionary Étude." *In Theory Only* 2.5 (1976): 3-12.

Smith, F. Joseph. "Musical Sound as a Model for Husslerian Intuition and Time." *J of Phenomenological Psychology* 4 (1973): 271-96.

Souvtchinsky, Pierre. "La notion du temps et la musique." *Revue musicale* 191 (1939): 309-20.

Sternberg, S., R. L. Knoll, and Paul Zukovsky. "Timing by Skilled Musicians." *The Psychology of Music*. Ed. Diana Deutsch. New York: Academic P, 1982. 181-239.

Stevens, J. *Words and Music in the Middle Ages: Song, Narrative, Dance, and Drama, 1050-1350*. Cambridge: Cambridge UP, 1986.

Stockhausen, Karlheinz. "Structure and Experiential Time." Trans. Leo Black. *Die Reihe* 2 (1959): 64-74.

Strohm, R. "Dramatic Time and Operatic Form in Wagner's *Tannhaeuser*." *Royal Musical Association Proceedings* 104 (1977-78): 1-10.

Tenney, J., and L. Polansky. "Temporal Gestalt Perception in Music." *J of Music Theory* 24.2 (Fall 1980): 205-41.

Tingley, G. P. "Metric Modulation and Elliott Carter's First String Quartet." *Indiana Theory Review* 4.3 (1981): 3-11.

Toch, Ernst. "Beginning and Ending." In his *The Shaping Forces in Music*. New York: Dover, 1977. 217-38.

Torkewitz, D. "Die 'entwickelte Zeit': zum *Intermezzo Op. 116, IV* von Johannes Brahms." *Musikforschung* 32.2 (1979): 135-40.

Underwood, J. "Time and Activity in Weber's *Opus 10*: A Composer's Viewpoint." *Indiana Theory Review* 3.2 (1980): 31-38.

Vanderwerf, Hendrik. "Concerning the Measurability of Medieval Music." *Current Musicology* 10 (1970): 69-73.

Velten, K. "Entwicklungsdenken und Zeiterfahrung in der Musik von

Johannes Brahms: das *Intermezzo Op. 117 Nr. 2.*" *Musikforschung* 34.1 (1981): 56-59.

Wallaschek, Richard. *Primitive Music*. London: Longmans, 1893.

Warren, R. M., and C. J. Obusek. "Identification of Temporal Order within Auditory Sequences." *Perception and Psychophysics* 12 (1972): 86-90.

Williams, B. M. "Time and the Structure of Stravinsky's *Symphony in C*." *Musical Quarterly* 59 (1973): 355-69.

Wiora, Walter. "Musik als Zeitkunst." *Musikforschung* 10 (1957): 15-28.

Zuckerkandl, Victor. "Time." In his *Sound and Symbol: Music and the External World*. Trans. W. R. Trask. New York: Pantheon, 1956. Chs. 11-13.

B. AFRICAN

Alkema, S. "A Study of African Rhythm." *Nada* 11 (1975): 221-26.

Augier, P. "La polyrythme dans les musiques du Sahara." *Libycor* 19 (1971): 217-33.

Brandel, Rose. *The Music of Central Africa*. The Hague: Martinus Nijhoff, 1961.

Chernoff, John M. *African Rhythm and African Sensibility*. Chicago: U of Chicago P, 1979.

Dauer, Alfons Michael. "Kinesis und Katharsis: Prolegomena zur Deutung afrikanischer Rhythmik." *Afrika Heute (Sonderbeilage)* 19 (October 1969): 1-12.

Gregoire, C. *Ägyptische Rhythmik: Rhythmen und Rhythmusinstrumente im heutigen Ägypten*. Strasbourg: Heitz, 1960.

Hornbostel, Erich M. von. "African Negro Music." *Africa* 1 (1928): 30-62.

Jones, Arthur M. "African Drumming." *Bantu Studies* 8 (1934): 1-16.

——. *African Rhythm*. London: International African Institute, 1955.

——. "Drums Down the Centuries." *African Music* 1.4 (1957): 4-10.

——. *Studies in African Music*. 2 vols. London: Oxford UP, 1959.

Kauffman, R. "African Rhythm: A Reassessment." *Ethnomusicology* 24 (1980): 393-415.

King, Anthony. "Employments of the 'Standard Pattern' in Yoruba Music." *African Music* 2.3 (1960): 51-54.

Knight, Roderic. "Mandinka Drumming." *African Arts* 7 (1974): 25-35.

Kubik, Gerhard. "The Phenomenon of Inherent Rhythms in East and Central African Instrumental Music." *African Music* 3.1 (1962):

33-42.

List, G. "African Influences in the Rhythmic and Metric Organization of Colombian Costeno Folksong and Folk Music." *Latin American Music Review* 1.1 (1980): 6-17.

Merriam, Alan P. "African Musical Rhythm and Concepts of Time-Reckoning." *Music East and West: Essays in Honor of Walter Kaufmann*. Ed. Thomas Noblitt. New York: Pendragon, 1981. 123-41.

Nketia, J. H. Kwabena. "Traditional Music of the Ga People." *African Music* 2.1 (1958): 21-27.

Pantaleoni, Hewitt. "Three Principles of Timing in Anlo Dance Drumming." *African Music* 5.4 (1972): 50-63.

Rouget, Gilbert. "Court Songs and Traditional History in the Ancient Kingdoms of Porto-Novo and Aomey." *Essays on Music and History in Africa*. Ed. Klaus P. Wachsmann. Evanston: Northwestern UP, 1971. 27-64.

Stone, Ruth M. "In Search of Time in African Music." *Music Theory Spectrum* 7 (1985): 139-48.

——. *Let the Inside Be Sweet: The Interpretation of Music Event among the Kpelle of Liberia*. Bloomington: Indiana UP, 1982.

——. "The Shape of Time in African Music." *Time, Science, and Society in China and the West: The Study of Time V*. Ed. J. T. Fraser, N. Lawrence, and F. C. Haber. Amherst: U of Massachusetts P, 1986. 113-25.

Waterman, Richard. "African Influence on the Music of the Americas." *Acculturation in the Americas*. Ed. Sol Tax. Chicago: U of Chicago P, 1952. 2:207-18.

C. EASTERN

Becker, Judith. "Hindu-Buddhist Time in Javanese Gamelan Music." *The Study of Time IV*. Ed. J. T. Fraser, N. Lawrence, and D. Park. New York: Springer, 1981. 161-72.

——. "Percussive Patterns in the Music of Mainland Southeast Asia." *Ethnomusicology* 12 (1968): 173-91.

——. "Time and Tune in Java." *The Imagination of Reality: Essays in Southeast Asian Coherence Systems*. Ed. A. Yengoyan, and A. L. Becker. Norwood: Alex Publishing, 1979. 197-210.

Ghosh, Nikhil. "Science and the Art of Keeping Time in Indian Music." *J of the Indian Musicological Society* 5.2 (1975): 27-32.

Hyeku, Lee. "Quintuple Meter in Korean Instrumental Music." *Asian Music* 13.1 (1981): 119-29.

Maceda, J. "A Concept of Time in Music of Southeast Asia (A Preliminary Account)." *Ethnomusicology* 30.1 (1986): 11-53.

Ohtsu, Yoko. "Hyôshi in Japanese Traditional Music and Time in Western Music" (in Japanese). *Bulletin of the Faculty of Music of Tokyo U, Fine Arts and Music* 4 (March 1979): 1-19.

Rowell, Lewis. "Abhinavagupta, Augustine, Time, and Music." *J of the Indian Musicological Society* 13.2 (1982): 18-36.

Saxena, S. K. "Aesthetic Theory and Hindustani Rhythm." *British J of Aesthetics* 16 (1976): 254-60.

Venkatarathnam, R. "Musicometric Dynamism of the Jatis." *Indian Musicological Society J* 13.1 (1982): 20-34.

Widdess, D. R. "Rhythm and Time Measurement in South Asian Art-Music: Some Observations on Tala." *Royal Musical Association Proceedings* 107 (1980-81): 132-39.

D. METRONOME

Creuzburg, H. "Beethovens Tempi und Metronomzahlen." *Das Orchester* 32 (October 1984): 848-49.

Forbes, W. "The Foundations of Music—The Beat." *The Strad* 90 (July 1979): 193-95.

Gleich, C. von. "Das Metronom und seine Deutung: Maelzel und die Folgen im 19. Jahrhundert." *Neue Zeitschrift für Musik* 1 (January 1986): 19-23.

——. "Die Theorie des variables Metronomgebrauchs." *Musikforschung* 41.1 (1988): 46-49.

Hess, W. "The Right Tempo: Beethoven and the Metronome." *Beethoven* 3.1 (1988): 16-17. Includes bibliography.

Howell, S. "Beethoven's Maelzel Canon—Another Schindler Forgery?" *The Musical Times* 120 (December 1979): 987-90.

Kolneder, W. "Zur Geschichte des Metronoms." *Hifi-Stereophonie* 19 (February 1980): 152-56.

Malloch, W. "Carl Czerny's Metronome Marks for Haydn and Mozart Symphonies." *Early Music* 16.1 (1988): 72-81.

Nauhaus, G. "Die Interpretation: Sind Robert Schumanns Metronomangaben richtig?" *Musik und Gesellschaft* 30 (April 1980): 228-30.

Rosenber, W. "Das verteufelte Metronom." *Hifi-Stereophonie* 19 (February 1980): 167-68.

Rosenblum, S. P. "Two sets of Unexplored Metronome Marks for Beethoven's Piano Sonatas." *Early Music* 16.1 (1988): 58-71.

Stadlen, P. "Beethoven and the Metronome." *Soundings* 9 (1982): 38-

56.
Stokes, J. L. "Beethoven's Endorsements of Maelzel's Metronome." *Studies in Music* (Canada) 1 (1976): 23-30.
Weinberg, N. "The Ultimate Metronome." *Instrument* 42 (April 1988): 56.

E. PSYCHOLOGY AND PHYSIOLOGY

Bengtsson, Ingmar, and Alf Gabrielsson. "Rhythm Research in Uppsala." In *Music, Room and Acoustics*. Stockholm: Royal Swedish Academy of Music, 1977. 19-56.
Bengtsson, Ingmar, Alf Gabrielsson, and Barbro Gabrielsson. "A Computer Program for Analysis of Rhythmic Performance." *Svensk Tidskrift för Musikforskning* 40.1 (1978): 15-24.
Bielawski, L. "The Zones of Time in Music and Human Activity." *The Study of Time IV*. Ed. J. T. Fraser, N. Lawrence, and D. Park. New York: Springer, 1981. 173-79.
Clynes, Manfred, and Janice Walker. "Neurobiologic Functions of Rhythm, Time, and Pulse in Music." *Music, Mind, and Brain*. Ed. Manfred Clynes. New York: Plenum, 1982. 171-216.
Deutsch, Diana. "Organizational Processes in Music." *Music, Mind, and Brain*. Ed. Manfred Clynes. New York: Plenum, 1982. 119-36.
——, ed. *The Psychology of Music*. New York: Academic P, 1982.
Fraisse, Paul. *The Psychology of Time*. New York: Harper and Row, 1963.
——. "Rhythm and Tempo." *The Psychology of Music*. Ed. Diana Deutsch. New York: Academic P, 1982. 149-80.
Gabrielsson, Alf. "Experimental Research on Rhythm." *Humanities Association Review* 30.1-2 (1979): 69-92.
——. "Perception and Performance of Musical Rhythm." *Music, Mind, and Brain*. Ed. Manfred Clynes. New York: Plenum, 1982. 159-69.
——. "Similarity Ratings and Dimension Analyses of Auditory Rhythm Patterns." *Scandinavian J of Psychology* 14 (1973): 138-60.
——. *Studies in Rhythm*. Uppsala: Almqvist and Wiksell, 1973.
Mainwaring, J. "Kinaesthetic Factors in the Recall of Music Experience." *British J of Psychology* 23 (1933): 284-307.
Mursell, James I. *The Psychology of Music*. New York: Norton, 1937.
Preusser, David. "The Effect of Structure and Rate on the Recognition and Description of Auditory Temporal Patterns." *Perception and Psychophysics* 11.3 (1972): 233-40.

Winckel, Fritz. "The Concept of Time." *Music, Sound and Sensation: A Modern Exposition*. Trans. Thomas Brinkley. New York: Dover, 1967. 75-86.

Yates, S. "Some Aspects of Time Difficulties and Their Relation to Music." *International J of Psychoanalysis* 16 (1935): 341-54.

F. RHYTHM AND METER

Aldrich, Putnam. *Rhythm in Seventeenth-Century Italian Monody*. New York: Norton, 1966.

Allen, W. Sidney. *Accent and Rhythm*. Cambridge: Cambridge UP, 1973.

Babbitt, Milton. "Twelve-Tone Rhythmic Structure and the Electronic Medium." *Perspectives of New Music* 1.1 (1962): 49-79.

Babitz, Sol. "A Problem of Rhythm in Baroque Music." *Musical Quarterly* 38 (1952): 533-65.

——. "The Restoration of Expressive Rhythm in Seventeenth- and Eighteenth-Century Music." *International Musician* 51 (February 1953): 24-25.

Barela, Margaret Mary. "Motion in Musical Time and Rhythm." 19.1 (1979): 78-92.

Bauer, W. "Antike Metren bei Beethoven." *Schweizerische Musikzeitung* 98 (1958): 249-52.

Becking, Gustav. *Der musikalische Rhythmus als Erkenntnisquelle*. Augsburg: Benno Filser, 1928.

Belotti, G. "L'asimmetria ritmica nella mazurca chopiniana." *Revista musicale italiana* 5 (1971): 657-58, 827-46.

Benary, Peter. "Metrum bei Mozart: zur metrischen Analyse seiner letzten drei Sinfonien." *Revue musicale de suisse romande* 114 (1974): 201-05.

Benjamin, W. E. "A Theory of Musical Meter." *Music Perception* 1.4 (1984): 355-413. Includes bibliography.

Berry, Wallace. "Metric and Rhythmic Articulation in Music." *Music Theory Spectrum* 7 (1985): 7-33.

——. "Rhythm and Meter." *Structural Functions in Music*. Englewood Cliffs: Prentice-Hall, 1976. 301-424.

Bresgen, Cesar. "J. S. Bach als Rhythmiker." *Musikerziehung* 29 (1975): 3-7, 57-61.

Brown, Samuel E., Jr. "New Evidence of Isomelic Design in Dufay's Isorhythmic Motets." *J of the American Musicological Society* 10 (1957): 7-13.

Broyles, M. E. "Rhythm, Metre and Beethoven." *Music Review* 33 (1972): 300-22.

Bruce, I. M. "A Note on Mozart's Bar-Rhythms." *Music Review* 17 (1956): 35-47.

Budde, Elmar. "Metrisch-rhythmische Probleme in Vokalwerk Weberns." In *Webern-Kongress Beiträge 1972-73*. Kassel: Bärenreiter, 1973. 52-60.

Caplin, William Earl. "Tonal Function and Metrical Accent: A Historical Perspective." *Music Theory Spectrum* 5 (1983): 1-14.

Cholopova, Valentina. "Russische Quellen der Rhythmik Strawinskys." *Musikforschung* 27 (1974): 435-46.

Cohen, Dalia. "Meter and Rhythm in Music and Poetry." *Israeli Studies in Musicology* 1 (1978): 99-141.

Collard, R., Peter G. Vos, and E. Leeuwenburg. "What Melody Tells about Metre in Music." *Zeitschrift für Psychologie* 189 (1981): 25-33.

Cone, Edward T. *Musical Form and Musical Performance*. New York: Norton, 1968.

Cooper, G. W., and L. B. Meyer. *The Rhythmic Structure of Music*. Chicago: U of Chicago P, 1960.

Court, R. "Rhythme, tempo, mesure." *Revue d'esthétique* 27.2 (1974): 143-59.

Crocker, Richard L. "'Musica Rhythmica' and 'Musica Metrica' in Antique and Medieval Theory." *J of Music Theory* 2.1 (April 1958): 2-23.

Dahlhaus, Carl. "Allegro frenetico: zum Problem des Rhythmus bei Berlioz." *Melos* 3 (1977): 212-24.

———. "Zur Rhythmik in Beethovens Diabelli Variationen." *Veröffentlichungen des Institut für Neue Musik und Musikerziehung Darmstadt* 6 (1967): 18-22.

Dietz, Hanns-Bertold. "Relations between Rhythm and Dynamics in Works of Beethoven." *Bericht über den internationalen musikwissenschaftlichen Kongress Bonn 1970*. Ed. Carl Dahlhaus, et al. Kassel: Bärenreiter, 1970. 47-53.

Dumesnil, René. *Rhythme musical*. 2nd ed. Paris: La Colombe, 1949.

Dunk, John L. *The Origin and Structure of Rhythm*. London: James Clarke, 1952.

Dürr, Walther. "Rhythm in Music: A Formal Scaffolding of Time." *The Voices of Time*. Ed. J. T. Fraser. 2nd ed. U of Massachusetts P, 1981. 180-200.

Dürr, Walther, Walter Gerstenberg, and Jonathan Harvey. "Rhythm." *The New Grove Dictionary of Music and Musicians*. Ed. Stanley Sadie. London: Macmillan, 1980. 15:804-24.

Egger, R. *Die Deklamationsrhythmik Hugo Wolfs in historischer Sicht.* Tutzing: Hans Schneider, 1963.

Eibner, F. "Zum Rhythmus der klassischen Musik." *Musikerziehung* 21.2 (1967): 67-72.

Einstein, Alfred. "Narrative Rhythm in Madrigal." Trans. Willis Wager. *Musical Quarterly* 29 (1943): 475-84.

Feil, Arnold. "Rhythm in Schubert: Some Practical Problems." Trans. L. W. Vyse. *Schubert Studies: Problems of Style and Chronology.* Ed. Eva Badura-Skoda, and Peter Branscombe. Cambridge: Cambridge UP, 1982. 327-45.

Forte, Allen. "Foreground Rhythm in Early Twentieth-Century Music." *Music Analysis* 2 (1983): 239-68.

Friedheim, Philip. "Berlioz and Rhythm." *Music Review* 37.1 (1976): 5-44.

Frisius, Rudolf. "Xenakis und der Rhythmus." *Neue Zeitschrift für Musik* 144.4 (1983): 13-17.

Gatty, Reginald. "Syncopation and Emphasis." *Musical Times* 53 (1912): 369-72, 442-46.

Gennrich, F. "Grundsätzliches zur Rhythmik der mittelalterlichen Monodie." *Musikforschung* 7 (1954): 150-76.

Georgiades, Thrasybulos Georgos. *Der griechische Rhythmus: Musik, Reigen, Vers und Sprache.* Tutzing: Hans Schneider, 1977.

Ghyka, Matila C. *Essai sur le rhythme.* Paris: Gallimard, 1938.

Gollingham, Bryan R. "Modal Rhythm and the Medieval Sequence." *Studies in Music* 1 (1976): 105-23.

Graebner, Eric. "Some Aspects of Rhythm in Berlioz." *Soundings* 2 (1971-72): 18-28.

Grave, F. K. "Rhythmic Harmony in Mozart." *Music Review* 41.2 (1980): 87-102.

Green, Douglass. "The Allegro Misterioso of Berg's *Lyric Suite*: Iso- and Retrorhythms." *J of the American Musicological Society* 30 (1977): 507-16.

Haimo, Ethan. "Rhythmic Theory and Analysis." *In Theory Only* 4.1 (1978): 18-35.

Handel, Stephen. "Using Polyrhythms to Study Rhythm." *Music Perception* 1 (1984): 465-84.

Handel, Stephen, and James S. Oshinsky. "The Meter of Syncopated Auditory Polyrhythms." *Perception and Psychophysics* 30 (1981): 1-9.

Hasty, Christopher F. "Rhythm in Post-Tonal Music: Preliminary Questions of Duration and Motion." *J of Music Theory* 25.2 (Fall 1981): 183-216.

Hauptmann, Moritz. *Die Natur der Harmonik und der Metrik.* Leipzig:

Breitkopf and Härtel, 1853.

Hlawiczka, Karol. "Musikalischer Rhythmus und Metrum." *Musikforschung* 24 (1971): 385-406.

Hori, Tsukiko. "Musical Rhythm: From the Point of View of Time" (in Japanese). *Ongaku gaku* 15.1 (1969): 66-76.

Houghton, Edward. "Rhythm and Meter in 15th-Century Polyphony." *J of Music Theory* 18.1 (Spring 1974): 190-212.

Houle, G. *Meter in Music, 1600-1800: Performance, Perception and Notation*. Bloomington: Indiana UP, 1987.

Hulse, S. H. "Processing of Rhythmic Sound Structures by Birds." *Annals of the New York Academy of Sciences* 423 (1984): 407-19.

Husmann, H. "Das System der modalen Rhythmik." *Archiv für Musikwissenschaft* 11 (1954): 1-38.

——. "Zur Rhythmik des Trouveregesanges." *Musikforschung* 5 (1952): 110-31.

Hyde, Martha M. "A Theory of Twelve-Tone Meter." *Music Theory Spectrum* 6 (1984): 14-51.

Imbrie, Andrew. "'Extra' Measures and Metrical Ambiguity in Beethoven." *Beethoven Studies*. Ed. Alan Tyson. New York: Norton, 1973. 44-66.

Isaacs, Elcanon. "The Nature of Rhythmic Experience." *Psychological Review* 27 (1920): 270-99.

Jammers, Ewald. "Gregorianischer Rhythmus, was ist das?" *Archiv für Musikwissenschaft* 31 (1974): 290-311.

Jander, Owen. "Rhythmic Symmetry in the Goldberg Variations." *Musical Quarterly* 52 (1966): 204-08.

Jarman, Douglas. "Some Rhythmic and Metric Techniques in Alban Berg's *Lulu*." *Musical Quarterly* 56 (1970): 349-66.

Jones, James Rives. "Some Aspects of Rhythm and Meter in Webern's *Opus 27*." *Perspectives of New Music* 7.1 (1968): 103-09.

Just, M. "Harmonischer Rhythmus." *Musikforschung* 29 (1976): 7-20.

Keller, Hans. "Rhythm: Gershwin and Stravinsky." *Score* 20 (1957): 19-31.

Kidd, G., et al. "Some Effects of Rhythmic Context on Melody Recognition." *American J of Psychology* 97.2 (Summer 1984): 153-73.

Klages, Ludwig. *Vom Wesen des Rhythmus*. Kampen: Kampmann, 1934.

Knapp, Janet. "Two Thirteenth-Century Treatises on Modal Rhythm and the Discant." *J of Music Theory* 6 (1962): 200-15.

Komar, Arthur. *Theory of Suspensions*. Princeton: Princeton UP, 1971.

Krastewa, Iwanka Stoianowa. "Le langage rhythmique d'Olivier Messiaen et la métrique ancienne grecque." *Schweitzerische Musikzeitung* 112.2 (1972): 79-86.

Krebs, H. "Dramatic Functions of Metrical Consonance and Dissonance in *Das Rheingold.*" *In Theory Only* 10.5 (1988): 5-21. Includes bibliography.

Kungle, S. "Self-Reliance for Rhythm and Tempo: Beyond Metronomic Dictates." *American Music TCR* 37.3 (1988): 24-25.

Kunst, Jaap. *Metre, Rhythm, Multi-Part Music.* Leiden: Brill, 1950.

Lester, J. *The Rhythms of Tonal Music.* Carbondale: Southern Illinois UP, 1986.

Lewin, David. "A Metrical Problem in Webern's *Opus 27.*" *J of Music Theory* 16 (1972): 124-32.

——. "On Harmony and Meter in Brahms's *Op. 76, No. 8.*" *19th Century Music* 4.3 (1981): 261-65.

——. "Vocal Meter in Schoenberg's Atonal Music, with a Note on a Serial Hauptstimme." *In Theory Only* 6.4 (1982): 12-36.

Marra, James R. "Interrelations between Pitch and Rhythmic Structure in Webern's *Opus 11.*" *In Theory Only* 7.2 (1983): 3-33.

Mathiesen, Thomas J. "Rhythm and Meter in Ancient Greek Music." *Music Theory Spectrum* 7 (1985): 159-80.

Meyers, C. S. "A Study of Rhythm in Primitive Music." *British J of Psychology, General Section* 1 (1966): 397-406.

Monelle, Raymond. "Levels of Rhythm in Vocal Music." *British J of Aesthetics* 22 (1982): 17-36.

Morgan, Robert P. "The Theory and Analysis of Tonal Rhythm." *Musical Quarterly* 64 (1978): 435-73.

Murray, G. "The Authentic Rhythm of Gregorian Chant." *Caecilia* 86 (Summer 1959): 57-71.

Newell, R. "Rhythmic Levels of Perception." *American Society of University Composers, Proceedings* 11-12 (1978-79): 66-77.

Ouvrard, J. P. "Les jeux du mètre et du sens dans la chanson polyphonique française du XVIe siècle (1528-1550)." *Revue de Musicologie* 67.1 (1981): 5-34.

Pike, Alfred. "The Time Set as a Rhythmic Agent for the Series." *Music Review* 24 (1963): 168-75.

Rahn, John. "On Pitch or Rhythm: Interpretation of Orderings of and in Pitch and Time." *Perspectives of New Music* 13.2 (1975): 182-203.

Rayburn, John. *Gregorian Chant Rhythm: A History of the Controversy Concerning Its Interpretation.* New York: Rayburn, 1964.

Reckow, Fritz. "Proprietas und perfectio: zur Geschichte des Rhythmus, seiner Aufzeichnung und Terminologie im 13. Jahrhundert." *Acta Musicologica* 39.3-4 (1967): 115-43.

Ricci, Robert. "The Division of the Pulse: Progressive and Retrogressive Rhythmic Forces." *In Theory Only* 1.8 (1975): 13-19.

Riemann, Hugo. *System der musikalischen Rhythmik und Metrik.* Leipzig: Breitkopf and Härtel, 1903.

Risler, J. "Le rhythme de la musique grecque." *Compte rendu du premier congrès du rhythme.* Ed. A. Pfrimer. Geneva: Institut Jacques-Dalcroze, 1926. 5-9.

Rosenthal, Felix. "Probleme der musikalischen Metrik." *Musikwissenschaft* 8 (1925-26): 262-88.

Rothschild, F. *The Lost Tradition of Music: Rhythm and Tempo in the Time of J. S. Bach.* London: Oxford UP, 1953.

Rothstein, William. *Phrase Rhythm in Tonal Music.* New York: Schirmer, 1989.

Rowell, Lewis. "Aristoxenus on Rhythm." *J of Music Theory* 23.1 (Spring 1979): 63-79.

——. "Time and Rhythm in Music." *Music Theory Spectrum* 7 (1985): 1-215.

Sachs, Curt. *Rhythm and Tempo.* New York: Norton, 1953.

Sanders, E. H. "Consonance and Rhythm in the Organum of the 12th and 13th Centuries." *American Musicological Society* 33.2 (1980): 264-86.

Schachter, Carl. "Rhythm and Linear Analysis: A Preliminary Study." *The Music Forum IV.* Ed. Felix Salzer, and Carl Schachter. New York: Columbia UP, 1976. 281-334.

——. "Rhythm and Linear Analysis: Durational Reduction." *The Music Forum V.* Ed. Felix Salzer, and Carl Schachter. New York: Columbia UP, 1980. 197-232.

——. "Rhythm and Linear Analysis: Aspects of Meter." *The Music Forum VI.* Ed. Felix Salzer, and Carl Schachter. New York: Columbia UP, 1987. 1-59.

Schenkman, Walter. "Rhythmic Patterns of the Baroque." *Bach* 5.3-4 (1974): 21-28, 9-16.

Schneider, Marius. "Le rhythme de la musique artistique espagnole du XVI siècle vu à travers la chanson populaire." *Studia Musicologica* 7 (1965): 125-40.

Seay, Albert. "Classical Metrics and Medieval Music." *Bulletin of the Rocky Mountain Language Association* 23.2 (1969): 59-67.

Seidel, Wilhelm. *Rhythmus: eine Begriffsbestimmung.* Darmstadt: Wissenschaftliche Buchgesellschaft, 1976.

——. *Über Rhythmustheorien der Neuzeit.* Bern: Francke, 1975.

Smith, Charles J. "Rhythm Restratified." *Perspectives of New Music* 16.1 (1977): 144-76.

Smither, Howard E. "The Rhythmic Analysis of Twentieth-Century Music." *J of Music Theory* 8 (1964): 54-88.

Sommer, Antonius. *Die Komplikationen des musikalischen Rhythmus in den Bünenwerken Richard Wagners.* Wilhelming: Katzbichler, 1971.

Stearns, R. H. "Rhythm, Meter, and Notation: Some Thoughts on Common Techniques from the Renaissance to the Present." *The Soundboard* 10.1 (1983): 46-52.

Steedman, M. J. "The Perception of Musical Rhythm and Metre." *Perception* 6 (1977): 555-71.

Steglich, Rudolph. *Die elementare Dynamik des musikalischen Rhythmus.* Leipzig: C. Merseburger, 1930.

Stockhausen, Karlheinz. "Kadenzrhythmik bei Mozart." *Darmstädter Beiträge zur Neuen Musik* 4 (1962): 38-72.

Sturges, P. T., and J. G. Martin. "Rhythmic Structure in Auditory Temporal Pattern Perception." *J of Experimental Psychology* 102 (1974): 377-83.

Suchla, B. R. "Zu Notation, Metrum und Rhythmus des altfranzösischen Liedes." *Archiv für Musikwissenschaft* 36.3 (1979): 159-82.

Talbot, M. "Harmony and Metre: A Study in Relationships." *Music Review* 33.1 (1972): 47-52.

Telle, K. *Tanzrhythmen in der Vokalmusik Georg Friedrich Händels.* Munich: Katzbichler, 1977.

Thigpen, Edmund. *Rhythmus: Analyse und Koordinationsgrundlage.* Munich: K. Maas, 1980.

Tischler, Hans. "A Propos Meter and Rhythm in the Ars Antiqua." *J of Music Theory* 26.2 (Fall 1982): 313-29.

——. "Metrum und Rhythmus in französischer Dichtung und Musik des 13. Jahrhunderts." *Archiv für Musikwissenschaft* 32 (1975): 72-80.

——. "Rhythm, Meter, and Melodic Organization in Medieval Song." *Revue belge de musicologie* 28-30 (1971-72): 5-23.

——. "Die rhythmische Interpretation eines Trouvereliedes." *Musikforschung* 32 (1979): 17-25.

——. "Some Rhythmic Features in Early Thirteenth-Century Motets." *Revue belge de musicologie* 21 (1967): 107-17.

Toorn, Pieter C. van den. "Rhythmic (or Metric) Invention." In *The Music of Igor Stravinsky.* New Haven: Yale UP, 1983. 204-51.

Vanderwerf, Hendrik. "Deklamatorischer Rhythmus in den Chansons der Trouvères." *Musikforschung* 20 (1967): 122-44.

Vollaerts, Jan W. A. *Rhythmic Proportions in Early Medieval Ecclesiastical Chant.* 2nd ed. Leiden: Brill, 1960.

Vos, Peter G. "Pattern Perception in the Measurement of Sound with Metrical-Organized Time Structures." *Nederlands Tijdschrift voor de Psychologie en haar Grensgebieden* 28.8 (1974): 565-80.

——. "Temporal Duration Factors in the Perception of Auditory Rhythmic Patterns." *Scientific Aesthetics* 1 (1977): 183-99.

Waite, William G. *The Rhythm of Twelfth-Century Polyphony: Its Theory and Practice*. New Haven: Yale UP, 1954.

Weaver, H. E. "Syncopation: A Study in Musical Rhythms." *J of General Psychology* 20 (1939): 409-29.

Westphal, Rudolf. *Allgemeine Theorie der musikalischen Rhythmik seit J. S. Bach*. Leipzig: Breitkopf and Härtel, 1880.

——. *Aristoxenos von Tarent, Melik und Rhythmik der classischen Helenentums*. 1883; reprint, Hildesheim: Georg Olms, 1965.

——. *Die Fragmente und die Lehrsätze der griechischen Rhythmiker*. Leipzig: B. G. Teubner, 1861.

Willems, E. *Le rhythme musical: rhythme, rhythmique, métrique*. Paris: Presses Universitaires de France, 1954.

Williams, J. K. "Rhythmic Structure in Selected Works of Binchois and Dufay." *Indiana Theory Review* 1.2 (1978): 31-48.

Winick, Steven D. *Rhythm: An Annotated Bibliography*. Metuchen: Scarecrow, 1974.

Winold, Allen. "Rhythm in Twentieth-Century Music." *Aspects of Twentieth-Century Music*. Ed. Gary E. Wittlich. Englewood Cliffs: Prentice-Hall, 1975. 208-69.

Wolf, R. Peter. "Metrical Relationships in French Recitative of the Seventeenth and Eighteenth Centuries." *Recherches* 18 (1978): 29-49.

Wolff, Hellmuth Christian. "Der Rhythmus bei J. S. Bach." *Bach Jahrbuch* 37 (1940-48): 83-121.

Woodrow, Herbert A. "A Quantitative Study of Rhythm." *Archives of Psychology* 14 (1909): 1-66.

——. "The Role of Pitch in Rhythm." *Psychological Review* 18 (1911): 54-77.

Yeston, Maury. *The Stratification of Musical Rhythm*. New Haven: Yale UP, 1976.

Youens, Susan. "Poetic Rhythm and Musical Metre in Schubert's *Winterreise*." *Music and Letters* 65.1 (1984): 28-40.

Zuckerkandl, Victor. "Meter and Rhythm." In *The Sense of Music*. Princeton: Princeton UP, 1959. 98-136.

G. TEMPO AND BEAT

Auda, Antoine. *Théorie et pratique du tactus*. Brussels, 1965.

Auhagen, W. "Chronometrische Tempoangaben im 18. und 19. Jahrhundert." *Archiv für Musikwissenschaft* 44.1 (1987): 40-57.

Badura-Skoda, Paul. "Fehlende und Uberzählige Takte bei Schubert und Beethoven." *Österreichische Musikzeitung* 33 (1978): 284-94.

Bank, Joannes Antonius. *Tactus, Tempo, and Notation in Mensural Music from the 13th to the 17th Century*. Amsterdam: Annie Bank, 1972.

Barthe, Englehard. *Takt und Tempo: Studien über Zusamenhänge*. Hamburg: Sikorski, 1960.

Behne, Klaus-Ernst. *Der Einfluss des Tempos auf die Beurteilung von Musik*. Cologne: A. Volk and H. Gerig, 1972.

——. "Tempo und Affekt: Ergebnisse einer experimentellen Studie." *Musik und Bildung* 6 (1974): 668-71.

Benary, Peter. "Tempo und Tempoänderung: zum Verhältnis von Notentext und Wiedergabe." *Neue Zeitschrift für Musik* 131 (1970): 566-68.

Boal, E. TeSelle. "Purcell's Clock Tempos and the Fantasias." *Viola da Gamba Society of America. J* 20 (1983): 24-39.

Booth, R. E. "Tempo Choice and Tempo Freedom in Weber's Piano Music." *J of the American Liszt Society* 19 (June 1986): 77-87.

Brandenburg, Sieghard. "Über die Bedeutung der Änderungen von Taktvorschriften in einingen Werken Beethovens." *Beiträge 1976-78: Beethoven-Kolloquium 1977*. Ed. Rudolf Klein. Kassel: Bärenreiter, 1978.

Brown, A. P. "Performance Tradition, Steady and Proportional Tempos, and the First Movements of Schubert's Symphonies." *J of Musicology* 5.2 (1987): 296-307.

Camus, R. F. "On the Cadence of the March." *J of Band Research* 16.2 (1981): 13-23.

Christenson, Dieter. "Inner Tempo and Melodic Tempo." Trans. Bruno Nettl. *Ethnomusicology* 4.1 (1960): 9-14.

Dahlhaus, Carl. "Über das Tempo in der Musik des späten 16. Jahrhunderts." *Musica* 13 (1959): 767-69.

——. "Zur Entstehung des modernen Taktsystems im 17. Jahrhundert." *Archiv für Musikwissenschaft* 18 (1961): 223-40.

——. "Zur Theorie des Taktus im 16. Jahrhundert." *Archiv für Musikwissenschaft* 17 (1960): 22-39.

Duke, R. A., et al. "Effect of Tempo on Pitch Perception." *J of Research in Music Education* 36.2 (1988): 108-25. Includes bibliography.

Epstein, David. "Tempo Relations: A Cross-Cultural Study." *Music Theory Spectrum* 7 (1985): 34-71.

Forte, Allen. "The Structural Origin of Exact Tempi in the Brahms-Haydn Variations." *Music Review* 18 (1957): 138-49.

Gelfand, Y. "On Tempo Indications Based on Beethoven's Music." *College Music Symposium* 25 (1985): 92-129.

Gersmann, F. "Klassisches Tempo für klassische Musik." *Guitar and Lute* 8.6 (1986): 37-46.

Gullo, Salvatore. *Das Tempo in der Musik des 13. und 14. Jahrhunderts.* Bern: Haupt, 1964.

Heckmann, H. "Bemerkungen zu Beethovens Tempi." *Beethoven Jahrbuch* (1955-56): 24-54.

Higgins, T. "Tempo and Character in Chopin." *Music Quarterly* 59 (1973): 106-20.

Hodgson, Walter. "Absolute Tempo: Its Existence, Extent, and Possible Explanation." *Proceedings of the Music Teachers National Association* 48 (1951): 158-69.

Hofmann, F. "Das Tempo des Gemeindegesanges." *Gottesdienst und Kirchenmusik* 2 (March-April 1980): 46-50.

Imberty, M. "Il senso del tempo e della morte nell'immaginario debussiano." *Rivista Italiana di Musicologia* 21.3 (1987): 383-409.

Johnson, P. "Tempo Relation in a Class of Keyboard Compositions from Sweelinck to Bach." *Bach* 18.4 (1987): 4-28.

Kolinski, Mieczyslaw. "The Evaluation of Tempo." *Ethnomusicology* 3.2 (1959): 45-57.

Kolisch, Rudolf. "Tempo and Character in Beethoven's Music." *Musical Quarterly* 29 (1943): 169-87, 291-312.

Kravitt, Edward F. "Tempo as an Expressive Element in the Late Romantic Lied." *Musical Quarterly* 59 (1973): 497-518.

Kuhn, T. L. "Discrimination of Modulated Beat Tempo by Professional Musicians." *J of Research in Music Education* 22 (1974): 270-77.

Kümmel, Werner Friedrich. "Zum Tempo in der italienischen Mensuralmusik des 15. Jahrhunderts." *Acta Musicologica* 42.3-4 (1970): 150-63.

Madsen, C. K. "Modulated Beat Discrimination among Musicians and Nonmusicians." *J of Research in Music Education* 27.2 (1979): 57-67.

Margulis, V. "Eine Formel für Tempoverwandtschaft." *Musica* 36.1 (1982): 31-33.

Marshner, F. "Zählheit, Tempo und Ausdruck bei Beethoven." In *Beethoven-Zentenarfeier: Internationaler musikhistorischer Kongress.* Vienna: Universal, 1972. 100-04.

Mauser, S. "Zum Verhältnis von Tempo- und Ausdrucksbezeichnungen in den späten Klaviersonaten Beethovens." *Österreichische Musikzeitschrift* 36 (December 1981): 617-22.

Mendel, Arthur. "A Note on Proportional Relationships in Bach

Tempi." *Musical Times* (December 1959): 683-84.

Mertin, Josef. "Tempoprobleme um den Tactus." *Österreichische Musikzeitung* 25 (1973): 564-74.

Nastasi, M. "Zur Tempofrage bei Frescobaldi." *Tibia: Magazin für Freunde alter und neuer Bläsermusik* 1 (1979): 217-21.

Newman, William S. "Freedom of Tempo in Schubert's Instrumental Music." *Musical Quarterly* 61 (1975): 528-45.

——. "Tempo in Beethoven's Instrumental Music: Its Choice and Its Flexibility." *Piano Quarterly* 116 (1981-82): 22-29; 117 (1982): 22-31.

Perl, H. "Beethoven-Beschwörung: zur Frage des Tempos in der älteren Musik." *Das Orchester* 33 (March 1985): 246-49.

Planchart, Alejandro E. "The Relative Speed of Tempora in the Period of Dufay." *Royal Musical Association Research Chronicle* 17 (1981): 33-51.

Raeburn, C. "Das Zeitmass in Mozarts Opern." *Österreichische Musikzeitung* 12 (September 1957): 329-33.

Schenbeck, L. "Tempo and Rhythm in Bach's Cantata Overtures." *Bach* 17.1 (1986): 3-17.

Schulze, W. *Temporelationen im symphonischen Werk von Beethoven, Schubert, und Brahms*. Bern: Verlag Kreis der Freunde um Hans Kayser, 1981.

Schwarzmaier, Ernst. *Die Takt- und Tonordnung Joseph Riepels: ein Beitrag zur Geschichte der Formenlehre im 18.* Jahrhundert (1936). Regensburg: Bosse, 1978.

Seay, A. "The Setting of Tempo by Proportion in the Sixteenth Century." *The Consort* 37 (1981): 394-98.

Sundberg, Johan, and Violet Verrillo. "On the Anatomy of the Retard: A Study of Timing in Music." *Speech Transmission Laboratory Quarterly Progress and Status Report* 2-3 (1977): 44-57.

Temperly, N. "Tempo and Repeats in the Early Nineteenth Century." *Music and Letters* 47 (1966): 323-36.

Wagner, Christoph. "Experimentelle Untersuchungen über das Tempo." *Österreichische Musikzeitung* 29 (1974): 589-604.

Wang, C. C. "Discrimination of Modulated Music Tempo by Music Majors." *J of Research in Music Education* 31.1 (1983): 49-55. Includes bibliography.

Wehmeyer, G. "Interpretation im klassischen Tempo." *Österreichische Musikzeitschrift* 40 (July-August 1985): 369-76.

17. NAVIGATION

Admiralty Navigation Manual, Vol. 1. London: His Majesty's Stationery Office, 1938.

Allen, Oliver E., et al. *The Seafarers: The Pacific Navigators*. Alexandria: Time-Life Books, 1980.

Bedini, Silvio A. *The Pulse of Time—Galileo Galilei, the Determination of Longitude and the Pendulum Clock*. Florence: Olschki, 1990.

Berthoud, Ferdinand. *Traité des horloges marines....* Paris, 1773.

Brown, Lloyd A. *The Story of Maps*. Boston: Little, Brown, 1949.

Cassini, [Jacques-Dominique], fils. *Voyage fait par ordre du Roi en 1768, pour éprouver les montres marines inventées par M. Le Roy*. Paris, 1770.

Congleton, Carol. *Navigational Applications of Plane and Spherical Trigonometry*. Centreville: Cornell Maritime P, 1980.

Cotter, Charles H. *A History of Nautical Astronomy*. Toronto: Hollis and Carter, 1968.

Ditisheim, Paul, et al. *Pierre Le Roy et la chronométrie*. Paris: Éditions Tardy, 1940.

Earnshaw, Thomas. *Longitude: An Appeal to the Public Stating Mr. Thomas Earnshaw's Claim to the Original Invention of the Improvements in His Timekeepers...and His Consequent Right to National Reward*. London: privately printed, 1808.

Exposé succint des travaux de MM. Harrison et Le Roy, dans la recherche des longitudes en mer, et des épreuves faites de leurs ouvrages. Paris, 1768.

Forbes, Eric G. *The Birth of Navigation: The Solving in the 18th Century of the Problem of Finding Longitude at Sea*. Greenwich: National Maritime Museum, 1974.

——. "The Foundation and Early Development of *The Nautical Almanac*." *J of the Institute of Navigation* 18 (1965): 391-401.

Gould, Rupert T. *John Harrison and His Timekeepers*. 4th ed. Greenwich: National Maritime Museum, 1978.

——. *The Marine Chronometer: Its History and Development*. 1923; reprint, London: J. D. Porter, 1960.

Grant, S. *Integration of Passive Ranging Loran-C, Satellite Navigation and Ship's Gyrocompass*. Dartford: Bedford Institute, 1976. Includes bibliography.

Guyou, E. "La méthode des distances lunaires." *Revue maritime* 153 (June 1902): 943-63.

Hall, John Scoville. *Radar Aids to Navigation*. Lexington: Boston

Technical Publishers, 1964.

Harrison, John. *Remarks on a Pamphlet Lately Published by the Rev. Mr. Maskelyne under the Authority of the Board of Longitude*. London, 1767.

Hewson, J. B. *A History of the Practice of Navigation*. Glasgow: Brown, Son and Ferguson, 1951.

Laycock, William S. *The Lost Science of John "Longitude" Harrison*. Ashford, Kent: Brant Wright Associates, 1976.

Le Roy, Pierre. *Précis des recherches faites en France depuis l'année 1730 pour la détermination des longitudes en mer par la mesure artificielle du temps*. Amsterdam, 1773.

——. *Suite du Précis sur les montres marines de France; avec un supplément au Mémoires sur la meilleure manière de mesurer le tems en mer*. Leiden, 1774.

Macey, Samuel L. "*The Compleat Plattmaker*: British Chart and Map Making, 1650-1750." *Eighteenth-Century Studies* 12 (Summer 1979): 527-37.

——. "The Early History of Chronometers: A Background Study Related to the Voyages of Cook, Bligh and Vancouver." *B.C. Studies* 48 (Summer 1978): 14-23.

Manual of Seamanship. Vol. 1. London: His Majesty's Stationery Office, 1937.

Marguet, F. *Histoire de la longitude à la mer au XVIIIe siècle, en France*. Paris: Auguste Challamel, 1917.

Maskelyne, Nevil. *An Account of the Going of Mr. Harrison's Watch....* London, 1767.

May, W. E. *A History of Marine Navigation*. New York: Norton, 1973.

Mercer, Tony. *Mercer Chronometers: Radical Tom Mercer and the House He Founded*. Ashford, Kent: Brant Wright Associates, 1978.

Mercer, Vaudrey. *John Arnold and Son, Chronometer Makers, 1762-1843*. London: Antiquarian Horological Society, 1972.

Mills, Henry Robert. *Positional Astronomy and Astro-Navigation Made Easy: A New Approach Using the Pocket Calculator*. New York: Wiley, 1978.

Moorehead, Alan. *The Fatal Impact: An Account of the Invasion of the South Pacific, 1767-1840*. Harmondsworth: Penguin, 1968.

Morison, Samuel Eliot. *Admiral of the Ocean Sea: A Life of Christopher Columbus*. Boston: Little, Brown, 1942.

Padfield, Peter. *Tide of Empires: Decisive Naval Campaigns in the Rise of the West, Vol 1, 1481-1654*. London: Routledge and Kegan Paul, 1979.

Parry, John H. *The Age of Reconnaisance*. London: Weidenfeld and

Nicolson, 1964.

——. *The Discovery of the Sea*. New York: Dial, 1974.

Perkins, P. *The Seaman's Tutor: Explaining Geometry, Cosmography, and Trigonometry*. London: Obadiah Blagrave, 1682.

Quill, Humphrey. *John Harrison: The Man Who Found Longitude*. London: John Baker, 1966.

Randall, Anthony. "L'oeuvre chronométrique de Ferdinand Berthoud de 1760 à 1787." *Association Nationale des Collectionneurs et Amateurs d'Horlogerie Ancienne* 30 (Spring 1981): 23-35.

Sadler, D. H. *Man Is Not Lost: A Record of Two Hundred Years of Astronomical Navigation with the Nautical Almanac, 1767-1967*. London: National Maritime Museum and Royal Greenwich Observatory, 1968.

Schurz, William Lytle. *The Manila Galleon*. 1931; reprint, New York: Dutton, 1959.

Taylor, Eva Germaine Rimington. *The Haven-Finding Art: A History of Navigation from Odysseus to Captain Cook*. London: Hollis and Carter, 1956.

Thrower, Norman J. W., ed. *The Compleat Plattmaker: Essays on Chart, Map, and Globe Making in England in the Seventeenth and Eighteenth Centuries*. Berkeley: U of California P, 1978.

Warner, Deborah Jean. "Astronomers, Artisans and Longitude." In *Transport Technology and Social Change*. Stockholm: Techniska Museet Symposium, 1980. 133-40.

Waters, David. *The Art of Navigation in England in Elizabethan and Early Stuart Times*. 2nd ed. 2 vols. Greenwich: National Maritime Museum, 1978.

18. PHILOSOPHY

A. GENERAL

Andrade, Almir de. *As duas faces do tempo*. Rio de Janeiro: Livraria J. Olympio Editôra, 1971.

Arguessy, H., et al. *Time and the Philosophies*. Paris: UNESCO P, 1977.

Benjamin, A. Cornelius. "Ideas of Time in the History of Philosophy." *The Voices of Time*. Ed. J. T. Fraser. 2nd ed. Amherst: U of Massachusetts P, 1981. 3-30.

Bergmann, Werner. "Selbstreferenz und Zeit: Die Dynamische Stabilität des Bewusstseins." *Husserl Studies* 6 (1989): 155-75.

Boas, George. "The Acceptance of Time." *U of California Publications in Philosophy* 16.12 (1950): 249-70.

Boodin, J. E. *Time and Reality*. New York: Macmillan, 1904.

Brandon, S. G. F. *Time and Mankind: An Historical and Philosophical Study of Mankind's Attitude to the Phenomena of Change*. London: Hutchinson, 1951.

Brumbaugh, Robert S. "Kinds of Time: An Excursion in Metaphysics." *Experience, Existence, and the Good: Essays in Honor of Paul Weiss*. Ed. Irwin C. Lieb. Carbondale: Southern Illinois UP, 1961. 119-25.

——. "Metaphysical Presuppositions and the Study of Time." *The Study of Time III*. Ed. J. T. Fraser, N. Lawrence, and D. Park. New York: Springer, 1978. 1-21.

——. *Unreality and Time*. New York: State U of New York P, 1984.

Buckley, Jerome Hamilton. *The Triumph of Time: A Study of the Victorian Concepts of Time, History, Progress and Decadence*. Cambridge: Belknap, 1966.

Bunge, M. "Physique et métaphysique du temps." *Proceedings of the XIVth International Congress of Philosophy, I*. Vienna: Herder, 1968.

Cacciari, M. *Zeit ohne Kronos*. Klagenfurt: Ritter, 1986.

Calder, N. *Timescale: An Atlas of the Fourth Dimension*. London: Chatto and Windus, 1984.

Castelli, Enrico. *Il tempo esaurito*. Padua: CEDAM, 1968.

——, ed. *Temporalità e alienazione*. Padua: CEDAM, 1975.

Chapman, T. *Time, a Philosophical Analysis*. Boston: Reidel, 1982.

Cleugh, Mary Frances. *Time and Its Importance in Modern Thought*. New

York: Russell and Russell, 1970.

Daniels, Charles B. "Seeing Through a Time-Gap." *Australasian J of Philosophy* 48.3 (December 1970): 354-59.

Delacre, Georges. *El tiempo en perspectiva: introducción una filosofia del tiempo*. Rio Piedras: Editorial Universitaria, Universidad de Puerto Rico, 1975.

Dewey, John. "Time and Individuality." *Time and Its Mysteries*. Ed. Harlow Shapley. New York: Collier Books, 1962.

Dreyfus, H. L. "Human Temporality." *The Study of Time II*. Ed. J. T. Fraser, and N. Lawrence. New York: Springer, 1975. 150-62.

Dummett, Michael A. E. *Truth and Other Enigmas*. London: Duckworth, 1978.

Dunne, John W. *An Experiment with Time*. 5th ed. London: Faber, 1938.

——. *The Serial Universe*. New York: Macmillan, 1938.

Felt, J. W. "Faces of Time." *Thought* 62 (December 1987): 414-22. Includes bibliography.

Freeman, E., and Wilfrid Sellars, eds. *Basic Issues in the Philosophy of Time*. Lasalle: Open Court, 1971.

Gale, R. M. *The Language of Time*. London: Routledge and Kegan Paul, 1968.

——, ed. *The Philosophy of Time*. London: Macmillan, 1968.

Garciá Astrada, Arturo. *Tiempo y eternidad*. Madrid: Gredos, 1971.

Gardies, Jean Louis. *La logique du temps*. Paris: Presses universitaires de France, 1975.

Gonseth, Ferdinand. *Time and Method: An Essay on the Methodology of Research*. Trans. Eva H. Guggenheimer. Springfield: Charles C. Thomas, 1972.

Gray, John Glenn. "The Idea of Death in Existentialism." *J of Philosophy* 48.5 (March 1951): 113-27.

Guitton, Jean. *Man in Time*. Trans. Adrienne Foulke. Notre Dame: U of Notre Dame P, 1966.

Gurvitch, Georges. *The Spectrum of Social Time*. Trans. and ed. Myrtle Korenbaum. Dordrecht: Reidel, 1964.

Harris, Errol E. *The Reality of Time*. Albany: State U of New York P, 1988.

Health, Louise Robinson. *The Concept of Time*. Chicago: U of Chicago P, 1936.

Herbert, Gundala Samuel. *Time: A Metaphysical Study*. Trivandrum: College Book House, 1978.

Hoffman, Piotr. *Doubt, Time, Violence*. Chicago: U of Chicago P, 1987.

Ingarden, Roman. *Time and Modes of Being*. Springfield: Charles C. Thomas, 1964.

Kern, Stephen. *The Culture of Time and Space: 1880-1918*. Cambridge: Harvard UP, 1983.

Lawrence, Nathaniel. "The Origins of Time." *Time, Science, and Society in China and the West: The Study of Time V*. Ed. J. T. Fraser, N. Lawrence, and F. C. Haber. Amherst: U of Massachusetts P, 1986. 23-38.

——. "Time Represented as Space." *The Monist* 53.3 (July 1969): 447-56.

Lovejoy, Arthur Oncken. *The Reason, the Understanding, and Time*. Baltimore: Johns Hopkins P, 1961.

McInerney, Peter K. *Time and Experience*. Philadelphia: Temple UP, 1990.

McKeon, Richard P. *Freedom and History*. New York: Noonday P, 1952.

McKerlie, D. "Equality and Time." *Ethics* 99 (April 1989): 475-91; discussion, 492-502.

Maier, Annaliese. "Die Subjectivisierung der Zeit." *Philosophia Naturalis* 1 (1950): 364-65.

Matsumoto, M. "Time: Being or Consciousness Alone?—A Realist View." *The Study of Time II*. Ed. J. T. Fraser, and N. Lawrence. New York: Springer, 1975. 206-15.

Newton-Smith, W. *The Structure of Time*. Boston: Routledge and Kegan Paul, 1980.

Orth, E. W., ed. *Studien zum Zeitproblem in der Philosophie des 20. Jahrhunderts*. Freiburg: K. Alber, 1982.

Palter, Robert M. *Whitehead's Philosophy of Science*. Chicago: U of Chicago P, 1960.

The Problem of Time. Berkeley: U of California Publications in Philosophy, 1935.

Reitmeister, Louis Aaron. *A Philosophy of Time*. Westport: Greenwood, 1974.

Ricoeur, Paul. "Narrated Time." Trans. R. Sweeney. *Philosophy Today* 29 (Winter 1985): 259-71. Includes bibliography.

——. *Time and Narrative*. Trans. Kathleen McLaughlin and David Pellauer. 3 vols. Chicago: U of Chicago P, 1984-88.

Salmon, Charles Ray. *The Book of Purpose: A Introduction to the Fourth Dimension*. Santa Maria: Cronus College P, 1972.

Schindler, Walter. *Die reflexive Struktur objektiver Erkenntnis: eine Untersuchung zum Zeitbegriff der Kritik der reinen Vernunft*. Munich: C. Hanser, 1979.

Seddon, Keith. *Time: A Philosophical Treatment.* New York: Routledge, Chapman and Hall, 1987.
Seorabje, Richard. *Time, Creation and the Continuum.* Ithaca: Cornell UP, 1983.
Sherover, Charles M. *Heidegger, Kant and Time.* Bloomington: Indiana UP, 1971.
——, ed. *The Human Experience of Time: The Development of Its Philosophic Meaning.* New York: New York UP, 1975.
——. "*Res Cogitans*, The Time of Mind." *Time and Mind: The Study of Time VI.* Ed. J. T. Fraser. Madison: International Universities P, 1989. 279-94.
——. "Time and Ethics: How Is Morality Possible?" *The Study of Time II.* Ed. J. T. Fraser, and N. Lawrence. New York: Springer, 1975. 216-30.
Sivadjian, Joseph. *Le temps.* Paris: Hermann, 1938.
Smith, J. E. "Time and Qualitative Time." *Review of Metaphysics* 40 (September 1986): 3-16.
Stallknecht, Newton P., and Robert S. Brumbaugh. *The Spirit of Western Philosophy.* New York: Longmans, 1950.
Stüttgen, Albert. *Heimkehr zum Rhythmus.* Munich: Pfeil, 1989.
Szalai, Sándor. "Time and Environment." *New Hungarian Quarterly* 19 (Summer 1978): 133-39.
Taylor, R. "Time and Life's Meaning." *Review of Metaphysics* 40 (June 1987): 675-86.
Toda, M. "The Boundaries of the Notion of Time." *The Study of Time III.* Ed. J. T. Fraser, N. Lawrence, and D. Park. New York: Springer, 1978. 370-88.
Trivers, Howard. *The Rhythm of Being: A Study of Temporality.* New York: Philosophical Library, 1985.
Weiss, Paul. *Modes of Being.* Carbondale: Southern Illinois UP, 1958.
Wood, David. *The Deconstruction of Time.* Atlantic Highlands: Humanities P International, 1989.
Yaker, Henri, et al. *The Future of Time.* New York: Doubleday, 1972.
Yamamoto, Makoto. "What Time Is Not." *The Study of Time II.* Ed. J. T. Fraser, and N. Lawrence. New York: Springer, 1975. 231-38.
Zalewski, S. *Time and Being* (in Polish). Warsaw: PAX, 1971.
Zawirski, Z. *L'évolution de la notion de temps.* Krakow: Gebethner and Wolff, 1936.
Zohar, D. *Through the Time Barrier.* London: Paladin, 1983.
Zwart, P. *About Time: A Philosophical Inquiry into the Origin and Nature of Time.* New York: Elsevier, 1975.

B. CLASSICAL

Aristotle. *The Complete Works of Aristotle: The Revised Oxford Translation*. Ed. Jonathan Barnes. 2 vols. Princeton: Princeton UP, 1984. See under "Time" in the General Index, and in particular *Physics*, Book 4, Sections 10-14.

———. *De Motu Animalium 7*. Trans. A. S. L. Farquharson. In *Works*. Ed. J. A. Smith, and W. D. Ross. Oxford: Clarendon P, 1912. Early mechanical metaphor.

Böhme, Gernot. *Zeit und Zahl: Studien zur Zeittheorie bei Platon, Aristotles, Leibniz und Kant*. Frankfurt: Klostermann, 1974.

Brumbaugh, Robert S. "Time Passes: Platonic Variations." *Review of Metaphysics* 33 (June 1980): 711-26.

Callahan, John F. *Four Views of Time in Ancient Philosophy*. Cambridge: Harvard UP, 1948.

Corish, Denis. "The Beginning of the Beginning in Western Thought." *The Study of Time IV*. Ed. J. T. Fraser, N. Lawrence, and D. Park. New York: Springer, 1981. 34-45.

———. "The Emergence of Time." *Time, Science, and Society in China and the West: The Study of Time V*. Ed. J. T. Fraser, N. Lawrence, and F. C. Haber. Amherst: U of Massachusetts P, 1986. 69-78.

Goldschmidt, Victor. *Le système stoïcien et l'idée de temps*. 3rd ed. Paris: J. Vrin, 1977.

Hendley, Brian P., ed. *Plato, Time and Education: Essays in Honor of Robert S. Brumbaugh*. Albany: State U of New York P, 1987.

Hintikka, Jaakko. *Time and Necessity: Studies in Aristotle's Theory of Modality*. Oxford: Clarendon P, 1973.

Lee, H. D. P. *Zeno of Elea, Text with Translation and Notes*. Cambridge: Cambridge UP, 1936.

Maula, Erkka. *On the Semantics of Time in Plato's Timaeus*. Abo, Finland: Abo Akademi, 1970. 37 pages.

Moreau, Joseph. *L'espace et le temps selon Aristote*. Padua: Antenore, 1965.

Plato. *The Collected Dialogues of Plato, Including the Letters*. Trans. Lane Cooper, et al. Ed. E. Hamilton, and H. Cairns. Princeton: Princeton UP, 1961. See in particular time as "moving image" in *Timaeus*, 28-39.

Sambursky, Samuel, and S. Pines. *The Concept of Time in Late Neoplatonism*. Jerusalem: Israel Academy of Sciences and Humanities, Section of Humanities, 1971.

Waterlow, S. "Aristotle's Now." *Philosophical Quarterly* 34 (April 1984): 104-28.

Vlastos, Gregory. "The Disorderly Motion in *Timaeus*." *Studies in Plato's Metaphysics*. Ed. R. E. Allen. New York: Humanities P, 1965. 397-98.

C. COSMOLOGY, SPACE, AND TIME

Alexander, Samuel. *Space, Time, and Deity*. 2 vols. Dover, 1966.

Brentano, Franz Clemens. *Philosophische Untersuchungen zu Raum, Zeit und Kontinuum*. Ed. Stephan Körner, and Roderick Chisholm. Hamburg: Meiner, 1976.

Brumbaugh, Robert S. "Cosmography." *Review of Metaphysics* 25 (1971): 140-48.

——. "Cosmography: The Problem of Modern Systems." *Review of Metaphysics* 26 (1973): 511-21.

——. "Preface to Cosmography." *Review of Metaphysics* 7 (1952): 529-34.

Grünbaum, Adolf. *Philosophical Problems of Space and Time*. New York: Knopf, 1963.

Hinckfuss, Ian. *The Existence of Space and Time*. Oxford: Clarendon P, 1975.

Schwarz, Gerhard. *Raum und Zeit als naturphilosophisches Problem*. Vienna: Herder, 1972.

Smart, J. J. C., ed. *Problems of Space and Time*. New York: Macmillan, 1964.

Van Fraassen, Bastiaan C. *An Introduction to the Philosophy of Time and Space*. New York: Random House, 1970.

Wiggins, David. *Identity and Spatio-Temporal Continuity*. Oxford: Blackwell, 1967.

D. EASTERN

Ahmad, Aziz. *Change, Time, and Causality: With Special Reference to Muslim Thought*. Lahore: Pakistan Philosophical Congress, 1974.

Balslev, Anindita Niyogi. "Reflections on Time in Indian Philosophy." *Time, Science, and Society in China and the West: The Study of Time V*. Ed. J. T. Fraser, N. Lawrence, and F. C. Haber. Amherst: U of Massachusetts P, 1986. 104-12.

——. *A Study of Time in Indian Philosophy*. Wiesbaden: Otto Harrassowitz, 1983.

Needham, Joseph. "Time and Knowledge in China and the West." *The*

Voices of Time. Ed. J. T. Fraser. 2nd ed. Amherst: U of Massachusetts P, 1981. 92-135.

Schayer, Stanislaw. *Contributions to the Problem of Time in Indian Philosophy*. Krakow: Polski Builetyn Orientalistyczny, 1938.

"Time and Temporality." *Philosophy East and West* 24.2 (1974): 119-225.

Wilhelm, Hellmut. "The Concept of Time in the Book of Changes [the *I Ching*]." *Man and Time: Papers from the Eranos Year Books*. Trans. Ralph Manheim. Ed. Joseph Campbell. New York: Pantheon, 1957. 212-32.

Yamamoto, Makoto. *The Possibility of Metaphysics* (in Japanese). Tokyo: Tokyo UP, 1977.

Yinzhi, Zhang. "Mohist Views of Time and Space." *Time, Science, and Society in China and the West: The Study of Time V*. Ed. J. T. Fraser, N. Lawrence, and F. C. Haber. Amherst: U of Massachusetts P, 1986. 206-10.

Yu-lan, Fung. *A History of Chinese Philosophy*. Trans. Derek Bodde. Beijing: Henri Vetch, 1937.

E. MEDIEVAL

Augustine. *The Confessions*. Trans. Rex Warner. New York: New American Library, 1963. Book 11.

Boros, L. *Les catégories de la temporalité chez Saint Augustin*. Paris: Archives de Philosophie, 1958.

Chaix-Ruy, J. *Saint Augustin, temps et histoire*. Paris: Études Augustiniennes, 1956.

Decloux, Simon. *Temps, Dieu, liberté dans les "Commentaires aristotéliciens" de saint Thomas d'Aquin, essai sur la pensée grecque et la pensée chrétienne*. Paris: Desclée, De Brouwer, 1967.

Flood, E. T. "The Narrative Structure of Augustine's *Confessions*: Time's Quest for Eternity." *International Philosophical Quarterly* 28 (June 1988): 141-62.

Guitton, J. *Le temps et l'éternité chez Plotin et Saint Augustin*. Paris: Boivin, 1933.

Kilwardby, Robert. *On Time and Imagination:* De tempore, *and* De spiritu fantastico. Ed. P. Osmund Lewry. Oxford: Oxford UP, 1987.

McEvoy, J. "St. Augustine's Account of Time and Wittgenstein's Criticisms." *Review of Metaphysics* 37 (March 1984): 547-77.

Marrou, H. I. *L'ambivalence du temps de l'histoire chez Saint Augustin*. Paris: J. Vrin, 1950.

Plotinus. *The Enneads*. Trans. S. MacKenna. Ed. B. S. Page. 4th ed.

London: Faber, 1969. First Ennead, v, and Third Ennead, vii.
Quinn, J. M. *The Doctrine of Time in St. Thomas*. Washington: Catholic UP, 1960.
Quispel, Gilles. "Time and History in Patristic Christianity." *Man and Time: Papers from the Eranos Yearbooks*. Trans. Ralph Manheim. Ed. Joseph Campbell. New York: Pantheon, 1957. 85-107.
Schmidt, Ernst A. *Zeit und Geschichte bei Augustin*. Heidelberg: Winter, 1985.
Thomas Aquinas, Saint. *Summa theologiae* (Latin text and English translation). 61 vols. New York: McGraw-Hill, 1963-81. See the extensive entry under "Time" in the Index in Vol. 61.
Verwiebe, Walter. *Welt und Zeit bei Augustin*. Leipzig: F. Meiner, 1933.

F. RENAISSANCE TO ENLIGHTENMENT

Alexander, Samuel. *Spinoza and Time*. London: Allen and Unwin, 1921.
Arthur, R. T. W. "Continuous Creation, Continuous Time: A Refutation of the Alleged Discontinuity of Cartesian Time." *J of the History of Philosophy* 26 (July 1988): 349-75.
Baker, John Tull. *An Historical and Critical Examination of English Space and Time Theories from Henry More to Bishop Berkeley*. Bronxville: Sarah Lawrence College, 1930.
Bougeant. *A Philosophical Amusement upon the Language of Beasts*. London: T. Cooper, 1739.
Čapek, Milič. "The Conflict between the Absolutist and the Relational Theory of Time before Newton." *J of the History of Ideas* 48 (October-December 1987): 595-608.
Daniel, S. H. "Seventeenth-Century Scholastic Treatments of Time." *J of the History of Ideas* 42 (October-December 1981): 587-606.
Descartes, René. *L'homme*, and *Un traité de la formation du foetus*. Paris: Charles Angot, 1644.
——. *René Descartes, Selected Philosophical Writings*. Trans. J. Cottingham, R. Stoothoff, and D. Murdoch. New York: Cambridge UP, 1988. See index under "Time" and "Mechanics." Descartes is responsible for the clockwork dog of the mechanistic philosophy.
Gilead, A. "Teleological Time: A Variation on a Kantian Theme." *Review of Metaphysics* 38.3 (March 1985): 529-62.
Gregory, Joshua C. "The Animate and Mechanical Models of Reality." *J of Philosophical Studies* (1927): 302.
Hallett, Harold Foster. *Aeternitas: A Spinozistic Study*. Oxford: Clar-

endon P, 1930.

Havet, Jacques. *Kant et le problème du temps*. Paris: Gallimard, 1946.

Hegel, G. W. F. *The Philosophy of Nature*. Trans. A. V. Miller. Oxford: Oxford UP, 1970. Sections 257-59.

Heidemann, Ingeborg. *Spontaneität und Zeitlichkeit. Ein Problem der Kritik der reinen Vernunft*. Cologne, 1958.

Heinrichs, Jürgen. *Das Problem der Zeit in der praktischen Philosophie Kants*. Bonn: Bouvier, 1968.

Hume, David. *Dialogues Concerning Natural Religion*. Ed. Norman Kemp Smith. 2nd ed. New York: Social Sciences Publishers, 1948. Juxtaposes clock and organic metaphors.

——. *A Treatise of Human Nature*. Ed. L. A. Selby-Bigge. 3 vols. Oxford: Clarendon P, 1888. Book 1, Part 2.

Jalabert, J. *La théorie leibnizienne de la substance et ses rapports avec la notion du temps*. Paris: Presses universitaires de France, 1947.

Kant, Immanuel. *The Critique of Pure Reason*. Trans. N. K. Smith. New York: St. Martin's P, 1968. See in particular "Transcendental Aesthetic," Section 2 "Of Time," and "Transcendental Analytic," Book 2, Chapter 1.

——. *Inaugural Dissertation of 1770 (Of the Form and Principles of the Sensible and Intelligible World)*. Trans. William J. Eckoff. New York, 1894. See in particular Section 3, Paragraph 14 "Of Time."

Kirkinen, Heikki. *Les origines de la conception moderne de l'homme-machine*. Helsinki, 1960.

Krantz, Emile. *Essai sur l'esthétique de Descartes: Rapports de la doctrine cartésienne avec la littérature classique française au XVIIe siècle*. 1898; reprint, Geneva: Slatkine Reprints, 1970.

La Mettrie, Julien Offray de. *Histoire naturelle de l'âme*. La Haye, 1745.

——. *Oeuvres philosophiques*. London [Berlin], 1751. See also his *L'homme machine* under Vartanian in this section.

Landsberg, P. T., and D. A. Evans. "Free Will in a Mechanistic Universe?" *British J of Philosophy of Science* 21 (1970): 343-58.

Laudan, Laurens. "The Clock Metaphor and Probabilism: The Impact of Descartes on English Methodological Thought, 1650-1665." *Annals of Science* 22 (June 1966): 73-104.

Leibniz, G., and S. Clarke. *The Leibniz-Clarke Correspondence: Together with Extracts from Newton's* Principia *and* Optics. Ed. H. G. Alexander. Manchester: Manchester UP, 1956. The classic debate concerning Newton's "absolute" versus Leibniz's "relational" theories of time.

Leibniz, G. *New Essays on Human Understanding*. Trans. Peter Remnant and Jonathan Bennett. Cambridge: Cambridge UP, 1981. Chapters 4 and 5 of Book 2 constitute Leibniz's response to Locke's specific discussions on time in his *Essay*.

Lenoble, Robert. *Mersenne; ou, La naissance du méchanisme*. Paris: J. Vrin, 1943.

Locke, John. *Essay Concerning the Human Understanding*. Ed. John W. Yolton. 2 vols. London: Dent, 1961. See in particular Chapters 14 and 15 of Book 2.

Lorenz, K. "Kants Lehre vom apriorischen in Lichte gegenwärtiger Biologie." *Blätter für Deutsche Philosophie* 15 (1941): 94-125.

Lotze, R. Hermann. *Metaphysic*. Trans. T. H. Green. Oxford: Oxford UP, 1887. Book 2, Chapter 3.

Morris, John. "Pattern Recognition in Descartes' Automata." *Isis* (Winter 1969): 451-55.

Nicolson, Marjorie. "The Early Stage of Cartesianism in England." *Studies in Philology* 26 (1929): 356-74.

Rosenfield, Leonora Cohen. *From Beast-Machine to Man-Machine*. Enlarged ed. New York: Octagon Books, 1968.

Schindler, David L., ed. *Beyond Mechanism: The Universe in Recent Physics and Catholic Thought*. Lanham: UP of America, 1986.

Sebba, G. "Time and the Modern Self: Descartes, Rousseau, Beckett." *The Study of Time I*. Ed. J. T. Fraser, F. C. Haber, and G. H. Müller. New York: Springer, 1972. 452-69.

Sorensen, R. A. "Time Travel, Parahistory and Hume." *Philosophy* 62 (April 1987): 227-36.

Temmer, Mark J. *Time in Rousseau and Kant: An Essay on French Preromanticism*. Geneva: E. Droz, 1958.

Uferman, K. *Untersuchungen über das Gesetz der Kontinuität bei Leibniz*. Berlin, 1927.

Vartanian, Aram, ed. *La Mettrie's L'Homme Machine* (critical edition with an introductory monograph). Princeton: Princeton UP, 1960.

Whitrow, G. J. "Kant and the Extra-Galactic Nebulae." *Quarterly J of the Royal Astronomical Society* 8 (1967): 48-56.

Wieland, Wolfgang. *Schellings Lehre von der Zeit: Grundlagen und Voraussetzungen der Weltalterphilosophie*. Heidelberg: C. Winter Universitätsverlag, 1956.

G. SPECIFIC RECENT AND CONTEMPORARY PHILOSOPHERS

Bergson, Henri Louis. *Duration and Simultaneity: With Reference to*

Einstein's Theory. Trans. L. Jacobson. Indianapolis: Bobbs-Merrill, 1965.

———. *Réflexions sur le temps, l'espace et la vie*. Paris: Payot, 1920.

———. *Time and Free Will*. Trans. R. L. Pogson. London: Allen and Unwin, 1950.

Bernet, Rudolf. "Endlichkeit und Unendlichkeit in Husserls Phänomenologie der Wahrnehmung." *Tijdschrift voor Filosofie* 40.2 (June 1978): 251-69.

———. "Die Frage nach dem Ursprung der Zeit bei Husserl und Heidegger." *Heidegger Studies*. Vols 3-4. Berlin: Duncker and Humblot, 1987-88. 89-104.

———. "Origine du temps et temps originaire chez Husserl et Heidegger." *Revue philosophique de Louvain* 85.4 (November 1987): 499-521.

Blaisdell, Chuck. "Heidegger's Structure of Time and Temporality." *Dialogue (Phi Sigma Tau)* 18 (April 1976): 44-53.

Bulhof, Ilse Nina. *Apollos Wiederkehr. Eine Untersuchung der Rolle des Kreises in Nietzsches Denken über Geschichte und Zeit*. The Hague: Martinus Nijhoff, 1969.

Camele, Anthony M. "Time in Merleau-Ponty and Heidegger." *Philosophy Today* 19.3/4 (Fall 1975): 256-68.

Caputo, John D. "Time and Being in Heidegger." *The Modern Schoolman* 50 (1973): 325-49.

Chahine, Osman E. *La durée chez Bergson*. Paris: H. Boucher, 1970.

Collingwood, Robin George. "Some Perplexities About Time: With an Attempted Solution." *Proceedings of the Aristotelian Society*, 1926.

Dauer, D. W. "Nietzsche and the Concept of Time." *The Study of Time II*. Ed. J. T. Fraser, and N. Lawrence. New York: Springer, 1975. 81-97.

Fell, Joseph P. "Heidegger's Notion of Two Beginnings." *Review of Metaphysics* 25.2 (December 1971): 213-37.

Feuerbach, Ludwig. *Principles of the Philosophy of Future*. Trans. M. H. Vogel. Indianapolis: Bobbs-Merril, 1966.

Giroux, Laurent. *Durée pure et temporalité: Bergson et Heidegger*. Montreal: Ballarmin, 1971.

Granel, Gérard. *Le sens du temps et de la perception chez E. Husserl*. Paris: Gallimard, 1968.

Gray, John Glenn. "Martin Heidegger: On Anticipating My Own Death." *Personalist* 46 (1965): 439-58.

Gupta, R. J. "What is Heidegger's Notion of Time?" *Revue Internationale de Philosophie* 14 (1960): 163-93.

Heidegger, Martin. *Being and Time*. Trans. John Macquarrie and

Edward Robinson. New York: Harper and Row, 1962.

———. *History of the Concept of Time: Prolegomena*. Trans. Theodore Kisiel. Bloomington: Indiana UP, 1985.

———. *On Time and Being*. Trans. Joan Stambaugh. New York: Harper and Row, 1972.

Heine, Steven. *Existential and Ontological Dimensions of Time in Heidegger and Dogen*. Albany: State U of New York P, 1985.

Held, Klaus. *Lebendige Gegenwart*. The Hague: Martinus Nijhoff, 1966. About Edmund Husserl.

Helm, Bertrand P. *Time and Reality in American Philosophy*. Amherst: U of Massachusetts P, 1985.

Hinners, Richard. "The Freedom and Finiteness of Existence in Heidegger." *The New Scholasticism* 33 (1959): 32-48.

Husserl, Edmund. *Cartesian Meditations: An Introduction to Phenomenology*. Trans. D. Cairns. The Hague: Martinus Nijhoff, 1931.

———. *The Phenomenology of Internal Time-Consciousness*. Trans. James S. Churchill. The Hague: Martinus Nijhoff, 1964.

———. *Texte zur Phänomenologie des inneren Zeitbewusstseins (1893-1917)*. Ed. Rudolf Bernet. Hamburg: Felix Meiner, 1985.

James, William. *The Principles of Psychology*. 2 vols. New York: Holt, 1890. See in particular Chapter 15.

McInerney, P. K. "What is Still Valuable in Husserl's Analyses of Inner Time-Consciousness." *J of Philosophy* 85.11 (November 1988): 605-16.

McTaggart, John, and Ellis McTaggart. *The Nature of Existence*. Cambridge: Cambridge UP, 1927. Volume 2, Chapter 33.

———. "The Unreality of Time." *Mind* New Series 68 (October 1908): 457-74.

Manchester, Peter B. "Time in Whitehead and Heidegger." *Process Studies* 5 (Summer 1975): 106-13.

Mason, David R. *Time and Providence: An Essay Based on an Analysis of the Concept of Time in Whitehead and Heidegger*. Washington: UP of America, 1982.

Mays, W. "Temporality and Time in Hegel and Marx." *The Study of Time II*. Ed. J. T. Fraser, and N. Lawrence. New York: Springer, 1975. 98-113.

———. "Whitehead and the Philosophy of Time." *The Study of Time I*. Ed. J. T. Fraser, F. C. Haber, and G. H. Müller. New York: Springer, 1972. 354-69.

Merlan, Philip. "Time-Consciousness in Husserl and Heidegger." *Philosophy and Phenomenological Research* 8 (1947-48): 23-54.

Merleau-Ponty, Maurice. "Temporality." In *Phenomenology of Percep-*

tion. Trans. C. Smith. New York: Humanities P, 1962.

Miller, Izchak. *Husserl, Perception, and Temporal Awareness*. Cambridge: MIT P, 1984.

Minkowski, Eugene. *Lived Time: Phenomenological and Psychopathological Studies*. Trans. Nancy Metzel. Evanston: Northwestern UP, 1970.

Oaklander, L. Nathan. "McTaggart, Schlesinger, and the Two-Dimensional Time Hypothesis." *Philosophical Quarterly* 33 (October 1983): 391-97. Includes bibliography.

——. *Temporal Relations and Temporal Becoming: A Defense of a Russellian Theory of Time*. Lanham: UP of America, 1984.

O'Mahoney, B. E. "Martin Heidegger's Existential of Death." *Philosophical Studies* (Ireland) 18 (1969): 58-75.

Orth, E. W., ed. *Zeit und Zeitlichkeit bei Husserl und Heidegger*. Freiburg: K. Alber, 1983.

Pears, David. "Time, Truth and Inference." *Essays in Conceptual Analysis*. Ed. A. Flew. London: Macmillan, 1963. 228-52.

Rapaport, Herman. *Heidegger and Derrida: Reflections on Time and Language*. Lincoln: Nebraska UP, 1989.

Royce, Josiah. "The Reality of the Temporal." *International J of Ethics* (April 1910): 157-71.

——. *The World and the Individual—Second Series*. New York: Macmillan, 1904. Lecture 3.

Russell, Bertrand. "On the Experience of Time." *The Monist* 25.2 (1915): 212-33.

Sallis, John C. "World, Finitude, Temporality in the Philosophy of Martin Heidegger." *Philosophy Today* 9 (1965): 40-52.

Santayana, George. "Pictorial Space and Sentimental Time." *The Realm of Matter*. New York: Scribners, 1920.

Sartre, Jean Paul. *Being and Nothingness: An Essay on Phenomenological Ontology*. Trans. Hazel Barnes. New York: Philosophical Library, 1956. Volume 2, Chapter 2.

Sheehan, Thomas J. "Heidegger: From Beingness to the Time-Being." *Listening* 8 (1973): 17-31.

Sherover, Charles M. *Heidegger, Kant and Time*. Bloomington: Indiana UP, 1971.

Shin, Gui Hyun. *Die Struktur des inneren Zeitbewusstseins: eine Studie über den Begriff der Protention in den veröffentlichten Schriften Edmund Husserls*. Las Vegas: P. Lang, 1978.

Slaatte, Howard Alexander. *Time, Existence and Destiny: Nicholas Berdyaev's Philosophy of Time*. New York: P. Lang, 1988.

Stambaugh, Joan. "Time and Dialectic in Hegel and Heidegger." *Re-*

search in Phenomenology 4 (1974): 87-97.

——. *Untersuchungen zum Problem der Zeit bei Nietzsche*. The Hague: Martinus Nijhoff, 1959.

Steiner, Rudolf. *Philosophy, Cosmology and Religion: Ten Lectures*. Ed. Stewart C. Easton, et al. Hudson: Anthroposophic P, 1984.

Wallack, F. Bradford. *The Epochal Nature of Process in Whitehead's Metaphysics*. Albany: State U of New York P, 1980.

Whitehead, Alfred North. *Process and Reality*. New York: Macmillan, 1929.

——. "Time." In *The Concept of Nature*. Cambridge: Cambridge UP, 1920.

Williams, J. R. "Heidegger, Death and God." *Studies in Religion* (Toronto) 1 (1972): 298-320.

Wohlfort, Günter. *Der Augenblick: Zeit und ästhetische Erfahrung bei Kant, Hegel, Nietzsche und Heidegger, mit einem Exkurs zu Proust*. Freiburg: K. Alber, 1982.

Wren, T. "Heidegger's Philosophy of History." *J of British Society of Phenomenology* 3 (1972): 111-25.

Zeit und Zeitlichkeit bei Husserl und Heidegger. Beiträge von Rudolf Bernet, et al. Freiburg: K. Alber, 1983.

H. TIME'S PASSAGE, PAST, PRESENT, FUTURE

Anscombe, Gertrude Elizabeth Margaret. *Times, Beginnings, and Causes*. London: Oxford UP, 1975. 20 pages.

Beer, M. "Temporal Indexicals and the Passage of Time." *Philosophical Quarterly* 38 (April 1988): 158-64.

Benthem, J. van. "Later than Late: On the Logical Origin of the Temporal Order." *Pacific Philosophical Quarterly* 63 (April 1982): 193-203.

Boschke, Friedrich L. *Und 1000 Jahre sind wie ein Tag. Die Zeit, das unverstandene Phänomen*. Munich: Bertelsmann, 1979.

Brandon, S. G. F. "Time and the Destiny of Man." *The Voices of Time*. Ed. J. T. Fraser. 2nd ed. Amherst: U of Massachusetts P, 1981. 140-57.

Brumbaugh, Robert S. "Applied Metaphysics: Truth and Passing Time." *Review of Metaphysics* 19.4 (June 1966): 647-66.

Čapek, Milič. "The Fiction of Instants." *The Study of Time I*. Ed. J. T. Fraser, F. C. Haber, and G. H. Müller. New York: Springer, 1972. 332-44.

Cassirer, E. "On the Reality of Becoming." *The Study of Time I*. Ed.

J. T. Fraser, F. C. Haber, and G. H. Müller. New York: Springer, 1972. 345-53.

Denbigh, K. G. "The Objectivity, or Otherwise, of the Present." *The Study of Time III*. Ed. J. T. Fraser, N. Lawrence, and D. Park. New York: Springer, 1978. 307-29.

Gale, R. M. "Why a Cause Cannot Be Later than Its Effect." *Review of Metaphysics* 19.2 (December 1965): 209-34.

Grünbaum, A. "The Status of Temporal Becoming." *Annals of the New York Academy of Sciences* 138 (1967): 374-95.

Hamblin, C. L. "Instants and Intervals." *The Study of Time I*. Ed. J. T. Fraser, F. C. Haber, and G. H. Müller. New York: Springer, 1972. 324-31.

Kümmel, Friedrich. "Time as Succession and the Problem of Duration." *The Voices of Time*. Ed. J. T. Fraser. 2nd ed. Amherst: U of Massachusetts P, 1981. 31-55.

Lavelle, Louis. *Du temps et de l'éternité*. Aubier: Éditions Montaigne, 1945.

Lawrence, Nathaniel. "Temporal Passage and Spatial Metaphor." *The Study of Time II*. Ed. J. T. Fraser, and N. Lawrence. New York: Springer, 1975. 196-205.

Merleau-Ponty, J. "Ideas of Beginnings and Endings in Cosmology." *The Study of Time III*. Ed J. T. Fraser, N. Lawrence, and D. Park. New York: Springer, 1978. 333-50.

Park, David. "The Past and the Future." *The Study of Time III*. Ed. J. T. Fraser, N. Lawrence, and D. Park. New York: Springer, 1978. 351-67.

Pellauer, D. "Time and Narrative and Theological Reflection." *Philosophy Today* 31.3/4 (Fall 1987): 262-86.

Plumer, G. "Expressions of Passage." *Philosophical Quarterly* 37 (October 1987): 341-54.

Prior, A. N. "The Notion of the Present." *The Study of Time I*. Ed. J. T. Fraser, F. C. Haber, and G. H. Müller. New York: Springer, 1972. 320-23.

Rucker, R. *Infinity and the Mind*. London: Paladin, 1984.

Rudavsky, T. M. "Creation Time and Infinity in Gersonides." *J of the History of Philosophy* 26 (January 1988): 25-44.

Schlesinger, G. N. "How Time Flies." *Mind* 91 (October 1982): 501-23.

Schuster, M. M. "Is the Flow of Time Subjective?" *Review of Metaphysics* 39 (June 1986): 695-714.

Sherover, Charles M. "Are We in Time?" *International Philosophical Quarterly* 26 (March 1986): 33-46.

——. "Experiential Time and the Religious Concern." *Zygon* 16 (December 1981): 323-44.

——. *The Human Experience of Time*. New York: New York UP, 1975.

——. "Perspectivity and the Principle of Continuity." *The Study of Time IV*. Ed. J. T. Fraser, N. Lawrence, and D. Park. New York: Springer, 1981. 136-46.

Shoham, Yoav. *Reasoning about Change: Time and Causation from the Standpoint of Artificial Intelligence*. Cambridge: MIT P, 1988.

Ushenko, Andrew Paul. *The Logic of Events: An Introduction to a Philosophy of Time*. Berkeley: U of California P, 1929.

Williams, Donald C. "The Myth of Passage." *The Philosophy of Time*. Ed. Richard M. Gale. Garden City: Doubleday, 1967. 98-116.

Wright, Georg Henrik von. *Time, Change and Contradiction*. London: Cambridge UP, 1969.

19. PHYSICS

A. GENERAL

Augustynek, Z. *The Nature of Time* (in Polish). Warsaw: PWN, 1975.

——. *The Properties of Time* (in Polish). Warsaw: PWN, 1970.

Barbera, G., ed. *Galileo Galilei and the Measure of Time*. Florence: Comitato Nazionale per Le Manifestazione Celebrative del IV Centario della Nascita di Galileo Galilei, 1967.

Barrow, J. D. *The World within the World*. Oxford: Clarendon P, 1988.

Böhme, Gernot. *Über die Zeitmode: Eine Untersuchung über das Verstehen von Zeit als Gegenwart, Vergangenheit und Zukunft mit besonderer Berücksichtigung der Beziehungen zum zweiten Haupsatz der Thermodynamik*. Göttingen: Vandenhoeck and Ruprecht, 1966.

Brans, Jean-Pierre, Isabelle Stengers, and Philippe Vincke, eds. *Temps et devenir: À partir de l'oeuvre d'Ilya Prigogine*. Geneva: Patiño, 1988.

Bunge, M. "Physical Time: The Objective and Relational Theory." *Philosophy of Science* 35 (1968): 355-88.

Costa de Beauregard, Olivier. *Le second Principe de la science du temps*. Paris: Éditions du Seuil, 1963.

——. "The Third Storm of the Twentieth Century: The Einstein Paradox." *The Study of Time III*. Ed. J. T. Fraser, N. Lawrence, and D. Park. New York: Springer, 1978. 53-70.

——. *Time, the Physical Magnitude*. Dordrecht: Reidel, 1987.

Davies, P. C. W. "Time and Reality." *Reduction, Time and Reality: Studies in the Philosophy of the Natural Sciences*. Ed. R. Healy. Cambridge: Cambridge UP, 1981. 63-78.

Denbigh, Kenneth George. *An Inventive Universe*. New York: Braziller, 1975.

——. "Physical Time and Mental Time." *Time and Mind: The Study of Time VI*. Ed. J. T. Fraser. Madison: International Universities P, 1989. 143-48.

Eddington, A. S. *Space, Time and Gravitation*. New York: Harper Torchbooks, 1959.

Fagg, Lawrence W. *Two Faces of Time*. Wheaton: Theosophical Publishing House, 1985.

Flood, Raymond, and Michael Lockwood, eds. *The Nature of Time*. Cambridge: Blackwell, 1987.

Fraser, J. T. *The Genesis and Evolution of Time: A Critique of Interpretation in Physics*. Amherst: U of Massachusetts P, 1982. Tokyo: Kodansha Scientific, 1984 (in Japanese).

———. "Revolution in Time," an essay review in *Foundations of Physics* 14 (1984): 799-804.

Hawking, Stephen W. *A Brief History of Time: From the Big Bang to Black Holes*. Toronto: Bantam, 1988.

Hinton, Charles Howard. *The Fourth Dimension*. London: Allen and Unwin, 1934.

Hund, F. "Zeit als physikalischer Begriff." *The Study of Time I*. Ed. J. T. Fraser, F. C. Haber, and G. H. Müller. New York: Springer, 1972. 39-52.

———. "Die Zeit in der Begriffswelt des Physikers." *Studium Generale* 8 (1955): 469-76.

Janich, Peter. *Protophysics of Time: Constructive Foundation and History of Time Measurement*. Dordrecht: Reidel, 1985.

Johnson, Conrad Dale. "Scientific Explanation and the Evolution of Time." *Time, Science, and Society in China and the West: The Study of Time V*. Ed. J. T. Fraser, N. Lawrence, and F. C. Haber. Amherst: U of Massachusetts P, 1986. 39-50.

Kamefuchi, S. "A Non-Causal Approach to Physical Time." *The Study of Time II*. Ed. J. T. Fraser, and N. Lawrence. New York: Springer, 1975. 239-48.

Kroes, Peter. *Time: Its Structure and Role in Physical Theories*. Boston: Reidel, 1984.

Landsberg, Peter T., ed. *The Enigma of Time*. Bristol: A. Hilger, 1982.

———. "The Physical Concept of Time in the 20th Century." *Physics in the Making*. Ed. A. Sarlemijn, and M. J. Sparnaay. Amsterdam: Elsevier, 1989. 131-65.

Laplace, P. S. *Celestial Mechanics*. Trans. N. Bowditch. 4 vols. 1829-39; reprint, New York: Chelsea Publishing, 1966. Translation of Laplace's *Mécanique céleste* (1799-1825).

Lowe, Victor. *Understanding Whitehead*. Baltimore: Johns Hopkins P, 1962.

Metz, André. *Temps, espace, relativité*. Paris: G. Beauchesne, 1928.

Misner, C. "Absolute Zero of Time." *Physical Review* 186.5B (October 25, 1969): 1328-33.

Mittelstaedt, Peter. *Der Zeitbegriff in der Physik*. 2nd ed. Mannheim: Bibliographisches Institut, 1980.

Morris, R. *Time's Arrows*. New York: Simon and Schuster, 1984.

Neal, Harry Edward. *The Mystery of Time*. New York: J. Messner, 1966.

North, J. D. *The Measure of the Universe*. New York: Oxford UP, 1965.

Park, David. *Classical Dynamics and Its Quantum Analogues*. New York: Springer, 1979.

———. *The How and the Why: An Essay on the Origins and Development of Physical Theory*. Princeton: Princeton UP, 1988.

———. *The Image of Eternity: Roots of Time in the Physical World*. Amherst: U of Massachusetts P, 1980.

———. "Laws of Physics and Ideas of Time." *The Study of Time II*. Ed. J. T. Fraser, and N. Lawrence. New York: Springer, 1975. 258-66.

———. "The Myth of the Passage of Time." *The Study of Time I*. Ed. J. T. Fraser, F. C. Haber, and G. H. Müller. New York: Springer, 1972. 110-21.

———. "Should Physicists Say That the Past Really Happened?" *Time and Mind: The Study of Time VI*. Ed. J. T. Fraser. Madison: International Universities P, 1989. 125-39.

Pedersen, O., and M. Pihl. *Early Physics and Astronomy: A Historical Introduction*. New York: Neale Watson, 1974.

Penrose, Roger. *The Emperor's New Mind: Concerning Computers, Minds, and the Laws of Physics*. Oxford: Oxford UP, 1989.

Popper, Karl. *Of Clouds and Clocks*. Washington: Washington UP, 1966.

Prigogine, Ilyia. *From Being to Becoming: Time and Complexity in the Physical Sciences*. San Francisco: Freeman, 1980.

Prigogine, Ilyia, and I. Stengers. *Order Out of Chaos*. London: Heinemann, 1984.

Raju, C. K. "Classical Time Symmetric Electrodynamics." *J of Physics A* 13 (1980): 3303-17.

———. "Quantum Mechanics and the Micro-Physical Structure of Time." *In Search of Quantum Reality*. Ed. R. Nair. Singapore: World Scientific, 1990.

Rigal, J. L., ed. *Le temps et la pensée physique contemporaine*. Paris: Dunod, 1968.

Rosnay, Joël de. *The Macroscope: A New World Scientific System*. Trans. Robert Edwards. New York: Harper and Row, 1979.

Salmon, Wesley C. *Space, Time, and Motion: A Philosophical Introduction*. Encino: Dickenson, 1975.

Schilpp, P., ed. *Albert Einstein: Philosopher-Scientist*. New York: Tudor, 1951.

Schlegel, Richard. *Time and the Physical World*. New York: Dover, 1968.

Stafleu, M. D. "Analysis of Time in Modern Physics." *Philosophia Reformata* 35 (1970): 119-31.

Taylor, J. G. "Time in Particle Physics." *The Study of Time I*. Ed. J. T. Fraser, F. C. Haber, and G. H. Müller. New York: Springer, 1972. 53-58.

Tolman, R. C. *Relativity, Thermodynamics, and Cosmology*. Oxford: Oxford UP, 1934.

Uritam, R. A. "Physics and the View of Nature in Traditional China." *American J of Physics* 43.2 (1975): 136-52.

Wagn, Klaus. *What Time Does*. Trans. Sophie Wilkins. Munich: Caann-Verlag, 1976.

Wallis, Robert. *Time: Fourth Dimension of the Mind*. Trans. B. B. Montgomery, and D. B. Montgomery. New York: Harcourt, Brace, 1968.

Watanabe, Michael Satosi. "Causality and Time." *The Study of Time II*. Ed. J. T. Fraser, and N. Lawrence. New York: Springer, 1975. 267-82.

——. *The History of Time Seen Through Physical Science* (in Japanese). Tokyo: Tosho, 1973.

——. "Time and the Probabilistic View of the World." *The Voices of Time*. Ed. J. T. Fraser. 2nd ed. Amherst: U of Massachusetts P, 1981. 527-63.

Wheeler, J. A. "Frontiers of Time." In *Proceedings of the International School of Physics, "Enrico Fermi," Course 72*. Amsterdam: North-Holland, 1979. 395-496.

Whitehead, Alfred North. *Process and Reality*. 1929; reprint, New York: Macmillan, 1969.

——. *Science in the Modern World*. 1925; reprint, New York: Macmillan, 1967.

Whiteman, Michael. *Philosophy of Space and Time and the Inner Constitution of Nature: A Phenomenological Study*. London: Allen and Unwin, 1967.

Whitrow, G. J. "Time and the Universe." *The Voices of Time*. Ed. J. T. Fraser. 2nd ed. Amherst: U of Massachusetts P, 1981. 564-81.

——. *What Is Time?* London: Thames and Hudson, 1972.

B. ASTRONOMY

Clube, Victor, ed. *Catastrophes and Evolution: Astronomical Foundations*. New York: Cambridge UP, 1989.

Cohen, Martin. *In Darkness Born: The Story of Star Formation*. New York: Cambridge UP, 1988.

Cornell, James, and Paul Gorenstein, eds. *Astronomy from Space: Sputnik to Space Telescope*. Cambridge: MIT P, 1985.

Cornell, James, and Alan P. Lightman, eds. *Revealing the Universe: Prediction and Proof in Astronomy*. Cambridge: MIT P, 1981.

Dyson, F. J. "Time Without End: Physics and Biology in an Open Universe." *Reviews of Modern Physics* 51 (1979): 447-60.

Fowler, W. A. "The Age of the Observable Universe." *Quarterly J, Royal Astronomical Society* 28 (1987): 87-108.

Grosser, Morton. *The Discovery of Neptune*. New York: Dover, 1979.

Heather, M. A. "The Reliability of Ancient Eclipse Records." *Geophysical J* 15.5 (1968): 549-50.

Hillier, Rodney R. *Gamma-Ray Astronomy*. New York: Oxford UP, 1984.

Hirsh, Richard. *Glimpsing an Invisible Universe: The Emergence of X-Ray Astronomy*. New York: Cambridge UP, 1983.

King, Henry C., and John R. Millburn. *Geared to the Stars: The Evolution of Planetariums, Orreries and Astronomical Clocks*. Toronto: U of Toronto P, 1978.

Kopal, Z., and J. Rahe. *Binary and Multiple Stars as Tracers of Stellar Evolution*. Norwell: Kluwer Academic, 1982.

Krupp, E. C., ed. *Archaeoastronomy and the Roots of Science*. Boulder: Westview P, 1984.

Mulholland, J. D. "Measures of Time in Astronomy." *Publications. Astronomical Society of the Pacific* 84 (1972): 357-64.

Newton, R. R. "The Application of Ancient Astronomy to the Study of Time." *Endeavour* 33 (1974): 34-39.

Sadler, D. H. "Astronomical Measures of Time." *Quarterly J of the Royal Astronomical Society* 9 (1968): 281-93.

Sandage, Allan. "The Redshift-Distance Relation—II." *Astrophysical J* 178 (1972): 1-24.

Spencer Jones, H. "The Rotation of the Earth and the Secular Accelerations of the Sun, Moon and Planets." *Monthly Notices, Royal Astronomical Society* 99.7 (1939): 541-58.

Verschuur, G. L., and K. I. Kellerman, eds. *Galactic and Extra-Galactic Radio-Astronomy*. New York: Springer, 1974.

Whipple, Fred L. *Orbiting the Sun: Planets and Satellites of the Solar System*. Cambridge: Harvard UP, 1981.

Wilkins, John. *Discourse concerning a New Planet; tending to prove, that it is probable our Earth is one of the Planets*. London: John Maynard, 1640.

Winkler, L. "Astronomically Determined Dates and Alignments." *American J of Physics* 40.1 (1972): 126-32.

Wolstencroft, R. D., and W. B. Burton, eds. *Millimetre and Submillimetre Astronomy*. Norwell: Kluwer Academic, 1988.

C. COSMOLOGY

Arnot, F. L. *Time and the Universe, a New Basis for Cosmology*. Sydney: Australasian Medical Publishing, 1941.

Barrow, J. D. "The Inflationary Universe: Modern Developments." *Quarterly J, Royal Astronomical Society* 29 (1988): 101-17.

——. *The World Within the World*. Oxford: Oxford UP, 1988.

Barrow, J. D., and F. J. Tipler. *The Anthropic Cosmological Principle*. Oxford: Clarendon P, 1986.

Bass, R. W., and L. Witten. "Remark on Cosmological Models." *Reviews of Modern Physics* 29 (1957): 452-53.

Berry, M. *Principles of Cosmology and Gravitation*. Cambridge: Cambridge UP, 1976.

Bondi, H. *Cosmology*. 2nd ed. Cambridge: Cambridge UP, 1960.

——. "Physics and Cosmology." *Observatory* 82 (1962): 133-43.

Davies, P. C. W. "Space-Time Singularities in Cosmology and Black Hole Evaporation." *The Study of Time III*. Ed J. T. Fraser, N. Lawrence, and D. Park. New York: Springer, 1978. 74-91.

Dicke, R. H., et al. "Cosmic Black Body Radiation." *Astrophysical J* 142 (1965): 414-29.

——. "Radiation from the Origin of Time." *Nature* 215 (1967): 1056-57.

Eddington, A. S. *The Expanding Universe*. Cambridge: Cambridge UP, 1933.

Einstein, Albert. "Cosmological Considerations in the General Theory of Relativity." In H. A. Lorenz, et al. *The Principle of Relativity*. Trans. W. Perrett, and G. B. Jeffrey. New York: Dover Reprint, n.d.

Fang, L. Z., and R. Ruffini, eds. *Quantum Cosmology*. Singapore: World Scientific, 1987.

Gal-Or, Benjamin. *Cosmology, Physics, Philosophy*. New York: Springer, 1981.

Gödel, K. "An Example of a New Type of Cosmological Solution of Einstein's Field Equations of Gravitation." *Reviews of Modern Physics* 21 (1949): 447-50.

Gott, J. R. "Computer Simulation of the Universe." *Comments on Astrophysics* 8.3 (1979): 55-67.

Gribbin, J. "Oscillating Universe Bounces Back." *Nature* 259 (1975): 15-16.

Harrison, E. R. *Cosmology: The Science of the Universe*. Cambridge: Cambridge UP, 1981.

——. "Cosmology without General Relativity." *Annals of Physics* 35 (1965): 437-46.

———. "The Early Universe." *Physics Today* 21 (June 1968): 31-39.

Hawking, S. W., and G. F. R. Ellis. *The Large Scale Structure of Space-Time*. Cambridge: Cambridge UP, 1973.

Heller, M. "Global Time Problem in Relativistic Cosmology." *Annales de la Société Scientifique de Bruxelles* 4 (89) (1975): 522-32.

———. "The Origins of Time." *The Study of Time IV*. Ed. J. T. Fraser, N. Lawrence, and D. Park. New York: Springer, 1981. 90-93.

Islam, Jumal N. *The Ultimate Fate of the Universe*. New York: Cambridge UP, 1983.

Johnson, M. C. *Time and the Universe for the Scientific Conscience*. Cambridge: Cambridge UP, 1952.

Kerszberg, Pierre. *The Invented Universe: The Einstein-De Sitter Controversy*. New York: Oxford UP, 1989.

Landsberg, P. T., and N. T. Bishop. "A Cosmological Deduction of the Order of Magnitude of an Elementary Particle Mass and of the Cosmological Coincidences." *Physics Letters* 53A (1975): 109-10.

———. "A Principle of Impotence Allowing for Newtonian Cosmologies with a Time-Dependent Gravitational Constant." *Monthly Notices, Royal Astronomical Society* 171 (1975): 279-86.

Landsberg, P. T., and R. K. Pathria. "Cosmological Parameters for a Restricted Class of Closed Big-Bang Universes." *Astrophysical J* 192 (1974): 577-79.

Layzer, D. "A Unified Approach to Cosmology." *Relativity and Astrophysics. Vol. 8*. Ed. J. Ehlers. Providence: American Mathematical Society P, 1967.

McCrea, W. H. "A Philosophy for Big Bang Cosmology." *Nature* 228 (1970): 21-24.

Merleau-Ponty, Jacques, and Bruno Morando. *The Rebirth of Cosmology*. New York: Knopf, 1976.

Meyerstein, F. Walter, ed. *Foundations of Big Bang Cosmology*. Singapore: World Scientific, 1989.

Mezzetti, M., et al., eds. *Large Scale Structure and Motions in the Universe*. Norwell: Kluwer Academic, 1989.

Milne, E. A. *Modern Cosmology and the Christian Idea of God*. Oxford: Oxford UP, 1952.

Park, David. "The Beginning and the End of Time in Physical Cosmology." *The Study of Time IV*. Ed. J. T. Fraser, N. Lawrence, and D. Park. New York: Springer, 1981. 103-18.

Reeves, Hubert. *Atoms of Silence: An Exploration of Cosmic Evolution*. Trans. Ruth Lewis and John Lewis. Cambridge: MIT P, 1983.

Robson, John M., ed. *Origin and Evolution of the Universe: Evidence for Design?* Toronto: U of Toronto P, 1987.

Rowan-Robertson, M. *The Cosmological Distance Ladder*. San Francisco: Freeman, 1985.

Schramm, D. N. "Cosmology and New Particles." *Particles and Fields —1977*. New York: American Institute of Physics, 1978.

Sciama, Dennis W. *Modern Cosmology*. New York: Cambridge UP, 1982.

Seeds, Michael A. *Exploring the Universe*. 3rd ed. Belmont: Wadsworth Publishing, 1989.

Seitter, W. C., et al., eds. *Large-Scale Structures in the Universe*. New York: Springer, 1988.

Shalom, Albert. "Time and Mind in the Constitution of the Universe: A Critique of the Participatory Anthropic Principle." *Time and Mind: The Study of Time VI*. Ed. J. T. Fraser. Madison: International Universities P, 1989. 151-76.

Shuter, W. L. *Kinematics, Dynamics and Structure of the Milky Way*. Norwell: Kluwer Academic, 1983.

Snow, Theodore P. *The Dynamic Universe: An Introduction to Astronomy*. 3rd ed. Saint Paul: West Publishing, 1988.

Tenreiro, R., and M. Quiros. *Introduction to Cosmology and Particle Physics*. Singapore: World Scientific, 1987.

Trefil, James S. *The Dark Side of the Universe: Searching for the Outer Limits of the Cosmos*. New York: Scribner, 1988.

Tucker, Wallace, and Ricardo Giacconi. *The X-Ray Universe*. Cambridge: Harvard UP, 1987.

Tucker, Wallace, and Karen Tucker. *The Dark Matter: Contemporary Science's Quest for the Mass Hidden in Our Universe*. New York: Morrow, 1988.

Unruh, W. G., ed. *The Early Universe*. Norwell: Kluwer Academic, 1987.

Unsold, A., and R. B. Baschek. *The New Cosmos*. Trans. R. C. Smith, et al. 3rd ed. New York: Springer, 1985.

Van Helden, Albert. *Measuring the Universe: Cosmic Dimensions from Aristarchus to Halley*. Chicago: U of Chicago P, 1986.

Wahlin, Lars. *The Collapsing Universe*. Boulder: Colutron Research, 1985.

Wald, Robert M. *Space, Time and Gravity: The Theory of the Big Bang and Black Holes*. Chicago: U of Chicago P, 1981.

Weinberg, Steven. *The First Three Minutes: A Modern View of the Origin of the Universe*. New York: Basic Books, 1976.

———. *Gravitation and Cosmology*. New York: Wiley, 1972.

Weinberg, Steven, and T. Piran, eds. *Intersection Between Elementary Particle Physics and Cosmology*. Teaneck: World Scientific Publish-

ing, 1986.
West, Richard M., ed. *Understanding the Universe: The Impact of Space Astronomy*. Norwell: Kluwer Academic, 1983.

D. QUANTUM THEORY

Bohm, David. *Quantum Theory*. Englewood Cliffs: Prentice-Hall, 1951.
Bohr, Niels. *Atomic Theory and the Description of Nature*. Cambridge: Cambridge UP, 1934.
Castell, L., et al., eds. *Quantum Theory and the Structure of Time and Space: Papers Presented at a Conference Held in Feldafing, July 1974*. Munich: C. Hanser, 1975.
De Witt, Bryce S. "The Everett-Wheeler Interpretation of Quantum Mechanics." *Batelle Rencontres: Lectures on Mathematics and Physics*. New York: W. A. Benjamin, 1968. 318-32.
Dirac, P. A. M. *Principles of Quantum Mechanics*. 4th ed. Oxford: Oxford UP, 1958.
Finkelstein, David. "Beneath Time: Explorations in Quantum Topology." *The Study of Time III*. Ed. J. T. Fraser, N. Lawrence, and D. Park. New York: Springer, 1978. 94-112.
Heisenberg, Werner. *Physical Principles of the Quantum Theory*. Chicago: U of Chicago P, 1930.
Landsberg, P. T. "Does Quantum Mechanics Exclude Life?" *Nature* 203 (1964): 928-30.
Merzbacher, Eugen. *Quantum Mechanics*. New York: Wiley, 1961.
Ono, Ken-ichi. "On the Origin of Indeterminacy." *The Study of Time II*. Ed. J. T. Fraser, and N. Lawrence. New York: Springer, 1975. 249-57.
Ono, S. "The Long Time Behaviour of Quantum Mechanical Systems." *Memoirs. Faculty of Engineering. Kyushu U* 11 (1949): 125-40.
Park, D. *Introduction to Quantum Theory*. 3rd ed. New York: McGraw-Hill, 1991.
——. "Time in Quantum Mechanics." *Fundamental Questions in Quantum Mechanics*. Ed. L. M. Roth, and A. Inomata. New York: Gordon and Breach, 1986.
Phipps, T. E., Jr. "Time Asymmetry and Quantum Equations of Motion." *Foundations of Physics* 3.4 (1973): 435-55.
Rae, A. I. M. *Quantum Mechanics*. New York: Wiley, 1981.
Salecker, H., and E. P. Wigner. "Quantum Limitations on the Measurement of Space-Time Distances." *Physical Review* 109.2A (January 15, 1958): 571-77.

Zimmerman, E. J. "Time and Quantum Theory." *The Voices of Time*. Ed. J. T. Fraser. 2nd ed. Amherst: U of Massachusetts P, 1981. 479-99.

E. RELATIVITY THEORY

Aoki, Shinko. "A Note on the Variability of Time Standards Due to Relativistic Effects." *Astronomical J* 69 (1964): 2213.

Baltay, C., et al. "Search for Uncharged Faster-than-Light Particles." *Physical Review D* 1.3 (February 1, 1970): 759-70.

Barbier, Osvaldo. *Tempo e relatività: il significato del tempo nella concezione relativistica*. Rome: Bizzari, 1976.

Benton, M. *The Clock Problem (Clock Paradox) in Relativity*. Bethesda: U.S. Naval Research Laboratory, 1959.

Bergson, Henri. *Durée et simultanéité*. 7th ed. Paris: Presses Universitaires de France, 1968.

Bilaniuk, O. M., and E. C. G. Sudarshan. "Particles Beyond the Light Barrier." *Physics Today* 22 (May 1969): 43-51.

Callahan, J. J. "The Curvature of Space in a Finite Universe." *Scientific American* 235 (August 1976): 90-100.

Čapek, Milič. "Time in Relativity Theory: Arguments for a Philosophy of Becoming." *The Voices of Time*. Ed. J. T. Fraser. 2nd ed. Amherst: U of Massachusetts P, 1981. 434-54.

Chaisson, Eric. *Relatively Speaking: Relativity, Black Holes, and the Fate of the Universe*. New York: Norton, 1988.

Costa de Beauregard, Olivier. "Time in Relativity Theory." *The Voices of Time*. Ed. J. T. Fraser. 2nd ed. Amherst: U of Massachusetts P, 1981. 417-33.

Costa de Beauregard, Olivier, and W. F. V. Rosser. *An Introduction to the Theory of Relativity*. London: Butterworths, 1964.

De Felice, F., and C. J. Clarke. *Relativity on Curved Manifolds*. New York: Cambridge UP, 1989.

Einstein, Albert. "The Foundations of the General Theory of Relativity." In H. A. Lorenz, et al. *The Principle of Relativity*. Trans. W. Perrett, and G. B. Jeffrey. New York: Dover Reprint, n.d.

———. "On the Electrodynamics of Moving Bodies." *Annalen der Physik* 17 (1905). Reprinted in H. A. Lorenz, et al. *The Principle of Relativity*. Trans. W. Perrett, and G. B. Jeffrey. New York: Dover Reprint, n.d.

———. "Prinzipielles zur allgemeinen Relativitätstheorie." *Annalen der Physik* 55 (1918): 241-44.

———. *Relativity: The Special and the General Theory*. Trans. R. W. Lawson. New York: Peter Smith, 1920.

———. *Sidelights on Relativity*. London: Methuen, 1922.

Ellis, George F., and Ruth M. Williams. *Flat and Curved Space-Times*. New York: Oxford UP, 1988.

Evett, Arthur. *Understanding the Space-Time Concepts of Special Relativity*. New York: Halsted, 1982.

Feinberg, G. "Possibility of Faster-than-Light Particles." *Physical Review* 159.5 (July 25, 1967): 1089-1105.

Fock, V. *The Theory of Space, Time and Gravitation*. Trans. N. Kemmer. New York: Pergamon, 1959.

Gardner, Martin. *The Relativity Explosion: A Lucid Account of Why Quasars, Pulsars, Black Holes and the New Atomic Clocks are Vindicating Einstein's Revolutionary Theory*. New York: Random House, 1976.

Howard, D., and J. Stachel. *Einstein and the History of General Relativity*. Cambridge: Birkhauser, 1989.

Hughston, L., and K. P. Tod. *Introduction to General Relativity*. New York: Cambridge UP, 1989.

Islam, J. N. *Rotating Fields in General Relativity*. New York: Cambridge UP, 1985.

Kaufman, W. J. III. *The Cosmic Frontiers of General Relativity*. Boston: Little, Brown, 1977.

Landsberg, Peter T. "Concepts in Special Relativistic Thermodynamics." *Essays in Physics* 2 (1970): 93-129.

———. "Time in Statistical Physics and Special Relativity." *The Study of Time I*. Ed. J. T. Fraser, F. C. Haber, and G. H. Müller. New York: Springer, 1972. 59-109.

Levi-Civita, Tullio. *The Absolute Differential Calculus: Calculus of Tensors*. Trans. Marjorie Long. Ed. Enrico Persico. London: Blackie, 1929.

Lorenz, H. A., A. Einstein, H. Minkowski, and H. Weyl. *The Principle of Relativity*. Trans. W. Perrett, and G. B. Jeffrey. New York: Dover Reprint, n.d. [1952].

McVittie, G. C. "General Relativity and Time in the Solar System." *The Study of Time I*. Ed. J. T. Fraser, F. C. Haber, and G. H. Müller. New York: Springer, 1972. 33-38.

———. "Interplanetary Radar Time-Delays in General Relativity." *Astronomy J* (1970): 287-96.

Marder, L. *Time and the Space Traveler*. London: Allen and Unwin, 1971.

Martin, J. L. *General Relativity: A Guide to Its Consequences for Gravity and Cosmology*. New York: Wiley, 1988.

Minkowski, Hermann. "Space and Time." In H. A. Lorenz, et al. *The Principle of Relativity*. Trans. W. Perrett, and G. B. Jeffrey. New York: Dover Reprint, n.d. This is Minkowski's address of 1908.

Misner, C. W., K. S. Thorne, and J. A. Wheeler. *Gravitation*. San Francisco: Freeman, 1973.

North, J. D. "The Time Coordinate in Einstein's Restricted Theory of Relativity." *The Study of Time I*. Ed. J. T. Fraser, F. C. Haber, and G. H. Müller. New York: Springer, 1972. 12-32.

Obanian, H. *Gravitation and Spacetime*. New York: Norton, 1976.

Pauli, W. *The Theory of Relativity*. London: Pergamon, 1958.

Russell, Bertrand. *The ABC of Relativity*. Revised ed. Ed. F. Pirani. New York: New American Library, 1958.

Taylor, E. F., and J. A. Wheeler. *Spacetime Physics*. San Francisco: W. H. Freeman, 1966.

Terletskii, Y. P. *Paradoxes in the Theory of Relativity*. New York: Plenum, 1968.

Weinberg, S. *Gravitation and Cosmology*. New York: Wiley, 1972.

Weyl, Hermann. *Space-Time-Matter*. Trans. H. L. Brose. London: Methuen, 1922.

F. THERMODYNAMICS AND ENTROPY

Blanch, Harvey W., and E. Terry Popoutsakis, eds. *Foundations of Biochemical Engineering: Kinetics and Thermodynamics in Biological Systems*. Washington: American Chemical Society, 1983.

Brittin, W., and G. Gamow. "Negative Entropy and Photosynthesis." *Proceedings of the National Academy of Sciences* 47.5 (1961): 724-27.

Bronstein, M., and L. Landau. "On the Second Law of Thermodynamics and Global Connections in the Cosmos" (in German). *Physikalische Zeitschrift der Sowjetunion* 4 (1933): 114-19.

Burghardt, M. D. *Engineering Thermodynamics with Applications*. 3rd ed. New York: Harper and Row, 1986.

Caldirola, P., ed. *Theories*. New York: Academic P, 1961.

Callen, Herbert B. *Thermodynamics and an Introduction to Thermostatistics*. 2nd ed. New York: Wiley, 1985.

Coe, L. "The Nature of Time." *American J of Physics* 37 (August 1969): 810-15.

Denbigh, K. G., and J. S. Denbigh. *Entropy in Relation to Incomplete Knowledge*. New York: Cambridge UP, 1985.

Ehrenfest, P., and T. Ehrenfest. *The Conceptual Foundations of the*

Statistical Approach in Mechanics. Trans. M. Moravcsik. Ithaca: Cornell UP, 1959.

Ellis, R. S. *Entropy, Large Deviations and Statistical Mechanics*. New York: Springer, 1985.

Fast, J. D. *Entropy*. 2nd ed. New York: Gordon and Breach, 1970.

Fenn, John B. *Engines, Energy, and Entropy: A Thermodynamics Primer*. New York: W. H. Freeman, 1982.

Ford, J. "The Transition from Analytical Dynamics to Statistical Mechanics." *Advances in Chemical Physics* 24 (1973): 155-86.

Hill, Terrell L. *An Introduction to Statistical Thermodynamics*. New York: Dover, 1986.

Holman, Jack P. *Thermodynamics*. 4th ed. New York: McGraw-Hill, 1988.

Honig, J. M. *Thermodynamics: Principles Characterizing Physical and Chemical Processes*. New York: Elsevier, 1982.

Huang, K. *Statistical Mechanics*. New York: Wiley, 1963.

James, Arthur M. *A Dictionary of Thermodynamics*. New York: Halsted P, 1976.

Jones, M. N., ed. *Biochemical Thermodynamics*. New York: Elsevier, 1988.

Justice, J. H., ed. *Maximum Entropy and Bayesian Methods in Applied Statistics*. New York: Cambridge UP, 1986.

Kac, M. *Probability and Related Topics in Physical Sciences*. London: Interscience, 1959.

Kestin, Joseph, ed. *The Second Law of Thermodynamics*. New York: Academic P, 1976.

Khinchin, A. I. *Mathematical Foundations of Statistical Mechanics*. Trans. G. Gamow. New York: Dover, 1944.

Kieffer, S. W., and A. Navrotsky, eds. *Microscopic to Macroscopic: Atomic Environments to Mineral Thermodynamics*. Washington: Mineralogical Society of America, 1985.

Kilmister, C. W., ed. *Disequilibrium and Self-Organisation*. Norwell: Kluwer Academic, 1986.

Kyle, B. G. *Chemical and Process Thermodynamics*. Englewood Cliffs: Prentice-Hall, 1984.

Lamprecht, I., and A. I. Zotin, eds. *Thermodynamics and Pattern Formation in Biology*. Hawthorne: De Gruyter, 1988.

——, eds. *Thermodynamics and Regulation of Biological Processes*. Hawthorne: De Gruyter, 1985.

Landsberg, Peter T. *Entropy and the Unity of Knowledge*. Cardiff: U of Wales P, 1961.

——. "Heat Death and Oscillation in Model Universes Containing Interacting Matter and Radiation." *Astrophysical J* 262 (1982): 432-41.

———. "Thermodynamics, Cosmology, and the Physical Constants." *The Study of Time III*. Ed. J. T. Fraser, N. Lawrence, and D. Park. New York: Springer, 1978. 116-38.

Landsberg, Peter T., and David Park. "Entropy in an Oscillating Universe." *Proceedings of the Royal Society* A346 (1975): 485-95.

Lauffer, M. A. *Entropy-Driven Processes in Biology: Polymerization of Tobacco Mosaic Virus Protein and Similar Reactions*. New York: Springer, 1975.

Lavenda, B. H. *Nonequilibrium Statistical Thermodynamics*. New York: Wiley, 1985.

Lindblad, Goran. *Non-Equilibrium Entropy and Irreversibility*. Norwell: Kluwer Academic, 1983.

Mayer, J. E. "Approach to Thermodynamic Equilibrium." *J of Chemical Engineers* 34 (1961): 1207-23.

Noll, W. *The Foundations of Mechanics and Thermodynamics: Selected Papers*. New York: Springer, 1974.

Oster, G. F., et al., eds. *Irreversible Thermodynamics and the Origin of Life*. New ed. New York: Gordon and Breach, 1974.

Owen, D. R. *A First Course in the Mathematical Foundations of Thermodynamics*. New York: Springer, 1983.

Penrose, O. "Foundations of Statistical Mechanics." *Reports on Progress in Physics* 42 (1979): 1937-2006.

Pippard, A. B. *Elements of Classical Thermodynamics*. New York: Cambridge UP, 1957.

Planck, Max. *Theory of Heat*. London: Macmillan, 1932.

Relf, F. *Fundamentals of Statistical and Thermal Physics*. New York: McGraw-Hill, 1965.

Riedi, P. C. *Thermal Physics: An Introduction to Thermodynamics, Statistical Mechanics, and Kinetic Theory*. 2nd ed. New York: Oxford UP, 1988.

Schlegel, Richard. "Time and Entropy." *Time in Science and Philosophy*. Ed. Jiri Zeman. Prague: Academia, 1971.

———. "Time and Thermodynamics." *The Voices of Time*. Ed. J. T. Fraser. 2nd ed. Amherst: U of Massachusetts P, 1981. 500-23.

Serrin, J., ed. *New Perspectives in Thermodynamics*. New York: Springer, 1986.

Smith, C. Ray, and W. T. Grandy, Jr., eds. *Maximum-Entropy and Bayesian Methods in Inverse Problems*. Norwell: Kluwer Academic, 1985.

Sonntag, Richard E., and Gordon J. Van Wylen. *Introduction to Thermodynamics: Classical and Statistical*. 2nd ed. New York: Wiley,

1988.
Ubbelohde, Alfred René. *Time and Thermodynamics*. London: Oxford UP, 1947.
Verboven, E. J. "Quantum Thermodynamics of an Infinite System of Harmonic Oscillators." *Statistical Mechanics*. Ed. T. A. Bak. New York: Benjamin, 1967. 49-54.

G. TIME, SPACE, AND SPACE-TIME

Abe, S. "Experimental Study on the Correlation between Time and Space." *Tohoku Psychologica Folia* 3 (1935): 53-68.
Akhundov, M. D. *Conceptions of Space and Time*. Trans. Charles Rougle. Cambridge: MIT P, 1986.
Benedicks, Carl Axel Fredrik. *Space and Time: An Experimental Physicist's Conception of These Ideas and of Their Alteration*. London: Methuen, 1924.
Blokhintsev, Dmitrii Ivanovich. *Space and Time in the Microworld*. Trans. Zdenka Smith. Dordrecht: Reidel, 1973.
Borel, Émile Félix Édouard Justin. *Space and Time*. Trans. Angelo S. Rappoport, and John Dougall. New York: Dover, 1960.
Costa de Beauregard, Olivier. *La notion de temps: Équivalence avec l'espace*. Paris: Hermann, 1963.
Davies, P. C. W. *Space and Time in the Modern Universe*. Cambridge: Cambridge UP, 1977.
Earman, John, et al., eds. *Foundations of Space-Time Theories*. Minneapolis: U of Minnesota P, 1977.
Esposito, Paul, and Louis Witten, eds. *Asymptotic Structures of Space-Time*. New York: Plenum, 1977.
Galli, Mario Giuseppe. *Spazio e tempo nella scienza moderna*. 2 vols. Florence: Baccini and Chiappi, 1965-67.
Grünbaum, A. *Philosophical Problems of Space and Time*. New York: Knopf, 1963.
Ingraham, R. L. "Stochastic Space-Time." *Nuovo Cimento* 34 (1964): 182-97.
Jammer, Max. *Concepts of Space: The History of Theories of Space in Physics*. Foreword by Albert Einstein. 2nd ed. Cambridge: Harvard UP, 1969.
Jones, R. V. "Time and Distance." *Horological Journal* 128 (December 1985): 6-8, 14; and 128 (February 1986): 10-11.
Mermin, N. David. *Space and Time in Special Relativity*. Prospect Heights: Waveland P, 1989.

Noll, W. "Space-Time Structures in Classical Mechanics." *Delaware Seminar in the Foundations of Physics*. Ed. M. Bunge. Berlin: Springer, 1967.

Ono, K. "On the Fundamental Laws of Nature and the Role of Time in Time-Space Manifold." *The Study of Time III*. Ed. J. T. Fraser, N. Lawrence, and D. Park. New York: Springer, 1978. 141-64.

Palágyi, Menyhért. *Neue Theorie des Raumes und der Zeit. Die Grundbegriffe einer Metageometrie*. 1901; reprint, Darmstadt: Wissenschaftliche Buchgesellschaft, 1967. 48 pages.

Reichenbach, Hans. *The Philosophy of Space and Time*. Trans. Maria Reichenbach, and John Freund. New York: Dover, 1958.

Robb, A. A. *The Absolute Relations of Time and Space*. Cambridge: Cambridge UP, 1921.

——. *Geometry of Time and Space*. Cambridge: Cambridge UP, 1936.

Sklar, Lawrence. *Space, Time and Spacetime*. Berkeley: U of California P, 1974.

Suppes, Patrick Colonel, ed. *Space, Time and Geometry*. Dordrecht: Reidel, 1973.

Szamosi, Gesa. *The Twin Dimensions: Inventing Time and Space*. New York: McGraw-Hill, 1986.

Trefil, James F. *Space, Time, Infinity*. Washington: Smithsonian Books, 1985.

Vilnitskij, M. *On the History of the Concepts of Space and Time in Classical Physics* (in Russian). Kiev: Nauka, 1955.

H. TIME'S ARROW

Aharony, A. "Time Reversal, Symmetry Violation and the H-Theorem." *Physics Letters* 37A (1971): 45-46.

Blum, Harold F. *Time's Arrow and Evolution*. 3rd ed. Princeton: Princeton UP, 1968.

Bunge, M. "Time Asymmetry, Time Reversal, and Irreversibility." *The Study of Time I*. Ed. J. T. Fraser, F. C. Haber, and G. H. Müller. New York: Springer, 1972. 122-30.

Cocke, W. J. "Statistical Time Symmetry and Two-Time Boundary Conditions in Physics and Cosmology." *Physical Review* 160 (1967): 1165-70.

Costa de Beauregard, Olivier. "Irreversibility Problems." *Logic, Methodology and Philosophy of Science*. Ed. Y. Bar-Hillel. Amsterdam: North-Holland, 1965. 313-42.

——. "No Paradox in the Theory of Time Anisotropy." *The Study of*

Time I. Ed. J. T. Fraser, F. C. Haber, and G. H. Müller. New York: Springer, 1972. 131-39.

Davies, P. C. W. "Inflation in the Universe and Time Asymmetry." *Nature* 312 (1984): 524-26.

——. *The Physics of Time Asymmetry.* Berkeley: U of California P, 1974.

Dako, M. "The Direction of Time." *Studium Generale* 22 (1969): 965-84.

Denbigh, K. G. "In Defence of *the* Direction of Time." *The Study of Time I.* Ed. J. T. Fraser, F. C. Haber, and G. H. Müller. New York: Springer, 1972. 148-58.

Earman, John. "The Anisotropy of Time." *Australasian J of Philosophy* 47 (December 1969): 273-95.

——. "Irreversibility and Temporal Asymmetry." *J of Philosophy* 64.18 (September 1967): 543-49.

Gal-Or, B. "The Crisis About the Origin of Irreversibility." *Science* 176 (1972): 11-17.

Gold, T. "The Arrow of Time." *American J of Physics* 30 (1962): 403-10.

Grünbaum, Adolf. "The Anisotropy of Time." *The Nature of Time.* Ed. Thomas Gold. Ithaca: Cornell UP, 1967.

Hobson, A. "Further Comments on Irreversibility." *Physics Letters* 28A (1968): 183-84.

——. "Irreversibility in Finite Quantum-Statistical-Mechanical Systems." *J of Chemical Physics* 46 (1967): 1365-72.

——. "Irreversibility in Simple Systems." *American J of Physics* 34.5 (May 1966): 411-16.

Horwich, P. *Asymmetries in Time: Problems in the Philosophy of Science.* Cambridge: MIT P, 1987.

Kampen, N. G. van. "Fundamental Problems in the Statistical Mechanics of Irreversible Processes." *Fundamental Problems in Statistical Mechanics I.* Ed. E. G. D. Cohen. Amsterdam: North-Holland, 1962. 173-202.

Landsberg, P. T. "Time Reversibility." *Solid State Theory: Methods and Applications.* Ed. P. T. Landsberg. New York: Wiley, 1969.

Layzer, D. "The Arrow of Time." *Scientific American* 233 (December 1975): 56-69.

Lewis, G. N. "The Symmetry of Time in Physics." *Science* 71 (1930): 569-77.

Mehlberg, Henryk. "Physical Laws and Time's Arrow." *Current Issues in the Philosophy of Science.* Ed. Herbert Feigl, and Grover Maxwell. New York: Holt, Rinehart, 1961.

Morris, Richard. *Time's Arrows: Scientific Attitudes Toward Time.* New

York: Simon and Schuster, 1984.

Morrison, P. "Time's Arrow and External Perturbations." *Preludes in Theoretical Physics*. Ed. A. de-Shalit, H. Feshbach, and L. van Hove. Amsterdam: North-Holland, 1966. 347-51.

Penrose, O., and I. C. Percival. "The Direction of Time." *Proceedings. Physical Society* 79 (1962): 605-16.

Pompe, W., and H.-G. Schöpf. "Kosmologische und statistische Überlegungen zur Irreversibilität." *Annalen der Physik* 21 (1968): 26-30.

Popper, K. "Time's Arrow and Entropy." *Nature* 207 (1965): 233-34.

Prigogine, I. *Introduction to Thermodynamics of Irreversible Processes*. New York: Wiley, 1967.

Prior, A. N. *Past, Present and Future*. Oxford: Oxford UP, 1967.

Reichenbach, Hans. *The Direction of Time*. Ed. Maria Reichenbach. Berkeley: U of California P, 1956.

Rosenfeld, L. "General Introduction to Irreversibility." *Irreversibility in the Many-Body Problem*. Ed. J. Biel, and J. Rae. New York: Plenum, 1972.

——. "Questions of Irreversibility and Ergodicity." *Proceedings of the International School of Physics "Enrico Fermi," XIV Course*. New York: Academic P, 1961.

Sachs, Robert G. *The Physics of Time Reversal*. Chicago: U of Chicago P, 1987.

Schwarze, S. C. "The Branch System Hypothesis: A Critique." *The Study of Time IV*. Ed. J. T. Fraser, N. Lawrence, and D. Park. New York: Springer, 1981. 213-22.

Stannard, R. F. "Symmetry of the Time Axis." *Nature* 211 (1966): 693-94.

Weizsäcker, C. F. von. "Der zweite Hauptsatz und der Unterschied von Vergangenheit und Zukunft." *Annalen der Physik* 36 (1939): 275-83.

Whyte, L. L. "Pierre Curie's Principle of One-Way Process." *The Study of Time I*. Ed. J. T. Fraser, F. C. Haber, and G. H. Müller. New York: Springer, 1972. 140-47.

Zeh, H. D. *The Physical Basis of the Direction of Time*. Berlin: Springer, 1989.

20. POLITICAL SCIENCE

Adelman, I., and C. Morris. *Society, Politics and Economic Development*. Baltimore: Johns Hopkins P, 1967.

Almond, Gabriel A., and Sidney Verba. *Civic Culture: Political Attitudes and Democracy in Five Nations*. Princeton: Princeton UP, 1963.

Aron, Raymond. *Politics and History*. Ed. Miriam B. Conant. New Brunswick: Transaction Publications, 1984.

Aronoff, Myron J., et al., eds. *State Formation and Political Legitimacy*. New Brunswick: Transaction Publications, 1987.

Baber, Walter F. *Organizing the Future: Matrix Models for the Postindustrial Polity*. Tuscaloosa: U of Alabama P, 1983.

Balbus, Isaac D. *The Dialectics of Legal Repression: Black Rebels Before the American Criminal Courts*. New York: Russell Sage, 1973.

Ball, Terence, ed. *Idioms of Inquiry: Critique and Renewal in Political Science*. Albany: State U of New York P, 1987.

Binder, Leonard, et al., eds. *Crises and Sequences in Political Development*. Princeton: Princeton UP, 1971.

Campbell, Angus. "Politics through the Life Cycle." *The Gerontologist* 11.2 Part 1 (Summer 1971): 112-17.

———. "Social and Psychological Determinants of Voting Behavior." *Politics of Age*. Ed. Wilma Donahue, and Clark Tibbitts. Ann Arbor: Division of Gerontology, U of Michigan, 1962. 87-100.

Carlie, Michael K. "The Politics of Age: Interest Group or Social Movement?" *The Gerontologist* 9.4 Part 1 (Winter 1969): 259-63.

Catanese, Anthony J. *Planners and Local Politics: Impossible Dreams*. Newbury Park: Sage, 1974.

Charlton, Sue E., et al., eds. *Women, the State, and Development*. Albany: State U of New York P, 1989.

Cohen, David. *The Political Process*. New York: American Jewish Committee, 1981.

Corning, P. A. *The Synergism Hypothesis: A Theory of Progressive Evolution*. New York: McGraw-Hill, 1983.

Cottrell, Fred. "Aging and the Political System." *Aging and Social Policy*. Ed. John C. McKinney, and Frank de Vyver. New York: Appleton-Century-Crofts, 1966. 77-113.

———. "Governmental Functions and the Politics of Age." *Handbook of Social Gerontology*. Ed. Clark Tibbitts. Chicago: U of Chicago P, 1960. 627-65.

Cox, Harold. "The Motivation and Political Alienation of Older Americans." *International J of Aging and Human Development* 11.1 (1980): 1-12.

Crittendon, John. "Aging and Party Affiliation." *Public Opinion Quarterly* 26.1 (1962): 648-57.

——. "Aging and Political Participation." *The Western Political Quarterly* 16.2 (1963): 323-31.

Cutler, Neal E., ed. *Aging and Public Policy*. Croton-on-Hudson: Policy Studies, 1984.

Cutler, Neal E., and Vern L. Bengtson. "Age and Political Alienation: Maturation, Generation and Period Effects." *The Annals of the American Academy of Political and Social Sciences* 415 (1974): 160-75.

Cutler, Neal E., and John R. Schmidhauser. "Age and Political Behavior." *Aging: Scientific Perspectives and Social Issues*. Ed. Diana S. Woodruff, and James E. Birren. New York: Van Nostrand, 1975. 374-403.

Cutler, Stephen J., and Robert L. Kaufman. "Cohort Changes in Political Attitudes: Tolerance of Ideological Nonconformity." *The Public Opinion Quarterly* 39.1 (1975): 69-81.

Cutler, Stephen J., et al. "Aging and Conservatism: Cohort Changes in Attitudes about Legalized Abortion." *J of Gerontology* 35.1 (1980): 115-23.

Darby, W. Thomas. *The Feast: Meditations on Politics and Time*. Toronto: U of Toronto P, 1982.

Dauenhauer, Bernard P. *The Politics of Hope*. New York: Routledge, Chapman and Hall, 1986.

Dickerson, M. O., et al., eds. *Problems of Change in Urban Government*. Atlantic Highlands: Humanities P, 1980.

Foner, Anne. "Age Stratification and Age Conflict in Political Life." *American Sociological Review* 39.2 (1974): 187-96.

Glenn, Norval D. "Aging and Conservatism." *The Annals of the American Academy of Political and Social Sciences* 415 (1974): 176-86.

Glenn, Norval D., and Michael Grimes. "Aging, Voting, and Political Interest." *American Sociological Review* 33.4 (1968): 563-75.

Gubrium, Jaber F. "Continuity in Social Support, Political Interest, and Voting in Old Age." *The Gerontologist* 12.4 (1972): 421-23.

Gunnell, John G. "The History of Political Philosophy and the Myth of the Tradition." *The Study of Time II*. Ed. J. T. Fraser, and N. Lawrence. New York: Springer, 1975. 283-94.

——. *Political Philosophy and Time*. Middletown: Wesleyan UP, 1968.

Hudson, Robert B., and Robert H. Binstock. "Political Systems and Aging." *Handbook of Aging and Social Sciences*. Ed. Robert H.

Binstock, and Ethel Shanas. New York: Van Nostrand, 1976. 369-400.

Jacob, Margaret C. *The Newtonians and the English Revolution 1689-1720*. Hassocks, England: Harvester P, 1976.

Jacob, Margaret C., and W. A. Lockwood. "Political Millenarianism and Burnet's *Sacred Theory*." *Science Studies* 2 (1972): 265-79.

Johnson, Gregory, and Lawrence J. Kamara. "Growing Up and Growing Old: The Politics of Age Exclusion." *International J of Aging and Human Development* 8.2 (1977-78): 99-110.

Le Goff, Jacques. "Is Politics Still the Backbone of History?" *Daedalus* 100 (1971): 1-13.

Lewis, Arthur M. *Evolution—Social and Organic: Linnaeus, Darwin, Kropotkin, Marx*. Chicago: Charles H. Kerr, 1984.

McIntyre, Angus, ed. *Aging and Political Leadership*. Albany: State U of New York P, 1988.

Michelson, William, ed. *Public Policy in Temporal Perspective*. New York: Mouton, 1978.

Moe, Terry M. *The Organization of Interests: Incentives and the Internal Dynamics of Political Interests Groups*. Chicago: U of Chicago P, 1980.

Morgan, Edmund S. *Inventing the People: The Rise of Popular Sovereignty in England and America*. New York: Norton, 1988.

Morrill, R. *Political Redistricting and Geographic Theory*. Washington: Association of American Geographers, 1981.

Mouritzen, Hans. *Finlandization: Towards a General Theory of Adaptive Politics*. Brookfield: Gower, 1988.

Muller, Virginia L. *The Idea of Perfectibility*. Lanham: UP of America, 1986.

Orr, Robert. "The Time Motif in Macchiavelli." *Macchiavelli and the Nature of Political Thought*. Ed. M. Fleisher. New York: Atheneum, 1972.

Petty, Sir William. *Another Essay in Political Arithmetic...1682*. In *The Economic Writings*. Ed. Charles Henry Hull. 1899; reprint, New York: Augustus M. Kelley, 1963. Vol. 1.

Pocock, J. G. A. *Politics, Language and Time: Essays on Political Thought and History*. New York: Atheneum, 1971.

Ragan, Pauline K., and James J. Dowd. "The Emerging Political Consciousness of the Aged: A General Interpretation." *J of Social Issues* 30.3 (1974): 137-58.

Read, Leonard E. *Seeds of Progress*. Irvington-on-Hudson: Foundations for Economic Education, 1980.

Reid, H. "The Politics of Time." *Human Context* 4 (1972): 456-508.

Rosen, Michael. *Hegel's Dialectic and Its Criticism*. New York: Cambridge UP, 1985.

Roskin, Michael G., et al. *Political Science: An Introduction*. 3rd ed. Englewood Cliffs: Prentice-Hall, 1988.

Saxonhouse, Arlene W. *Women in the History of Political Thought: Ancient Greece to Machiavelli*. New York: Praeger, 1985.

Schaefer, David L., ed. *The New Egalitarianism: Questions and Challenges*. New York: Associated Faculty P, 1979.

Schluchter, Wolfgang. *The Rise of Western Rationalism: Max Weber's Developmental History*. Trans. Günther Roth. Berkeley: U of California P, 1981.

Schmidt, Helmut. *A Grand Strategy for the West: The Anachronism of National Strategies in an Interdependent World*. New Haven: Yale UP, 1985.

Schneider, Gerald. *How Far Do Governments Look Ahead?—A Comparative Analysis of the Factors Contributing to the Variance in Time Horizons of 40 States*. Frankfurt: Haag and Herchen, 1989.

Sherover, Charles M. "The Temporality of the Common Good: Futurity and Freedom." *Review of Metaphysics* 37.3 (March 1984): 475-97.

——. *Time, Freedom, and the Common Good: An Essay in Public Philosophy*. Albany: State U of New York P, 1989.

Smith, T. Alexander. *Time and Public Policy*. Knoxville: U of Tennessee P, 1989.

Stabile, Donald. *Prophets of Order: The Rise of the New Class, Technocracy and Socialism in America*. Boston: South End P, 1984.

Stohl, Michael, and George A. Lopez, eds. *The State As Terrorist: The Dynamics of Governmental Violence and Repression*. Westport: Greenwood, 1984.

Strauss, Leo. *History of Political Philosophy*. Chicago: Rand McNally, 1972.

——. *Natural Right and History*. Chicago: U of Chicago P, 1953.

Strauss, Leo, and Joseph Cropsey, eds. *History of Political Philosophy*. 3rd ed. Chicago: U of Chicago P, 1987.

Tokar, Brian. *The Green Alternative: Creating an Ecological Future*. San Pedro: R. and E. Miles, 1987.

Truman, David B. *The Governmental Process: Political Interests and Public Opinion*. 1951; reprint, Westport: Greenwood, 1981.

Wendorff, Rudolf. *Dritte Welt und westliche Zivilisation. Grundprobleme der Entwicklungspolitik*. Opladen: Westdeutscher Verlag, 1984.

21. PSYCHOLOGY

A. GENERAL

Aaronson, B. S. "Time, Time Stance, and Existence." *The Study of Time I*. Ed. J. T. Fraser, F. C. Haber, and G. H. Müller. New York: Springer, 1972. 293-311.

Blatt, S. J., and P. Quinlan. "The Psychological Effects of Rapid Shifts in Temporal Referents." *The Study of Time I*. Ed. J. T. Fraser, F. C. Haber, and G. H. Müller. New York: Springer, 1972. 506-22.

Block, Richard A., ed. *Cognitive Models of Psychological Time*. Hillsdale: Erlbaum, 1990.

——. "Models of Psychological Time." *Cognitive Models of Psychological Time*. Ed. Richard A. Block. Hillsdale: Erlbaum, 1990. 1-36.

Block, R. A., J. L. Saggau, and L. H. Nickol. "Temporal Inventory on Meaning and Experience: A Structure of Time." *Imagination, Cognition and Personality* 3 (1983-84): 203-25.

Bolton, T. L. "Rhythm." *American J of Psychology* 6 (1894): 145-238.

Boring, L. D., and E. G. Boring. "Temporal Judgments after Sleep." *Studies in Psychology*. Titchener Commemorative Volume. Worcester: Wilson, 1917. 255-79.

Boulter, L. R., and M. H. Appley. "Time and Effort as Determiners of Time-Production Error." *J of Experimental Psychology* 75.4 (1967): 447-52.

Brecher, G. A. "Die Entstehung und biologische Bedeutung der subjektiven Zeitenheit—des Momentes." *Zeitschrift für Vergleichende Psychologie* 18 (1932): 204-43.

Bruneau, Tom. "Personal Time and Self Identity." *Tempo e identità*. Ed. P. Reale. Milan: F. Angeli, 1988. 102-115.

Campbell, Scott S. "Circadian Rhythms and Human Temporal Experience." *Cognitive Models of Psychological Time*. Ed. Richard A. Block. Hillsdale: Erlbaum, 1990. 101-18.

——. "Duration and Placement of Sleep in a 'Disentrained' Environment." *Psychophysiology* 21 (1984): 106-13.

Chiselin, M. T. "Darwin and Evolutionary Psychology." *Science* 179 (1973): 964-68.

Cohen, J. *Psychological Time in Health and Disease*. Springfield: Charles C. Thomas, 1967.

Cottle, Thomas J., and S. L. Klineberg. *The Present of Things Future*. New York: Free P, 1974.

De Long, A. J. "Phenomenological Space-Time: Toward an Experiential Relativity." *Science* 213 (1981): 681-83.

Dewey, J. *Time and Its Mysteries*. New York: Collier, 1962.

Dietze, G. "Untersuchungen über den Umfang des Bewusstseins bei regelmässig aufeinanderfolgenden Schalleindrücken." *Philosophische Studien* 2 (1885): 362-93.

Dimond, S. J. "The Structural Basis of Timing." *Psychological Bulletin* 62 (1964): 348-50.

Dobbs, H. A. C. "The Dimensions of the Sensible Present." *The Study of Time I*. Ed. J. T. Fraser, F. C. Haber, and G. H. Müller. New York: Springer, 1972. 274-92.

Eisler, H., et al. "A Complementary Bibliography on the Psychology of Time." *JSAS: Catalog of Selected Documents in Psychology* 10, 55 (1980). Ms. No. 2101.

Fraisse, P. "Elements de chronopsychologie." *Le travail humain* 43 (1980): 353-72.

——. *Psychologie du temps*. 2nd ed. Paris: Presses Universitaires de France, 1967.

——. *The Psychology of Time*. Trans. Jennifer Leith. New York: Harper and Row, 1963.

——. "Rhythm and Tempo." *The Psychology of Music*. Ed. D. Deutsch. New York: Academic P, 1982. 149-80.

——. *Les structures rhythmiques*. Louvain: Studia Psychologica, 1956.

——. "Time and Rhythm Perception." *Handbook of Perception*. Ed. E. C. Carterette, and M. P. Friedman. New York: Academic P, 1978. 8:203-54.

——. "Time in Psychology." *Time and the Sciences*. Ed. F. Greenaway. Paris: UNESCO, 1979. 71-84.

Fraisse, Paul, et al. *Du temps biologique au temps psychologique*. Paris: Presses Universitaires de France, 1979.

——. "Vergleich der Zeitschätzungsmethoden." *Zeitschrift für Psychologie* 167 (1962): 268-77.

Friedman, W. J. *About Time: Inventing the Fourth Dimension*. Cambridge: MIT P, 1990.

Fröhlich, F. W. *Die Empfindungszeit: Ein Beitrag zur Lehre von der Zeit, Raum- und Bewegungsempfindung*. Jena: Gustav Fischer, 1929.

Fröhlich, W. D. "Sozialisation und kognitive Stile: Einige Denkmöglichkeiten und Befunde." *Handbuch der Psychologie, Vol. 7*. Ed. C. F. Graumann. Göttingen: Hogrefe, 1972.

Gooddy, William. *Time and the Nervous System*. New York: Praeger, 1988.

Gorman, B. S., and A. E. Wessman. "Images, Values, and Concepts of Time in Psychological Research." *The Personal Experience of*

Time. Ed. B. S. Gorman, and A. E. Wessman. New York: Plenum, 1977. 217-63.

——. *The Personal Experience of Time*. New York: Plenum, 1977.

Gorman, B. S., et al. "Linear Representation of Temporal Location and Stevens's Law." *Memory and Cognition* 1 (1973): 169-71.

Green, J., et al. "The Effect of Desynchronization on Meal Patterns of Humans Living in Time Isolation." *Physiology and Behavior* 39.2 (1987): 203-09.

Guyau, Jean Marie. *La genèse de l'idée de temps*. 2nd ed. Paris: Félix Alcan, 1902. Also in J. A. Michon, et al., eds. *Guyau and the Idea of Time*, 1988. See below.

Halberg, F. "Physiologic Considerations Underlying Rhythmometry, with Special Reference to Emotional Illness." *Symposium Bell-Air III*. Geneva: Masson, 1968. 73-126.

Hall, E. T. *The Dance of Life: The Other Dimension of Time*. Garden City: Anchor P, 1983.

Hall, G. S., and J. Jastrow. "Studies of Rhythm." *Mind* 11 (1886): 55-62.

Helson, H. "The Tau-Effect—An Example of Psychological Relativity." *Science* 71 (1930): 536-37.

Hoagland, H. "Consciousness and the Chemistry of Time." *Problems of Consciousness*. Ed. H. A. Abramson. New York: Josiah Macy, 1951. 164-78.

——. *Pacemakers in Relation to Aspects of Behavior*. New York: Macmillan, 1935.

Jackson, J. L. "Is the Processing of Temporal Information Automatic or Controlled?" *Time, Mind, and Behavior*. Ed. J. A. Michon, and J. L. Jackson. Heidelberg: Springer, 1985. 179-90.

James, W. *The Principles of Psychology*. Vol. 1. New York: Henry Holt, 1890. Includes a useful chapter on time.

Jones, M. R. "Some Thoughts on the Relevance of Bergson to Contemporary Psychology." *The Legacy of Henri Bergson: Towards a Unification of the Sciences*. Ed. A. C. Papanicolacu, and P. A. Y. Gunter. New York: Gordon and Breach, 1987. 250-70.

Jones, M. R., L. Summerell, and E. Marshburn. "Recognizing Melodies: A Dynamic Interpretation." *Quarterly J of Experimental Psychology* 39 (1987): 89-121.

Killeen, Peter R. "Incentive Theory III: Adaptive Clocks." *Annals of the New York Academy of Sciences* 423 (1984): 515-27.

Koffka, K. "Experimental-Untersuchungen zur Lehre des Rhythmus." *Zeitschrift für Psychologie* 52 (1909): 1-109.

Köhler, W. "Zur Theorie des Sukzessivvergleichs und der Zeitfehler."

Psychologische Forschung 4 (1923): 115-75.
Krauss, H. H. "Anxiety: The Dread of a Future Event." *J of Individual Psychology* 23 (1967): 88-93.
Lauth, Reinbord. *Die Konstitution der Zeit im Bewusstsein*. Hamburg: Meiner, 1981.
Lévy, J. C. *Le temps psychologique*. Paris: Dunod, 1969.
Lichtenstein, M. "Phenomenal Simultaneity with Irregular Timing of Components of the Visual Stimulus." *Perceptual and Motor Skills* 12 (1961): 47-60.
Macar, F. *Le temps—perspectives psychophysiologiques*. Brussels: Mardaga, 1980.
McGrath, Joseph E., ed. *Research Toward a Psychology of Time*. Beverly Hills: Sage, 1986.
Mach, E. *Die Analyse der Empfindungen und das Verhältnis des Physischen zum Psychischen*. Jena: G. Fischer, 1919.
Mann, H., M. Siegler, and H. Osmond. "The Many Worlds of Time." *J of Analytical Psychology* 13.1 (1968): 33-56.
Melges, F. T. "Guyau on the Illusions of Time: Normal and Pathological." *Guyau and the Idea of Time*. Ed. J. A. Michon, V. Pouthas, and J. L. Jackson. Amsterdam: North-Holland, 1988. 213-31.
——. "Time, Timing, and the Future." *Readings: A Journal of Review and Commentary in Mental Health* 3 (1988): 13-17.
Meredith, P. "The Psychophysical Structure of Temporal Information." *The Study of Time I*. Ed. J. T. Fraser, F. C. Haber, and G. H. Müller. New York: Springer, 1972. 259-73.
Michon, John A. "The Compleat Time Experiencer." *Time, Mind, and Behavior*. Ed. J. A. Michon, and J. L. Jackson. Heidelberg: Springer, 1985.
——. "Contemporary Approaches to Human Time Experience." *Teorie e Modelli* 2, Supplement 1 (1985): 29-47.
——. "Holes in the Fabric of Subjective Time: Figure Ground Relations in Event Sequences." *Acta Psychologica* 41 (1977): 191-203.
——. "Implicit and Explicit Representations of Time." *Cognitive Models of Psychological Time*. Ed. Richard A. Block. Hillsdale: Erlbaum, 1990. 37-58.
——. "A Natural History of Time" (in Dutch). *Groniek (Gronings Historisch Tijdschrift)* 103-04 (1989): 69-80.
——. "Temporality and Metaphor." *Time, Mind, and Behavior*. Ed. J. A. Michon, and J. L. Jackson. Berlin: Springer, 1985. 288-95.
——. *Timing in Temporal Tracking*. Assen: Van Gorcum, 1967.
——. "Timing Your Mind and Minding Your Time." *Time and Mind: The Study of Time VI*. Ed. J. T. Fraser. Madison: International

Universities P, 1989. 17-39.

——. "Le traitement de l'information temporelle." *De temps biologique au temps psychologique*. Ed. P. Fraisse, et al. Paris: Presses Universitaires de France, 1977. 255-87.

Michon, John A., and J. L. Jackson, eds. *Time, Mind, and Behavior*. Berlin: Springer, 1985.

Michon, J. A., V. Pouthas, and J. L. Jackson, eds. *Guyau and the Idea of Time*. Amsterdam: North-Holland, 1988.

Minkowski, Eugene. *Lived Time*. Trans. Nancy Metzel. Evanston: Northwestern UP, 1970.

Moenks, F. "Zeitperspektive als psychologische Variable." *Archiv für die gesamte Psychologie* 119 (1967): 131-61.

Moulyn, Adrian C. *The Meaning of Suffering: An Interpretation of Human Existence from the Viewpoint of Time*. Westport: Greenwood, 1982.

Mucciarelli, G., ed. *Psychology of Time*. Bologna: Pitagora, 1985.

Nichols, H. "The Psychology of Time." *American J of Psychology* 3 (1891): 453-529.

Pöppel, Ernst. "Excitability Cycles in Central Intermittency." *Psychologische Forschung* 34 (1970): 1-9.

——. *Mindworks: Time and Conscious Experience*. Trans. Tom Artin. Boston: Harcourt Brace, 1988.

——. "Tagesperiodische Veränderungen der akustischen Adaptation und des psychomotorischen Tempos mit und ohne Nachtruhe." *Pflügers Archiv* 300 (1968): R 11-R 12.

Pöppel, Ernst, J. C. Aschoff, and H. Giedke. "Tagesperiodische Veränderungen der Reaktionszeit bei Wahlreaktionen." *Zeitschrift für Experimentelle und Angewandte Psychologie* 17 (1970): 537-52.

Postman, L., and G. A. Miller. "Anchoring in Temporal Judgments." *American J of Psychology* 58 (1945): 43-53.

Quandt, J. "Das Problem des Zeitbewusstseins." *Archiv für die gesamte Psychologie* 8 (1906): 143-89.

Reale, Paola. *La psicologia del tempo*. Turin: Boringhieri, 1982.

Rizzi, Pietro. *I percorsi del tempo: Sulla psicogenesi della temporalità*. Milan: Uniccopli, 1988.

Rosenthal, Leslie, and Margot C. Howe. "A Comparison of Temporal Adaptation Among Day Versus Night Shift Workers." *Occupational Therapy in Mental Health* 4.2 (1984): 59-78.

Rotter, G. S. "Clock Speed as an Independent Variable in Psychological Research." *J of General Psychology* 81.1 (1969): 45-52.

Salmon, P. *Living in Time*. London: Dent, 1985.

Schmidt, R. A. "Anticipation and Timing in Human Motor Performance." *Psychological Bulletin* 70 (1968): 631-46.

——. "Movement Time as a Determiner of Timing Accuracy." *J of Experimental Psychology* 79 (1969): 43-47.

Scholz, W. von. *Der Zufall und das Schicksal*. Munich: P. List, 1959.

Senden, M. von. *Raum- und Gestaltauffassung bei operierten Blindgeborenen vor und nach der Operation*. Leipzig: Barth, 1932.

Shallice, T. "The Detection of Change and the Perceptual Moment Hypothesis." *British J of Statistical Psychology* 17 (1964): 113-35.

Siffre, M. *Beyond Time*. London: Chatto and Windus, 1965.

Sixtl, F. "Der Zeitfehler [time-order error] beim Schätzen der Reizzeit und als Funktion der Reizlänge der Intervallzeit und der Versuchswiederholung." *Zeitschrift für Experimentelle und Angewandte Psychologie* 10 (1963): 209-25.

Slife, Brent D. "Psychology's Reliance on Linear Time: A Reformation." *J of Mind and Behavior* 2.1 (1981): 27-46.

Stelmach, George E., Patricia A. Mullins, and Hans-Leo Teulings. "Motor Programming and Temporal Patterns in Handwriting." *Annals of the New York Academy of Sciences* 423 (1984): 144-57.

Stephenson, William. "William James, Neils Bohr, and Complementarity: IV. The Significance of Time." *Psychological Record* 38.1 (1988): 19-35.

Stern, W. "Psychische Präsenzzeit." *Zeitschrift für Psychologie* 13 (1897): 325-49.

Stevens, S. S. "On the Psychophysical Law." *Psychological Review* 64.3 (May 1957): 153-81.

Straus, Erwin Walter Maximilian. *Man, Time and World*. Trans. Donald Moss. Pittsburgh: Duquesne UP, 1982.

Temperley, N. M. "Personal Tempo and Subjective Accentuation." *J of General Psychology* 68 (1963): 267-87.

Tillich, Hannah. *From Time to Time*. New York: Stein and Day, 1973.

Treisman, Michel. "Temporal Rhythms and Cerebral Rhythms." *Annals of the New York Academy of Sciences* 423 (1984): 542-65.

Underwood, B. J., and R. A. Malmi. "An Evaluation of the Measures Used in Studying Temporal Codes for Words within a List." *J of Verbal Learning and Verbal Behavior* 17 (1978): 279-93.

Westman, Alida S. "Development of Time Concepts: Differentiating Clock and Calendar from Apparent Durations." *J of Genetic Psychology* 148.3 (1987): 259-70.

Wolff, P. "Trace Quality in the Temporal Ordering of Events." *Perceptual and Motor Skills* 22 (1966): 283-86.

Woodrow, H. "The Temporal Indifference Interval Determined by the Method of Mean Error." *J of Experimental Psychology* 17 (1934):

167-88.

Yaker, Henri M., et al., eds. *The Future of Time: Man's Temporal Environment*. Garden City: Doubleday, 1971.

B. BEHAVIOR

Aaronson, B. S. "Behavior and the Place Names of Time." *American J of Clinical Hypnosis* 9.1 (1966): 1-17.

Agarwal, Adesh, and Manju Srivastava. "Significance of Time Perspective in Reflection-Impulsivity." *Psychological Studies* 26.2 (1981): 100-03.

Agarwal, Adesh, and Sushma Tiwari. "Future Orientation: A Mediator in Temporal Coding." *International J of Psychology* 23.2 (1988): 151-63.

Agarwal, Adesh, and Kaushal K. Tripathi. "Influence of Prolonged Deprivation, Age and Culture on the Development of Future Orientation." *European J of Social Psychology* 14.4 (1984): 451-53.

Agarwal, Adesh, Kaushal K. Tripathi, and Manju Srivastava. "Social Roots and Psychological Implications of Time Perspective." *International J of Psychology* 18.5 (1983): 367-80.

Agarwal, Adesh, and L. B. Tripathi. "Time Perspective in Achievement Motivation." *Psychologia: An International J of Psychology in the Orient* 23.1 (1980): 50-62.

Bingham, D., and B. J. Hailey. "The Time-Urgency Component of the Type A Behavior Pattern: Time Pressure and Performance." *J of Applied Psychology* 19 (April 1989): 425-32. Includes bibliography.

Blatt, S. J., and P. Quinlan. "Punctual and Procrastinating Students: A Study of Temporal Organization." *J of Consulting Psychology* 31.2 (1967): 169-74.

Bouffard, Leandre, Willy Lens, and Joseph R. Nuttin. "Extension of Future Time Perspective in Relation to Frustration." *International J of Psychology* 18.5 (1983): 429-42.

Church, R. M., and M. Z. Deluty. "Bisection of Temporal Intervals." *J of Experimental Psychology: Animal Behavior Processes* 3 (1977): 216-28.

Cromer, Gerald. "Time Perspectives and Time Usage in a Basic Training Course in the Israeli Defense Forces." *Small Group Behavior* 19.1 (1988): 67-78.

Cummings, L. L. "Organizational Behavior." *Annual Review of Psychology* 33 (1982): 541-79.

Feather, N. T., and R. E. Volkmer. "Preference for Situations Involving Effort, Time Pressure, and Feedback in Relation to Type A Behavior, Locus of Control, and Test Anxiety." *J of Personality and Social Psychology* 55.2 (August 1988): 266-71. Includes bibliography.

Fuhrman, Robert W., and Robert S. Wyer. "Event Memory: Temporal-Order Judgments of Personal Life Experiences." *J of Personality and Social Psychology* 54.3 (1988): 365-84.

Gibbon, J. "Two Kinds of Ambiguity in the Study of Psychological Time." *Quantitative Analysis of Behavior*. Ed. M. Commons, and J. A. Nevin. Cambridge: Ballinger, 1981. 1:157-89.

Gibbon, J., and R. M. Church. "Time Left: Linear Versus Logarithmic Subjective Time." *J of Experimental Psychology: Animal Behavior Processes* 7 (1981): 87-108.

Helson, R., et al. "Personality and Patterns of Adherence and Non-Adherence to the Social Clock." *J of Personality and Social Psychology* 46.5 (1984): 1079-96.

Hulstijn, W., and G. P. Van Galen. "Programming in Handwriting: Reaction Time and Movement Time as a Function of Sequence Length." *Acta Psychologica* 54 (1983): 23-49.

Killeen, P. R., and J. G. Fetterman. "A Behavioral Theory of Timing." *Psychological Review* 95.2 (April 1988): 274-95. Includes bibliography.

Kirmeyer, S. L. "Coping with Competing Demands: Interruption and the Type A Pattern." *J of Applied Psychology* 73 (November 1988): 621-29. Includes bibliography.

Laties, V. G., B. Weiss, and A. B. Weiss. "Further Observations on Overt 'Mediating' Behavior and the Discrimination of Time." *J of the Experimental Analysis of Behavior* 12.1 (1969): 43-57.

Michon, John A. "A Model of Some Temporal Relations in Human Behavior." *Psychologische Forschung* 31 (1968): 287-98.

——. "Psychological and Physiological Aspects of the Temporal Organization of Behavior." *Time, Quality of Life, and Social Development*. Ed. C. A. Mallmann, and O. Nudler. San Carlos de Bariloche: Fundacion Bariloche, 1982.

Orme, John Edward. *Time, Experience and Behaviour*. New York: Elsevier, 1969.

Richelle, Marc, and Helga Lejeune. *Time in Animal Behavior*. New York: Pergamon, 1980.

Richelle, M., et al. "From Biotemporality to Nootemporality: Toward an Integrative and Comparative View of Time in Behavior." *Time, Mind, and Behavior*. Ed. J. A. Michon, and J. L. Jackson. Berlin: Springer, 1985. 75-99.

Stern, M. H. "Thyroid Function and Activity, Speed, and Timing Aspects of Behaviour." *Canadian J of Psychology* 13 (1959): 43-48.

Yarnold, P. R., et al. "Time Urgency Among Coronary-Prone Individuals." *J of Abnormal Psychology* 91.3 (June 1982): 175-77.

C. CHILDREN AND DEVELOPMENT

Aram, D. M., et al. "Auditory Temporal Perception of Children with Left or Right Brain Lesions." *Neuropsychologia* 26.6 (1988): 931-35.

Arlin, Marshall. "The Effects of Physical Work, Mental Work, and Quantity on Children's Time Perception." *Perception and Psychophysics* 45.3 (1989): 209-14.

Ball, S., et al. "The Tyranny of the 'Devil's Mill': Time and Task at School." *Readings on Interaction in the Classroom*. Ed. S. Delamont. London: Methuen, 1984. 41-57.

Bauer, Patricia J., and Jean M. Mandler. "One Thing Follows Another: Effects of Temporal Structure on 1- to 2-Year-Olds' Recall of Events." *Developmental Psychology* 25.2 (1989): 197-206.

Bentley, A. M. "Swazi Children's Understanding of Time Concepts: A Piagetian Study." *J of Genetic Psychology* 148 (December 1987): 443-52. Includes bibliography.

Bowers, R. L., and R. W. Richards. "Children's Time Perception as Affected by Order of Stimulus Presentation." *Perceptual and Motor Skills* 65 (October 1987): 587-92. Includes bibliography.

Brown, A. L. "Judgments of Recency for Long Sequences of Pictures: The Absence of a Developmental Trend." *J of Experimental Child Psychology* 15 (1973): 473-80.

Burd, Larry, John M. Dodd, and Wayne W. Fisher. "Making Functional Time Estimates: Comparison of School-Aged, Learning Disabled and Other Children." *Perceptual and Motor Skills* 59.3 (1984): 967-73.

Clark, H. H. "Space, Time, Semantics, and the Child." *Cognitive Development and the Acquisition of Language*. Ed. T. E. Moore. New York: Academic P, 1973. 27-63.

Colarusso, C. A. "The Development of Time Sense in Adolescence." *Psychoanalytic Study of the Child* 43 (1988): 179-97.

Dodd, John M., et al. "A Comparison of Learning Disabled and Other Children on the Ability to Make Functional Time Estimates." *Child Study J* 15.3 (1985): 189-97.

Dufresne, A., and A. Kobasigawa. "Developmental Differences in Children's Spontaneous Allocation of Study Time." *J of Genetic*

Psychology 149 (March 1988): 87-92.
Fivush, Robyn, and Jean M. Mandler. "Developmental Changes in the Understanding of Temporal Sequence." *Child Development* 56 (1985): 1437-46.
Flavell, J. *The Developmental Psychology of Jean Piaget*. New York: Van Nostrand, 1963.
Fraisse, P. "The Adaptation of the Child to Time." *The Developmental Psychology of Time*. Ed. W. J. Friedman. New York: Academic P, 1982. 113-40.
Friedman, W. J. "Conventional Time Concepts and Children's Structuring of Time." *The Developmental Psychology of Time*. New York: Academic P, 1982. 171-208.
——. "The Development of Children's Understanding of Cyclic Aspects of Time." *Child Development* 48 (1977): 1593-99.
——. "The Development of Children's Knowledge of Temporal Structure." *Child Development* 57.6 (1986): 1386-1400.
——, ed. *The Developmental Psychology of Time*. New York: Academic P, 1982.
Friedman, William J., and F. Laycock. "Children's Analog and Digital Clock Knowledge." *Child Development* 60 (April 1989): 357-71.
Goldstone, S., W. K. Boardman, and W. T. Lhamon. "Kinesthetic Cues in the Development of Time Concepts." *J of Genetic Psychology* 93 (1958): 185-90.
Greene, A. L. "Future-Time Perspective in Adolescence: The Present of Things Future Revisited." *J of Youth and Adolescence* 15.2 (1986): 99-113.
Harner, Lorraine. "Children Talk About the Time and Aspect of Actions." *Child Development* 52 (1981): 498-506.
Hayes, Donald S., and B. Suzanne Kelly. "Young Children's Processing of Television: Modality Differences in the Retention of Temporal Relations." *J of Experimental Child Psychology* 38.3 (1984): 505-14.
Hicks, R. E., et al. "A Developmental Study of Judgements." *J of Genetic Psychology* 144 (March 1984): 31-38.
Hirsch, Bennett Z., and Janet M. Kulberg. "Television and Temporal Development." *J of Early Adolescence* 7.3 (1987): 331-44.
Inhelder, B., and J. Piaget. *The Early Growth of Logic in the Child*. New York: Harper and Row, 1964.
Irwin, R. J., et al. "The Development of Auditory Temporal Acuity in Children." *Child Development* 56 (June 1985): 614-20. Includes bibliography.
Kagan, J., and H. A. Moss. "The Stability of Passive and Dependent

Behavior from Childhood through Adulthood." *Child Development* 31 (1960): 577-91.

Levin, I. "The Nature and Development of Time Concepts in Children: The Effects of Interfering Cues." *The Developmental Psychology of Time*. Ed. William J. Friedman. New York: Academic P, 1982. 47-85.

——. "Principles Underlying Time Measurement: The Development of Children's Constraints on Counting Time." *Time and Human Cognition: A Life-Span Perspective*. Ed. I. Levin, and D. Zakay. Amsterdam: North-Holland, 1989. 145-83.

Levin, I., and T. Globerson. "The Development of Time Concepts among Advantaged and Disadvantaged Israeli Children." *J of Genetic Pychology* 145 (September 1984): 117-25.

Levin, Iris, Friedrich Wilkening, and Dembo Yoram. "Development of Time Quantification: Integration and Nonintegration of Beginnings and Endings in Comparing Durations." *Child Development* 55.6 (1984): 2160-72.

Levin, I., et al. "The Role of Memory and Integration in Early Time Concepts." *J of Experimental Child Psychology* 37.2 (April 1984): 262-70.

Lipsitt, L. P. "The Concepts of Development and Learning in Child Behavior." *Brain Function, Vol. 4*. Ed. D. B. Lindsley, and A. A. Lumsdaine. Berkeley: U of California P, 1967.

——. "Learning in the Human Infant." *Developmental Approaches*. Ed. H. W. Stevenson. New York: Wiley, 1967. 225-47.

Macar, F. "Temporal Regulations in Children of 3 to 5 Years Old." *Cahiers de Psychologie Cognitive* 8.1 (1988): 39-51.

Macar, F., and S. Grondin. "Temporal Regulation as a Function of Muscular Parameters in 5-Year-Old Children." *J of Experimental Child Psychology* 45.2 (April 1988): 159-74.

Malrieu, Philippe. *Les origines de la conscience du temps. Les attitudes temporelles de l'enfant*. Paris: Presses Universitaires de France, 1953.

Matsuda, Fumiko, and Michihiko Matsuda. "A Longitudinal Study of Learning Process of Duration Estimation in Young Children." *Japanese Psychological Research* 25.3 (1983): 119-29.

Montangero, J. "The Development of Temporal Inferences and Meanings in 5- to 8-Year Old Children." *Time, Mind, and Behavior*. Ed. J. A. Michon, and J. L. Jackson. Berlin: Springer, 1985. 279-87.

——. *La notion de durée chez l'enfant de 5 à 9 ans*. Paris: Presses Universitaires de France, 1977.

Montare, Alberto. "Learning Effects of Knowledge of Results Upon Time Estimation." *Perceptual and Motor Skills* 60.3 (1985): 871-77.

——. "Temporal Correlates of Children's Discrimination Learning." *J of Psychology* 115.2 (1983): 137-44.

Piaget, J. *Die Bildung des Zeitbegriffs beim Kinde*. Zürich: Rascher, 1955.

——. "The Child and Modern Physics." *Scientific American* 196 (March 1957): 46-51.

——. *The Child's Conception of Time*. Trans. A. J. Pomerans. New York: Basic Books, 1970.

——. *Le développement de la notion de temps chez l'enfant*. Paris: Presses Universitaires de France, 1946.

——. "Time Perception in Children." *The Voices of Time*. Ed. J. T. Fraser. 2nd ed. Amherst: U. of Massachusetts P, 1981. 202-16.

Pouthas, V. "Timing Behavior in Young Children: A Developmental Approach to Conditioned Spaced Responding." *Time, Mind, and Behavior*. Ed. J. A. Michon, and J. L. Jackson. Berlin: Springer, 1985. 100-09.

Pouthas, Viviane, and Anne-Yvonne Jacquet. "A Developmental Study of Timing Behavior in 4-1/2- and 7-Year Old Children." *J of Experimental Child Psychology* 43.2 (1987): 282-99.

Raundalen, Magne, and O. J. Finney. "Children's and Teen-Agers' Views of the Future." *J of Mental Health* 15.1-3 (1986): 114-25.

Richards, D. D. "Children's Time Concepts: Going the Distance." *The Developmental Psychology of Time*. Ed. W. J. Friedman. New York: Academic P, 1982. 13-19.

Rizzo, A. E. "The Time Moratorium." *Adolescence* 2.8 (1967-68): 469-80.

Rosseel, E. "The Impact of Attitudes Toward the Personal Future on Study Motivation and Work Orientations of Nonworking Adolescents." *Adolescence* 24 (Spring 1989): 73-93. Includes bibliography.

Sabbadini, Andrea. "How the Infant Develops a Sense of Time." *British J of Psychotherapy* 5.4 (1989): 475-84.

Seth, Usha. "Future Time Orientation of Boys and Girls of Different Educational Faculties." *Indian J of Current Psychological Research* 1.2 (1986): 107-11.

Singer, J. L., H. Wilensky, and V. G. McGraven. "Delaying Capacity, Fantasy and Planning Ability." *J of Consulting Psychology* 20 (1956): 375-83.

Smith, C. "The Acquisition of Time Talk: Relations Between Child and Adult Grammars." *J of Child Language* 7 (1980): 263-78.

Smith, P. Hull. "Five-Month-Old Infant Recall and Utilization of

Temporal Organization." *J of Experimental Child Psychology* 38.3 (1984): 400-14.

Smith, P. H., et al. "Expectancies and Memory for Spaciotemporal Events in 5-Month-Old Infants." *J of Experimental Child Psychology* 47.2 (April 1989): 210-35.

Springer, D. "Development in Young Children of an Understanding of Time and the Clock." *J of Genetic Psychology* 67 (1946): 97-106.

Sundberg, Norman D., Millicent E. Poole, and Leona E. Tyler. "Adolescents' Expectations of Future Events: A Cross-Cultural Study of Australians, Americans, and Indians." *International J of Psychology* 18.5 (1983): 415-27.

Tismer, K. G. "Psychological Aspects of Temporal Dominance during Adolescence." *Psychological Reports* 61 (October 1987): 647-54. Includes bibliography.

Verstraeten, Damielle. "Level of Realism in Adolescent Future Time Perspective." *Human Development* 23.3 (1980): 177-91.

Wilkening, F., et al. "Children's Counting Strategies for Time Quantification and Integration." *Developmental Psychology* 23 (November 1987): 823-31. Includes bibliography.

Zern, David. "The Influence of Certain Child-Rearing Factors upon the Development of a Structured and Salient Sense of Time." *Genetic Psychology Monographs* 81.2 (1970): 197-254.

D. MEMORY, LEARNING, AND COGNITION

Aiken, L. R. "Learning and Retention in the Estimation of Short Time Intervals: A Circuit and a Study." *Perceptual and Motor Skills* 20 (1965): 509-17.

Albert, S. "Time, Memory, and Affect: Experimental Studies of the Subjective Past." *The Study of Time III*. Ed. J. T. Fraser, N. Lawrence, and D. Park. New York: Springer, 1978. 269-90.

Block, Richard A., ed. *Cognitive Models of Psychological Time*. Hillsdale: Erlbaum, 1990.

——. "Contextual Coding in Memory: Studies of Remembered Duration." *Time, Mind, and Behavior*. Ed. J. A. Michon, and J. L. Jackson. Heidelberg: Springer, 1985. 169-78.

——. "A Contextualistic View of Time and Mind." *Time and Mind: The Study of Time VI*. Ed. J. T. Fraser. Madison: International Universities P, 1989. 61-79.

——. "Experiencing and Remembering Time: Affordances, Context, and Cognition." *Time and Human Cognition; A Life-Span Perspec-*

tive. Ed. I. Levin, and D. Zakay. Amsterdam: North-Holland, 1989. 333-63.

——. "Models of Psychological Time." *Cognitive Models of Psychological Time*. Ed. R. A. Block. Hillsdale: Erlbaum, 1990. 1-35.

——. "Remembered Duration: Imagery Processes and Contextual Encoding." *Acta Psychologica* 62 (1986): 103-22.

Brown, N. R., et al. "The Subjective Dates of Natural Events in Very-Long-Term Memory." *Cognitive Psychology* 17 (April 1985): 139-77. Includes bibliography.

Church, R. M. "The Internal Clock." *Cognitive Processes in Animal Behavior*. Ed. S. H. Hulse, H. Fowler, and W. H. Honig. Hillsdale: Erlbaum, 1978. 277-310.

——. "Short-Term Memory for Time Intervals." *Learning and Motivation* 11 (1980): 208-19.

Clauser, G. *Die Kopfuhr. Das automatische Erwachen*. Stuttgart: Ferdinand Enke, 1954.

Cooper, L. F., and M. H. Erickson. *Time Distortion in Hypnosis*. Baltimore: Williams and Wilkins, 1959.

Culbertson, James T. *Sensations, Memories, and the Flow of Time*. Santa Margarita: Cromwell P, 1976.

Cunningham, T. F., A. F. Healy, and D. M. Williams. "Effects of Repetition on Short-Term Retention of Order Information." *J of Experimental Psychology: Learning, Memory, and Cognition* 10 (1984): 575-97.

Deutsch, Diana. "Dislocation of Tones in a Musical Sequence: A Memory Illusion." *Nature* 226 (1970): 286.

——. "Grouping Mechanisms in Music." *The Psychology of Music*. Ed. Diana Deutsch. New York: Academic P, 1982. 99-134.

De Volder, Maurice L., and Willy Lens. "Academic Achievement and Future Time Perspective as a Cognitive-Motivational Concept." *J of Personality and Social Psychology* 42.3 (1982): 566-71.

Estes, William K. "Is Human Memory Obsolete?" *American Scientist* 68 (1980): 62-69.

——. "Memory for Temporal Information." *Time, Mind, and Behavior*. Ed. J. A. Michon, and J. L. Jackson. Berlin: Springer, 1985. 151-68.

Fisher, J., and R. A. Hinde. "The Opening of Milk Bottles by Birds." *British Birds* 42 (1949): 347-57.

Fraisse, P. "Cognition of Time in Human Activity." *Cognition in Human Motivation and Learning*. Ed. G. d'Ydewalle, and W. Lens. Hillsdale: Erlbaum, 1981. 233-59.

Frake, Charles O. "Cognitive Maps of Time and Tide Among Medieval

Seafarers." *Man* 20 N.5 (1985): 254-70.

Frederickson, Charles G. "Temporal Experience: A Two Component Model." *Perceptual and Motor Skills* 66.1 (1988): 63-68.

Friedman, W. J. "A Follow-up to 'Scale Effects in Memory for the Time of Events': The Earthquake Study." *Memory and Cognition* 15 (1987): 518-20.

——. "Image and Verbal Processes in Reasoning about the Months of the Year." *J of Experimental Psychology. Learning, Memory, and Cognition* 9 (October 1983): 650-66. Includes bibliography.

Friedman, W. J., and A. J. Wilkins. "Scale Effects in Memory for the Time of Events." *Memory and Cognition* 13 (1985): 168-75.

Galinat, Withold H., and Ingwer Borg. "On Symbolic Temporal Information: Beliefs About the Experience of Duration." *Memory and Cognition* 15.4 (1987): 308-17.

Geissler, H.-G. "The Temporal Architecture of Central Information Processing: Evidence for a Tentative Time-Quantum Model." *Psychological Research* 49 (1987): 99-106.

Gibbon, J. "On the Form and Location of the Psychometric Bisection Function for Time." *J of Mathematical Psychology* 24 (1981): 58-87.

——. "Scalar Expectancy Theory and Weber's Law in Animal Timing." *Psychological Review* 84 (1977): 279-325.

——. "The Structure of Subjective Time: How Time Flies." *The Psychology of Learning and Motivation* 20 (1986): 105-35.

Guay, M. "Long-Term Retention of Temporal Information." *Perceptual and Motor Skills* 54 (1982): 843-49.

Guay, M., A. Salmoni, and E. Buckolz. "Vierordt's Law for the Reproduction of Temporal Information." *J of Human Movement Studies* 15 (1988): 245-54.

Guay, M., and R. B. Wilberg. "Immediate Reproduction of Temporal Information under Four Cognitive Strategies." *Perceptual and Motor Skills* 56 (April 1983): 375-81.

——. "Short-Term Retention of Temporal Visual Information." *Perceptual and Motor Skills* 56 (February 1983): 139-46.

Guay, M., et al. "Effect of Interpolated Temporal Activity on Temporal Short-Term Memory." *Perceptual and Motor Skills* 67.3 (1988): 743-48.

Guenther, R. K., and M. Linton. "Mechanisms of Temporal Coding." *J of Experimental Psychology: Human Learning and Memory* 1 (1975): 182-87.

Heath, Richard A. "Memory for the Order of Sequential Visual Events." *Australian J of Psychology* 34.3 (1982): 267-79.

Hellström, Å. "Factors Producing and Factors Not Producing Time

Errors: An Experiment with Loudness Comparisons." *Perception and Psychophysics* 23 (1978): 433-44.

——. "Time Errors and Differential Sensation Weighting." *J of Experimental Psychology* 5 (1979): 460-77.

——. "The Time-Order Error and Its Relatives: Mirrors of Cognitive Processes in Comparing." *Psychological Bulletin* 97.1 (1985): 35-61.

Herman, James F., Stephen F. Roth, and Lawra M. Norton. "Time and Distance in Spatial Cognition Development." *International J of Behavioral Development* 7.1 (1984): 35-51.

Hinrichs, J. V. "A Two-Process Memory-Strength Theory for Judgment of Recency." *Psychological Review* 77.3 (May 1970): 223-33.

Hintzman, D. L., and R. A. Block. "Memory for the Spacing of Repetitions." *J of Experimental Psychology* 99 (1973): 70-74.

——. "Repetition and Memory: Evidence for a Multiple-Trace Hypothesis." *J of Experimental Psychology* 88 (1971): 297-306.

Hintzman, D. L., R. A. Block, and J. J. Summers. "Contextual Associations and Memory for Serial Position." *J of Experimental Psychology* 97 (1973): 220-29.

Hintzman, D. L., J. J. Summers, and R. A. Block. "Spacing Judgments as an Index of Study-Phase Retrieval." *J of Experimental Psychology: Human Learning and Memory* 1 (1975): 31-40.

Hirsh, I. J. "Auditory Perception of Temporal Order." *J of the Acoustical Society of America* 31 (1959): 759-67.

Hirsh, I. J., and C. E. Sherrick. "Perceived Order in Different Sense Modalities." *J of Experimental Psychology* 62 (1961): 423-32.

Huttenlocher, J., L. Hedges, and V. Prohaska. "Hierarchical Organization in Ordered Domains: Estimating the Dates of Events." *Psychological Review* 95 (1988): 471-84.

Jackson, Janet L. "The Processing of Temporal Information: Do We Indeed Time Our Minds?" *Time and Mind: The Study of Time VI*. Ed. J. T. Fraser. Madison: International Universities P, 1989. 43-57.

Janet, P. *L'évolution de la mémoire et de la notion de temps*. Paris: Chahine, 1928.

Jones, M. R. "Cognitive Representations of Serial Patterns." *Human Information Processing: Tutorials in Performance and Cognition*. Ed. B. Kantowitz. Hillsdale: Erlbaum, 1974. 187-229.

——. "Structural Organization of Events in Time." *Time, Mind, and Behavior*. Ed. J. A. Michon, and J. L. Jackson. Berlin: Springer, 1985. 192-214.

Jones, M. R., and M. Boltz. "Dynamic Attending and Responses to Time." *Psychological Review* 96 (1989): 459-91.

Jones, M. R., D. J. Maser, and G. R. Kidd. "Rate and Structure in Memory for Auditory Patterns." *Memory and Cognition* 6 (1978): 246-58.

Kausler, D. H., et al. "Temporal Memory over the Adult Lifespan." *American J of Psychology* 101 (Summer 1988): 207-15. Includes bibliography.

Kawai, Masao. "Newly Acquired Pre-cultural Behaviour of Japanese Monkeys on Koshiu Islet." *Primates* 6 (1965): 1-30.

Keele, S. W., and R. I. Ivry. "Modular Analysis of Timing in Motor Skill." *The Psychology of Learning and Motivation*. Ed. G. Bower. New York: Academic P, 1987. 21:183-228.

Loftus, E. F., and W. Marburger. "Since the Eruption of Mount St. Helens, Has Anyone Beaten You Up? Improving the Accuracy of Retrospective Reports with Landmark Events." *Memory and Cognition* 11 (1983): 114-20.

Loftus, E. F., et al. "Time Went by So Slowly: Overestimation of Event Duration by Males and Females." *Applied Cognitive Psychology* 1 (1987): 3-13.

Macar, F. "Time Psychophysics and Related Models." *Time, Mind, and Behavior*. Ed. J. A Michon, and J. L. Jackson. Berlin: Springer, 1985. 112-30.

Meck, W. H. "Attentional Bias Between Modalities: Effect on the Internal Clock, Memory, and Decision Stages Used in Animal Time Discrimination." *Timing and Time Perception*. Ed. J. Gibbon, and L. Allen. New York: New York Academy of Sciences, 1984. 528-41.

———. "Hippocampal Function Is Required for Feedback Control of an Internal Clock's Criterion." *Behavioral Neuroscience* 102 (1988): 54-60.

———. "Selective Adjustment of the Speed of Internal Clock and Memory Processes." *J of Experimental Psychology* 9 (April 1983): 171-201.

Meck, W. H., and R. M. Church. "Mode Control Model of Counting and Timing Processes." *J of Experimental Psychology* 9 (July 1983): 320-34.

———. "Nutrients that Modify the Speed of Internal Clock and Memory Storage Processes." *Behavioral Neuroscience* 101 (1987): 465-75.

Meck, Warren H., Russell M. Church, and David S. Olton. "Hippocampus, Time, and Memory." *Behavioral Neuroscience* 98.1 (1984): 3-22.

Michon, John A. "Implicit and Explicit Representations of Time." *Cognitive Models of Psychological Time*. Ed. R. A. Block. Hillsdale:

Erlbaum, 1990. 37-58.

———. "J. T. Fraser's 'Levels of Temporality' as Cognitive Representations." *Time, Science, and Society in China and the West: The Study of Time V.* Ed. J. T. Fraser, N. Lawrence, and F. C. Haber. Amherst: U of Massachusetts P, 1986. 51-66.

———. "Processing of Temporal Information and the Cognitive Theory of Time Experience." *The Study of Time I.* Ed. J. T. Fraser, F. C. Haber, and G. H. Müller. New York: Springer, 1972. 242-58.

———. "Programs and 'Programs' for Sequential Patterns in Motor Behavior." *Brain Research* 71 (1974): 413-24.

———. "Time Experience and Memory Processes." *The Study of Time II.* Ed. J. T. Fraser, and N. Lawrence. New York: Springer, 1975. 302-13.

Michon, John A., and J. L. Jackson. "Attentional Effort and Cognitive Strategies in the Processing of Temporal Information." *Timing and Time Perception.* Ed. J. Gibbon, and L. Allen. *Annals of the New York Academy of Sciences* 423 (1984): 298-321.

Morgan, R. *Training the Time Sense: Hypnotic and Conditioning Approaches.* Fresno: California School of Professional Psychology, 1983.

Naveh-Benjamin, M. "Coding of Temporal Order Information: An Automatic Process?" *J of Experimental Psychology: Learning, Memory, and Cognition* 16 (1990): 117-26.

Nuttin, J. R. "The Future Time Perspective in Human Motivation and Learning." *Proceedings of the 17th International Congress of Psychology.* Amsterdam: North-Holland, 1974. 60-82.

Ornstein, Robert E. *On the Experience of Time.* Harmondsworth: Penguin, 1969.

Park, D. "Brain Time and Mind Time." *Time, Mind, and Behavior.* Ed. J. A. Michon, and J. L. Jackson. Berlin: Springer, 1985. 53-64.

Pellionisz, A., et al. "Space-Time Representation in the Brain: The Cerebellum as a Predictive Space-Time Metric Tensor." *Neuroscience* 7.12 (1982): 2949-70.

Peterson, L. R. "Search and Judgment in Memory." *Concepts and the Structure of Memory.* Ed. B. Kleinmuntz. New York: Wiley, 1967.

Petterson, M. "Diffusion of a New Habit among Green Finches." *Nature* 117 (1956): 705-10.

Pöppel, E., et al. "A Hypothesis Concerning Timing in the Brain." *Synergetics of Cognition.* Ed. H. Haken, and M. Stadler. Heidelberg: Springer, 1990. 144-49.

Poynter, W. D. "Duration Judgment and the Segmentation of Experience." *Memory and Cognition* 11 (1983): 77-82.

———. "Inferring Time's Passage." *Time and Human Cognition: A Life-Span Perspective*. Ed. I. Levin, and D. Zakay. Amsterdam: North-Holland, 1989. 305-31.

Poynter, W. D., and D. Homa. "Duration Judgment and the Experience of Change." *Perception and Psychophysics* 33 (1983): 548-60.

Roberts, S. "Cross-Modal Use of an Internal Clock." *J of Experimental Psychology: Animal Behavior Processes* 8 (1982): 2-22.

———. "Isolation of an Internal Clock." *J of Experimental Psychology: Animal Behavior Processes* 7 (1981): 242-68.

———. "Properties and Function of an Internal Clock." *Animal Cognition and Behavior*. Ed. R. L. Mellgren. Amsterdam: North-Holland, 1983. 345-98.

Roberts, S., and R. M. Church. "Control of an Internal Clock." *J of Experimental Psychology: Animal Behavior Processes* 4 (1978): 318-37.

Roberts, S., and M. D. Holder. "What Starts an Internal Clock?" *J of Experimental Psychology* 10 (July 1984): 273-96.

Robinson, G. M. "The Delayed Auditory Feedback Effect Is a Function of Speech Rate." *J of Experimental Psychology* 95 (1972): 1-5.

Roitblat, H. L., B. Pologe, and R. A. Scopatz. "The Representation of Items in Serial Position." *Animal Learning and Behavior* 11 (1983): 489-98.

Roitblat, H. L., R. A. Scopatz, and T. G. Bever. "The Hierarchical Representation of Three-Item Sequences." *Animal Learning and Behavior* 15 (1987): 179-92.

Rubin, D. C., and A. D. Baddeley. "Telescoping Is Not Time Compression: A Model of the Dating of Autobiographical Events." *Memory and Cognition* 17 (1989): 653-61.

Rumiati, Rino, and Sergio Roncato. "Time Representation in Sentence Comprehension." *Acta Psychologica* 59.3 (1985): 271-83.

Shaffer, L. H. "Rhythm and Timing in Skill." *Psychological Review* 89 (1982): 109-22.

———. "Timing in Action." *Time, Mind, and Behavior*. Ed. J. A. Michon, and J. L. Jackson. Berlin: Springer, 1985. 226-41.

———. "Timing in Musical Performance." *Timing and Time Perception*. Ed. J. Gibbon, and L. Allan. New York: New York Academy of Sciences, 1984. 420-28.

———. "Timing in the Motor Programming of Typing." *Quarterly J of Experimental Psychology* 30 (1978): 333-45.

Shaffer, L. H., E. F. Clarke, and N. P. Todd. "Metre and Rhythm in Piano Playing." *Cognition* 20 (1985): 61-77.

Staddon, J. E. "Time and Memory." *Annals of the New York Academy of Sciences* 423 (1984): 322-34.

Thomas, E. A. C., and N. E. Cantor. "Interdependence between the Processing of Temporal and Non-Temporal Information." *Attention and Performance VII*. Ed. J. Requin. Hillsdale: Erlbaum, 1978. 43-62.

Thomas, E. A. C., and W. B. Weaver. "Cognitive Processing and Time Perception." *Perception and Psychophysics* 17 (1975): 363-67.

Tilley, A., and P. Warren. "Retrieval from Semantic Memory at Different Times of Day." *J of Experimental Psychology* 9.4 (October 1983): 718-24.

Toda, Masanao. "Future Time Perspective and Human Cognition: An Evolutional View." *International J of Psychology* 18.5 (1983): 351-65.

——. "Time and the Structure of Human Cognition." *The Study of Time II*. Ed. J. T. Fraser, and N. Lawrence. New York: Springer, 1975. 314-24.

Tzeng, O. J. L., and B. Cotton. "A Study-Phase Retrieval Model of Temporal Coding." *J of Experimental Psychology: Human Learning and Memory* 6 (1980): 705-16.

Tzeng, O. J. L., A. T. Lee, and C. D. Wetzel. "Temporal Coding in Verbal Information Processing." *J of Experimental Psychology: Human Learning and Memory* 5 (1979): 52-64.

Underwood, Benton J. "Attributes of Memory." *Psychological Review* 76 (1969): 559-74.

——. *Temporal Codes for Memories: Issues and Problems*. Hillsdale: Erlbaum, 1977.

Vroon, P. A., H. Timmers, and S. Tempelaars. "On the Hemispheric Representation of Time." *Attention and Performance VI*. Ed. S. Dornic. Hillsdale: Erlbaum, 1977. 231-45.

Wilkening, Friedrich. "Integrating Velocity, Time, and Distance Information: A Developmental Study." *Cognitive Psychology* 13 (1981): 231-47.

Winograd, Eugene, and Robert M. Soloway. "Reminding as a Basis of Temporal Judgments." *J of Experimental Psychology: Learning, Memory, and Cognition* 11.2 (1985): 262-71.

Yates, Francis A. *The Art of Memory*. Chicago: U of Chicago P, 1966.

Yntema, D. B., and F. P. Trask. "Recall as a Search Process." *J of Verbal Learning and Verbal Behavior* 2 (1963): 65-74.

Zakay, D., and E. Fallach. "Immediate and Remote Time Estimation: A Comparison." *Acta Psychologica* 57 (1984): 69-81.

Zung, W. W. K., and W. P. Wilson. "Time Estimation During Sleep." *Biological Psychiatry* 3 (1971): 159-64.

E. PERSONALITY AND SOCIETY

Bond, M. J., and N. T. Feather. "Some Correlates of Structure and Purpose in the Use of Time." *J of Personality and Social Psychology* 55 (August 1988): 321-29. Includes bibliography.

Calabresi, R., and J. Cohen. "Personality and Time Attitudes." *J of Abnormal Psychology* 73.5 (1968): 431-39.

Cottle, Thomas J. "The Circles Test: An Investigation of Perceptions of Temporal Relatedness and Dominance." *J of Projective Techiques and Personality Assessment* 31.5 (1967): 58-71.

——. *Perceiving Time: A Psychological Investigation with Men and Women*. New York: Wiley, 1976.

Green, H. B., and R. H. Knapp. "Time Judgment, Aesthetic Preference, and Need for Achievement." *J of Abnormal and Social Psychology* 58 (1959): 140-42.

Knapp, R. H. "Attitudes Toward Time and Aesthetic Choice." *J of Social Psychology* 56 (1962): 79-87.

——. "Personality and the Psychology of Time." *The Study of Time I*. Ed. J. T. Fraser, F. C. Haber, and G. H. Müller. New York: Springer, 1972. 311-19.

Knapp, R. H., and J. T. Garbutt. "Variation in Time Descriptions and Need Achievement." *J of Social Psychology* 67 (1965): 269-72.

Knapp, R. H., and H. B. Green. "Time Imagery and the Achievement Motive." *J of Personality* 25 (1958): 426-34.

Levine, R. V. "The Pace of Life." *American Scientist* 78 (1990): 450-59.

——. "The Pace of Life." *Psychology Today* (October 1989): 42-46.

——. "The Pace of Life Across Cultures." *The Social Psychology of Time: New Perspectives*. Ed. J. McGrath. Newbury Park: Sage, 1988. 39-60.

——. "The Waiting Game: Power Has Its Privilege." *Psychology Today* (April 1987): 24-33.

Levine, R. V., and K. Bartlett. "Pace of Life, Punctuality, and Coronary Heart Disease in Six Countries." *J of Cross-Cultural Psychology* 15 (1984): 233-55.

McGrath, Joseph E., ed. *The Social Psychology of Time: New Perspectives*. Newbury Park: Sage, 1988.

McGrath, Joseph E., and Janice R. Kelly. *Time and Human Interaction: Toward a Social Psychology of Time*. New York: Guilford P, 1986.

McGrath, J. E., J. R. Kelly, and D. E. Machatka. "The Social Psychology of Time: Entrainment of Behavior in Social and Organizational Settings." *Applied Social Psychology, Vol. 5: Applications in Organizational Settings*. Beverly Hills: Sage, 1984. 21-44.

McGrath, J. J., and J. O'Hanlon. "Temporal Orientation and Vigilance Performance." *Attention and Performance*. Ed. A. F. Sanders. Amsterdam: North-Holland, 1967. 410-19.

Mass, Jeannette P. "Beliefs Relating to Temporal Dimensions Held by Three Ethnic Groups in the Fiji Islands." *Pychological Studies* 30.1 (1985): 53-65.

Burke, R. J. "Beliefs and Fears Underlying Type A Behavior: Correlates of Time Urgency and Hostility." *J of General Psychology* 112 (April 1985): 133-45.

Cottle, T. J. "Temporal Correlates of Dogmatism." *J of Consulting and Clinical Psychology* 36.1 (1971): 70-81.

——. "Temporal Correlates of the Achievement Value and Manifest Anxiety." *J of Consulting and Clinical Psychology* 33.5 (1969): 541-50.

Dapkus, Marilyn A. "A Thematic Analysis of the Experience of Time." *J of Personality and Social Psychology* 49.2 (1985): 407-19.

Koothur, Joicy P., and S. Kanekar. "Type A Behavior Pattern and Academic Performance of College Students in India." *J of Social Psychology* 123.2 (1984): 275-76.

Lee, C., and J. M. Innes. "Type A Behavior Pattern, Time Urgency and Arrival Time: A Replication." *Perceptual and Motor Skills* 56 (February 1983): 177-78.

Schachter, S., and L. P. Gross. "Manipulated Time and Eating Behavior." *J of Personality and Social Psychology* 10.2 (October 1968): 98-106.

Weinstein, Neil D. "Unrealistic Optimism About Future Life Events." *J of Personality and Social Psychology* 39 (1980): 806-20.

Wessman, A. E., and B. S. Gorman. "The Emergence of Human Awareness and Concepts of Time." *The Personal Experience of Time*. Ed. B. S. Gorman, and A. E. Wessman. New York: Plenum, 1977. 3-55.

Zakay, D., J. Lomranz, and M. Kaziniz. "Extraversion-Introversion and Time Perception." *Personality and Individual Differences* 5 (1984): 237-39.

F. PSYCHOANALYSIS AND PSYCHIATRY (see Medicine: Psychiatry)

Aziz, Robert. *C. G. Jung's Psychology of Religion and Synchronicity*. Albany: State U of New York P, 1990.

Board, C., et al. "Time Perspective and Intimacy." *Archives of General Psychiatry* 1 (1959): 425-33.

Bonaparte, Marie. "Time and the Unconscious." *International J of Psychoanalysis* 21 (1940): 427-68.

Buckhout, R., P. Fox, and M. Rabinowitz. "Estimating the Duration of an Earthquake: Some Shaky Field Observations." *Bulletin of the Psychonomic Society* 27 (1989): 375-78.

Eigen, M. "On Time and Dreams." *Psychoanalytical Review* 70.2 (Summer 1983): 211-20.

Eisenbud, J. "Time and the Oedipus." *Psychoanalytic Quarterly* 25 (1956): 363-84.

Fraser, J. T. "Temporal Levels and Reality Testing." *International J of Psycho-Analysis* 62 (1981): 3-26.

Freeman, A. M., and F. T. Melges. "Depersonalization and Temporal Disintegration in Acute Mental Illness." *American J of Psychiatry* 134 (1977): 679-81.

——. "Temporal Disorganization, Depersonalization, and Persecutory Ideation in Acute Mental Illness." *American J of Psychiatry* 135 (1978): 123-24.

Freud, Sigmund. *The Standard Edition of the Complete Psychological Works*. 24 vols. London: Hogarth P, 1953-74. See "Time" under Subject Index in Vol. 24.

Grof, S. *Beyond the Brain: Birth, Death and Transcendence in Psychotherapy*. Albany: State U of New York P, 1985.

Jung, C. G. "On Synchronicity." *Man and Time: Papers from the Eranos Yearbooks*. Trans. R. F. C. Hull. Ed. Joseph Campbell. New York: Pantheon, 1957. 201-11.

——. *Structure and Dynamics of the Psyche*. Trans R. F. Hull. 2nd ed. *The Collected Works*. Princeton: Princeton UP, 1968.

Krolick, Sanford. *Recollective Resolve: A Phenomenological Understanding of Time and Myth*. Macon: Mercer UP, 1987.

Kurtz, S. A. "The Psychoanalysis of Time." *J of the American Psychoanalytic Association* 36.4 (1988): 985-1004.

Lewis, W. C. "Structural Aspects of the Psychoanalytic Theory of Instinctual Drives, Affects, and Time." *Psychoanalysis and Current Biological Thought*. Ed. N. S. Greenfield, and W. C. Lewis. Madison: U of Wisconsin P, 1965. 151-79.

Progoff, Ira. *Jung, Synchronicity, and Human Destiny: C. G. Jung's Theory of Meaningful Coincidence*. New York: Crown, 1987.

Rabin, Albert I. "Future Time Perspective and Ego Strength." *The Study of Time III*. Ed. J. T. Fraser, N. Lawrence, and D. Park. New York: Springer, 1978. 294-305.

Ruiz, R. A., and H. H. Krauss. "Anxiety, Temporal Perspective and Item Content of the Incomplete Thoughts Test (ITT)." *J of Clinical*

Psychology 24.1 (1968): 70-72.
Sabbadini, Andrea. "Boundaries of Timelessness: Some Thoughts about the Temporal Dimension of the Psychoanalytic Space." *International J of Psycho-Analysis* 70.2 (1989): 305-313.
——. "Tempo e identità: alcune considerazioni psicoanalitiche." *Tempo e identità*. Ed. P. Reale. Milan: Franco Angeli, 1988. 116-27.
——., ed. *Il tempo in psicoanalisi*. Milan: Feltrinelli, 1979.
——. "Unconscious Timelessness and Infantile Omnipresent in Psychoanalytic Theory." *J of the Arbours Association* 3 (1985): 15-28.
——. "The Year 2000: A Psychoanalytic Perspective on the Fantasy of the New Millennium." *Free Associations* 9 (1987): 56-71.
Schiffer, Irvine. *The Trauma of Time: A Psychoanalytic Investigation*. New York: International Universities P, 1978.
Sonnet, André. *The Twilight Zone of Dreams*. Trans. J. T. Fraser. Philadelphia: Chilton Company, 1961.
Squyres, E. M., et al. "Sex Differences in Time Orientation: A Jungian Perspective." *J of Analytical Psychology* 27.1 (January 1982): 71-81.
Von Franz, Marie-Louise. *Number and Time*. Evanston: Northwestern UP, 1974.
——. *Rhythm and Repose*. London: Thames and Hudson, 1978.
——. "Time and Synchronicity in Analytic Psychology." *The Voices of Time*. Ed. J. T. Fraser. 2nd ed. Amherst: U of Massachusetts P, 1981. 218-32.
Wood, Douglas K. *Men Against Time: Nicolas Berdyaev, T. S. Eliot, Aldous Huxley, and C. G. Jung*. Lawrence: UP of Kansas, 1982.

G. TIME PERCEPTION (TEMPORAL JUDGMENT)

Aaronson, B. S. "Hypnosis, Time Rate Perception, and Personality." *J of Schizophrenia* 2.1 (1968): 11-41.
——. "Hypnotic Alterations of Space and Time." *International J of Parapsychology* 10.1 (1968): 5-36.
Adam, N. "Mechanisms of Time Perception." *T.-I.-T. J of Life Sciences* 1 (1971): 41-52.
Ahmadi, Kate S. "Effects of Social Influences and Waiting on Time Judgment." *Perceptual and Motor Skills* 59.3 (1984): 771-76.
Aitken, R. C. B., and J. L. Gedye. "A Study of Two Factors which Affect Arousal Level and the Apparent Duration of a Ten-Minute Interval." *British J of Psychology* 59.3 (1968): 253-63.
Allan, L. G. "Comments on Current Ratio-Setting Models for Time Perception." *Perception and Psychophysics* 24 (1978): 444-50.

——. "Contingent Aftereffects in Duration Judgments." *Annals of the New York Academy of Sciences* 423 (1984): 116-130.

——. "Magnitude Estimation of Temporal Intervals." *Perception and Psychophysics* 33 (1983): 29-42.

——. "The Perception of Time." *Perception and Psychophysics* 26 (1979): 340-54.

——. "The Time-Order Error in Judgments of Duration." *Canadian J of Psychology* 31 (1977): 24-31.

Allan, L. G., and A. B. Kristofferson. "Judgments about the Duration of Brief Stimuli." *Perception and Psychophysics* 15 (1974): 434-40.

——. "Psychophysical Theories of Duration Discrimination." *Perception and Psychophysics* 16 (1974): 26-34.

Allen, D. A. "Filling Time Versus Affective Response to the Activity that Fills Time: Independent Effects on Time Judgment?" *Perceptual and Motor Skills* 51 (1980): 723-27.

Allen, D. A., and C. A. Clark. "Effect of Information Feedback on Temporal Discrimination in Concurrent and Post-Feedback Measures." *Perceptual and Motor Skills* 49 (1979): 935-39.

Allport, D. A. "Phenomenal Simultaneity and the Perceptual Moment Hypothesis." *British J of Psychology* 59 (1968): 395-406.

Arlin, M. "The Effects of Quantity and Depth of Processing on Children's Time Perception." *J of Experimental Child Psychology* 42 (1986): 84-98.

——. "The Effects of Quantity, Complexity, and Attentional Demand on Children's Time Perception." *Perception and Psychophysics* 40 (1986): 177-82.

Aschoff, J. "On the Perception of Time During Prolonged Temporal Isolation." *Human Neurobiology* 4 (1985): 41-52.

Axel, R. "Estimation of Time." *Archives of Psychology* 74 (1924): 1-77.

Baddeley, A. D. "Time Estimation at Reduced Body-Temperature." *American J of Psychology* 79 (1966): 475-79.

Barber, T. X., and D. S. Calverley. "Toward a Theory of 'Hypnotic' Behavior: An Experimental Study of 'Hypnotic Time Distortion.'" *Archives of General Psychiatry* 10 (1963): 209-16.

Baugh, Bernard T., and Susan E. Mason. "Need for Cognition Related to Time Perception." *Perceptual and Motor Skills* 62.2 (1986): 540-42.

Behar, I., and W. Bevan. "The Perceived Duration of Auditory and Visual Intervals: Cross-Modal Comparison and Interaction." *American J of Psychology* 74 (1961): 17-26.

Bell, C. R. "Effects of Lowered Temperature on Time Estimation." *Quarterly J of Experimental Psychology* 27 (1975): 531-38.

——. "Time Estimation and Increases in Body Temperature." *J of Experimental Psychology* 70 (1965): 232-34.

Bell, C. R., and K. A. Provins. "The Relation between Physiological Responses to Environmental Beat and Time Judgments." *J of Experimental Psychology* 66 (1963): 572-79.

Bell, C. R., and N. Watts. "Personality and Judgments of Temporal Intervals." *British J of Psychology* 57.1-2 (May 1966): 155-59.

Ben-Dov, G., and A. Carmon. "On Time Space and the Cerebral Hemispheres: A Theoretical Note." *International J of Neuroscience* 7 (1976): 29-33.

——. "Rhythm Length and Hemispheric Asymmetry." *Brain and Cognition* 3 (1984): 35-41.

Benussi, V. *Psychologie der Zeitauffassung.* Heidelberg: C. Winter, 1913.

——. "Zur experimentellen Analyse des Zeitvergleichs." *Archiv für Geschichte der Psychologie* 9 (1907): 366-449.

Bergler, E., and G. Roheim. "Psychology of Time Perception." *Psychoanalytic Quarterly* 15 (1946): 190-206.

Berglund, B., et al. "The Influence of Auditory Stimulus Intensity on Apparent Duration." *Scandinavian J of Psychology* 10.1 (1969): 21-26.

Bilting, M., et al. "Estimation of Time as a Measure of Pain Magnitude." *J of Psychosomatic Research* 27.6 (1983): 493-97.

Bindra, D., and H. Waksberg. "Methods and Terminology in Studies of Time Estimation." *Psychological Bulletin* 53 (1956): 155-59.

Björkman, M., and O. Holmkvist. "The Time-Order Error in the Construction of a Subjective Time Scale." *Scandinavian J of Psychology* 1 (1960): 7-13.

Block, R. A. "Memory and the Experience of Duration in Retrospect." *Memory and Cognition* 2 (1974): 153-60.

——. "Temporal Judgments and Contextual Change." *J of Experimental Psychology: Learning, Memory, and Cognition* 8 (1982): 530-44.

Block, Richard A., E. G. George, and M. A. Reed. "A Watched Pot Sometimes Boils: A Study of Duration Experience." *Acta Psychologica* 46 (1980): 81-94.

Block, R. A., J. L. Saggau, and L. H. Nickol. "Temporal Inventory on Meaning and Experience: A Structure of Time." *Imagination, Cognition and Personality* 3 (1983-84): 203-25.

Bokander, I. "Time Estimation as an Indicator of Attention-Arousal When Perceiving Complex and Meaningful Stimulus Material." *Perceptual and Motor Skills* 21 (1965): 323-28.

Boltz, M. "Time Judgments of Musical Endings: Effects of Expectancies on the 'Filled Interval Effect.'" *Perception and Psychophysics* 46

(1989): 409-18.

Borg, I., and W. H. Galinat. "Ratios in People's Beliefs about Experienced Duration." *Perceptual and Motor Skills* 64 (1987): 603-08.

Boring, E. G. "Temporal Perception and Operationism." *American J of Psychology* 49 (1936): 519-52.

Bowers, K. S. "Hypnosis and the Perception of Time." *International J of Clinical and Experimental Hypnosis* 27 (1979): 29-41.

——. "Time Distortion and Hypnotic Ability: Underestimating the Duration of Hypnosis." *J of Abnormal Psychology* 88 (1979): 435-39.

Brown, D. R., and L. Hitchcock. "Time Estimation: Dependence and Independence of Modality-Specific Effects." *Perceptual and Motor Skills* 21 (1965): 727-34.

Brown, Norman R., Lance J. Rips, and Steven K. Shevell. "The Subjective Dates of Natural Events in Very-Long-Term Memory." *Cognitive Psychology* 17.2 (1985): 139-77.

Brown, Scott W. "Time Perception and Attention: The Effects of Prospective Versus Retrospective Paradigms and Task Demands on Perceived Duration." *Perception and Psychophysics* 38.2 (1985): 115-24.

Brown, S. W., et al. "The Psychophysics of Retrospective and Prospective Timing." *Perception* 17.3 (1988): 297-310.

Bünning, E. *The Psychological Clock*. New York: Academic P, 1964.

Burnside, W. "Judgment of Short Time Intervals While Performing Mathematical Tasks." *Perception and Psychophysics* 9 (1971): 404-06.

Cahoon, D., and E. M. Edmonds. "The Watched Pot Still Won't Boil: Expectancy as a Variable in Estimating the Passage of Time." *Bulletin of the Psychonomic Society* 16 (1980): 115-16.

Cahoon, R. L. "Effect of Acute Exposure to Altitude on Time Estimation." *J of Psychology* 66.2 (1967): 321-24.

——. "Physiological Arousal and Time Estimation." *Perceptual and Motor Skills* 28.1 (1969): 259-68.

Campbell, S. S. "Estimation of Empty Time." *Human Neurobiology* 5 (1986): 205-07.

Carlson, V. R., and I. Feinberg. "Individual Variations in Time Judgment and the Concept of an Internal Clock." *J of Experimental Psychology* 77.4 (1968): 631-40.

——. "Time Judgment as a Function of Method, Practice, and Sex." *J of Experimental Psychology* 85.2 (1970): 171-80.

Church, R. M. "Properties of the Internal Clock." *Timing and Time Perception*. Ed. J. Gibbon and L. Allan. New York: New York Academy of Sciences, 1984. 566-82.

Cohen, J. "The Experience of Time." *Acta Psychologica* 10 (1954): 207-19.

——. "Preliminary Investigation on Time Perception and Autonomic Self-Regulation." *Biofeedback and Self Regulation* 6.3 (September 1981): 289-93.

——. *Psychological Time in Health and Disease*. Springfield: C. C. Thomas, 1967.

——. "Subjective Time." *The Voices of Time*. Ed. J. T. Fraser. 2nd ed. Amherst: U of Massachusetts P, 1981. 257-75.

Cohen, J., C. E. M. Hansel, and J. Sylvester. "An Experimental Study of Comparative Judgments of Time." *British J of Psychology* 45 (1954): 108-14.

Craik, K. H., and T. R. Sarbin. "The Effect of Covert Alterations of Clock Rate Upon Time Estimates and Personal Tempo." *Perceptual and Motor Skills* 16 (1963): 597-610.

Creelman, C. D. "Human Discrimination of Auditory Duration." *J of the Acoustical Society of America* 34 (1962): 582-93.

Curton, E. D., and D. S. Lordahl. "Effects of Attentional Focus and Arousal on Time Estimation." *J of Experimental Psychology* 103 (1974): 861-67.

Day, W. F. "Serial Non-Randomness in Auditory Differential-Thresholds as a Function of Interstimulus Interval." *American J of Psychology* 69 (1956): 387-94.

Delay, Eugene R., and Mianne E. Mathey. "Effects of Ambient Noise on Time Estimation by Humans." *Perceptual and Motor Skills* 61.2 (1985): 415-19.

Deppe, A. H. "Overload and Sensory Deprivation: Time Estimation in Novice Divers." *Perceptual and Motor Skills* 29.2 (1969): 481-82.

Deutsch, Diana. "The Processing of Structured and Unstructured Tonal Sequences." *Perception and Psychophysics* 28 (1980): 381-89.

Dewolfe, R. K. S., and C. P. Duncan. "Time Estimation as a Function of Level of Behavior of Successive Tasks." *J of Experimental Psychology* 58 (1959): 153-58.

Dupreez, P. D. "Relation between Verbal Estimation and Reproduction of a Short-Time Interval: Preliminary Study." *Perceptual and Motor Skills* 17 (1963): 45-46.

——. "Reproduction of Time Intervals after Short Periods of Delay." *J of General Psychology* 76 (1967): 59-71.

Edgell, B., and A. W. Waller. "On Time Judgments." *American J of Psychology* 14 (1903): 154-74.

Edmonds, E. M., D. Cahoon, and B. Bridges. "The Estimation of Time as a Function of Positive, Neutral, or Negative Expectancies." *Bulletin of the Psychonomic Society* 17 (1981): 259-60.

Efron, R. "Conservation of Temporal Information by Perceptual Systems." *Perception and Psychophysics* 14 (1973): 518-30.

——. "The Duration of the Present." *Annals of the New York Academy of Sciences* 138.2 (1967): 713-29.

——. "The Effect of Handedness on the Perception of Simultaneity and Temporal Order." *Brain* 86.2 (1963): 261-84.

——. "Effect of Stimulus Duration on Perceptual Onset and Offset Latencies." *Perception and Psychophysics* 8 (1970): 231-34.

——. "The Effect of Stimulus Intensity on the Perception of Simultaneity in Right- and Left-Handed Subjects." *Brain* 86.2 (1963): 285-94.

——. "An Invariant Characteristic of Perceptual Systems in the Time Domain." *Attention and Performance IV*. Ed. S. Kornblum. New York: Academic P, 1973. 713-36.

——. "The Measurement of Perceptual Durations." *The Study of Time I*. Ed. J. T. Fraser, F. C. Haber, and G. H. Müller. New York: Springer, 1972. 207-18.

——. "The Minimum Duration of a Perception." *Neuropsychologia* 8 (1970): 57-63.

——. "The Relationship between the Duration of a Stimulus and the Duration of a Perception." *Neuropsychologia* 8 (1970): 37-55.

——. "Temporal Perception, Aphasia and Déjà Vu." *Brain* 86 (1963): 403-24.

Efron, R., and D. N. Lee. "The Visual Persistence of a Moving Stroboscopically Illuminated Object." *American J of Psychology* 84 (1971): 365-75.

Eisler, H. "Applicability of the Parallel-Clock Model to Duration Discrimination." *Perception and Psychophysics* 29 (1981): 225-33.

——. "Attention to Visually and Auditorily Presented Durations." *Attention and Performance VI*. Ed. S. Dornic. Hillsdale: Erlbaum, 1977. 247-59.

——. "Experiments on Subjective Duration 1868-1975: A Collection of Power Function Exponents." *Psychological Bulletin* 83 (1976): 1154-71.

——. "On the Nature of Subjective Scales." *Scandinavian J of Psychology* 23 (1982): 161-71.

——. "Subjective Duration and Psychophysics." *Psychological Review* 82 (1975): 429-50.

Ekman, G., et al. "Apparent Duration as a Function of Intensity of Vibrotactile Stimulation." *Perceptual and Motor Skills* 28.1 (1969): 151-56.

Elliott, Didby, and Ruth Jones. "Short-Term Memory for Duration of

Movement." *Perceptual and Motor Skills* 58.1 (1984): 331-34.

Emley, G. S., C. R. Schuster, and B. R. Lucchesi. "Trends Observed in the Time Estimation of Three Stimulus Intervals Within and Across Sessions." *Perceptual and Motor Skills* 26.2 (1968): 391-98.

Eson, M. E., and J. S. Kafka. "Diagnostic Implications of a Study in Time Perception." *J of General Psychology* 46 (1952): 169-83.

Falk, J. L., and D. Bindra. "Judgment of Time as a Function of Serial Position and Stress." *J of Experimental Psychology* 47 (1954): 279-82.

Ferguson, Robert P., and Martin Parthena. "Long-Term Temporal Estimation in Humans." *Perception and Psychophysics* 33.6 (1983): 585-92.

Filer, R. J., and D. W. Meals. "The Effect of Motivating Conditions on the Estimation of Time." *J of Experimental Psychology* 39 (1949): 327-31.

Fox, G. J. "The Effects of Interval and Duration of Visual Stimuli upon the Time Error." *Psychological Newsletter* 45 (1952): 1-17.

Fox, R. H., et al. "Time Judgment and Body Temperature." *J of Experimental Psychology* 75.1 (1967): 88-96.

Fraisse, Paul. "Perception and Estimation of Time." *Annual Review of Psychology* 35 (1984): 1-36.

Fraisse, Paul, and C. Voillaume. "Conditionnement temporel du rhythme alpha et estimation du temps." *L'Année Psychologique* 69 (1969): 7-15.

François, M. "Contribution à l'étude du sens du temps. La température interne comme facteur de variation de l'appréciation subjective des durées." *L'Année Psychologique* 28 (1927): 186-204.

——. "Influence de la température interne sur notre appréciation du temps." *Comptes Rendus des Séances, Société de Biologie* 98 (1928): 201-03.

Frankenhaeuser, M. *Estimation of Time.* Stockholm: Almquist and Wiksell, 1959.

Gibbon, John, and Lorraine Allan, eds. *Timing and Time Perception.* New York: New York Academy of Sciences, 1984.

Gibson, J. J. "Events Are Perceivable But Time Is Not." *The Study of Time II.* Ed. J. T. Fraser, and N. Lawrence. New York: Springer, 1975. 295-301.

——. "The Problem of Temporal Order in Stimulation and Perception." *J of Psychology* 62 (1966): 141-49.

Gilliland, A. R., J. Hofeld, and G. Eckstrand. "Studies in Time Perception." *Psychological Bulletin* 43 (1946): 162-76.

Gilliland, A. R., and R. Martin. "Some Factors in Estimating Short Time Intervals." *J of Experimental Psychology* 27 (1940): 243-55.

Giulio, L. F. "From Simultaneity to Successiveness: Psychophysiological Characteristics of the Transition Period." *Estratto Dagli* (Turin) 121.5-6 (1987): 104-14.

Goldstone, S. "The Human Clock: A Framework for the Study of Healthy and Deviant Time Perception." *Annals of the New York Academy of Sciences* 138.2 (1967): 767-83.

Goldstone, S., W. K. Boardman, and W. T. Lhamon. "Effect of Quinal Barbitone, Dextroamphetamine, and Placebo on Apparent Duration." *British J of Psychology* 49.4 (November 1958): 324-28.

——. "The Time Sense: Anchor Effects and Apparent Duration." *J of Psychology* 44 (1957): 145-53.

Gomez, L. M., and L. C. Robertson. "The Filled-Duration Illusion: The Effect of Multi-Dimensional Stimuli and Temporal Discrimination." *Perception and Psychophysics* 27 (1980): 111-16.

——. "The Filled-Duration Illusion: The Function of Temporal and Nontemporal Set." *Perception and Psychophysics* 25 (1979): 432-38.

Gray, C. "Duration Differences: Attentional Demand or Time Error?" *Perception* 11 (1982): 97-102.

Groos, G., and S. Daan. "The Use of Biological Clocks in Time Perception." *Time, Mind, and Behavior*. Ed. J. A. Michon, and J. L. Jackson. Berlin: Springer, 1985. 65-74.

Guay, M. "Human Time Estimation: Bibliography." *Perceptual and Motor Skills* 44 (1977): 1087-92.

Guay, Michel, and Alan W. Salmoni. "Human Time Estimation: Procedural Effects." *Bulletin of the Psychonomic Society* 26.1 (1988): 19-22.

Gulliksen, H. "The Influence of Occupation upon the Perception of Time." *J of Experimental Psychology* 10 (1927): 52-59.

Handel, Stephen. "Space Is to Time as Vision Is to Audition: Seductive but Misleading." *J of Experimental Psychology: Human Perception and Performance* 14.2 (1988): 315-17.

Hartocollis, Peter. *Time and Timelessness, or the Varieties of Temporal Experience*. New York: International Universities P, 1983.

Harton, J. J. "The Influence of the Degree of Unity of Organization on the Estimation of Time." *J of General Psychology* 21 (1939): 25-49.

——. "The Influence of the Difficulty of Activity on the Estimation of Time." *J of Experimental Psychology* 23 (1938): 270-87, 428-33.

——. "An Investigation of the Influence of Success and Failure on the Estimation of Time." *J of General Psychology* 21 (1939): 51-62.

——. "Time Estimation in Relation to Goal Organization and Difficulty of Tasks." *J of General Psychology* 27 (1942): 63-69.

Hawkes, G. R., R. J. T. Joy, and W. O. Evans. "Autonomic Effects on Estimates of Time: Evidence for a Physiological Correlate of Temporal Experience." *J of Psychology* 53 (1962): 183-91.

Heath, Richard A. "Response Time and Temporal Order Judgment in Vision." *Australian J of Psychology* 36.1 (1984): 21-34.

Hellström, Å. "Time Errors Are Perceptual: An Experimental Investigation of Duration and a Quantitative Successive-Comparison Model." *Psychological Research* 39 (1977): 345-88.

Hicks, R. E. "Effect of Information Feedback upon Intertrial Consistency of Time Judgment." *Acta Psychologica* 40 (1976): 265-70.

Hicks, R. E., and D. A. Allen. "Counting Eliminates the Repetition Effect in Judgments of Temporal Duration." *Acta Psychologica* 43 (1979): 361-66.

———. "The Repetition Effect in Judgments of Temporal Duration Across Minutes, Days, and Months." *American J of Psychology* 92 (1979): 323-33.

Hicks, Robert E., Deborah A. Allen, and James P. Mayo. "A Developmental Study of Temporal Duration Judgments." *J of Genetic Psychology* 144.1 (1984): 31-38.

Hicks, R. E., and R. M. Brundige. "Judgments of Temporal Duration while Processing Verbal and Physiognomic Stimuli." *Acta Psychologica* 36 (1974): 447-53.

Hicks, R. E., and G. W. Miller. "Transfer of Time Judgments as a Function of Feedback." *American J of Psychology* 89 (1976): 303-10.

Hicks, R. E., G. W. Miller, and M. Kingsbourne. "Prospective and Retrospective Judgments of Time as a Function of Amount of Information Processed." *American J of Psychology* 89 (1976): 719-30.

Hicks, R. E., et al. "Cannabis, Atropine, and Temporal Information Processing." *Neuropsychobiology* 12.4 (1984): 229-37.

———. "Concurrent Processing Demands and the Experience of Time-in-Passing." *American J of Psychology* 90 (1977): 431-46.

———. "Effects of Response Language and Stimulus Context upon Judgments of Visual and Temporal Extent." *American J of Psychology* 96 (1983): 365-75.

Hinrichs, J. V., and H. Buschke. "Judgment of Recency Under Steady-State Conditions." *J of Experimental Psychology* 78.4 Part 1 (1968): 574-79.

Hirano, Shin. "Effect of Pause Between Discrete Events on Time Estimation." *J of Child Development* 17 (1981): 35-39.

Hoagland, H. "The Physiological Control of Judgments of Duration: Evidence for a Chemical Clock." *J of General Psychology* 9 (1933): 267-87.

Hogan, H. W. "Time Perception and Stimulus Preference as a Function of Stimulus Complexity." *J of Personality and Social Psychology* 31 (1975): 32-35.

Hoopen, Gert Ten, J. Vos, and J. Dispa. "Interaural and Nonaural Clicks and Clocks: Tempo Difference Versus Attention Switching." *J of Experimental Psychology: Human Perception and Performance* 8 (1982): 422-34.

Hornstein, A. D., and G. S. Rotter. "Research Methodology in Temporal Perception." *J of Experimental Psychology* 79.3 Part 1 (1969): 561-64.

Hulbert, R. J., et al. "Time Perspective, Time Attitude, and Time Orientation in Alcoholism: A Review." *International J of the Addictions* 23.3 (March 1988): 279-98.

Ihle, R. C., and W. E. Wilsoncroft. "Filled-Duration Illusion: Limits of Duration of Interval and Auditory Fillers." *Perceptual and Motor Skills* 56 (April 1983): 655-60.

Jackson, Janet L. "A Cognitive Approach to Temporal Information Processing." *Cognitive Models of Psychological Time*. Ed. Richard A. Block. Hillsdale: Erlbaum, 1990. 153-80.

Jackson, J. L., and J. A. Michon. "Effects of Item Concreteness on Temporal Coding." *Acta Psychologica* 57 (1984): 83-95.

Jackson, J. L., J. A. Michon, and A. Vermeeren. "The Processing of Temporal Information." *Timing and Time Perception*. Ed. J. Gibbon and L. Allan. *Annals of the New York Academy of Sciences* 423 (1984): 603-04.

Jackson, J. L., et al. "The Effect of Depth of Processing on Temporal Judgment Tasks." *Acta Psychologica* 62 (1986): 199-210.

Jones, M. R. "Attentional Rhythmicity in Human Perception." *Rhythm in Psychological, Linguistic, and Music Processes*. Ed. J. R. Evans, and M. Clynes. Springfield: Charles C. Thomas, 1986. 13-40.

——. "Controlled Attending as a Function of Melodic and Temporal Context." *Perception and Psychophysics* 32 (1982): 211-18.

——. "The Patterning of Time and Its Effect on Perceiving." *Annals of the New York Academy of Sciences* 423 (1984): 158-67.

——. "Time, Our Lost Dimension: Toward a New Theory of Perception, Attention, and Memory." *Psychological Review* 83.5 (September 1976): 323-55.

Josenhans, W. "Versuch einer Objektivierung von Änderungen im Zeiterleben und Zeitempfinden." *Naturwissenschaften* 46 (1959): 538-39.

Joubert, Charles E. "Structured Time and Subjective Acceleration of Time." *Perceptual and Motor Skills* 59.1 (1984): 335-36.

Juhnke, R., et al. "Psychology of Computer Use: V. Computer Use and the Experience of Time." *Perceptual and Motor Skills* 67.3

(December 1988): 863-70.

Kastenbaum, R. "The Dimensions of Future Time Perspective: An Experimental Analysis." *The J of General Psychology* 65 (1961): 203-18.

——. "The Structure and Function of Time Perspective." *J of Psychological Researches* (India) 8 (1954): 1-11.

Katz, D. "Experimentelle Beiträge zur Psychologie des Vergleichs im Gebiete des Zeitsinns." *Zeitschrift für Psychologie und Physiologie der Sinnesorgane* 42 (1906): 302-40, 424-50.

Kirkcaldy, Bruce D. "Individual Differences in Time Estimation." *International J of Sport Psychology* 15.1 (1984): 11-24.

Kleber, R. J., W. T. Lhamon, and S. Goldstone. "Hyperthermia, Hyperthyroidism, and Time Judgment." *J of Comparative and Physiological Psychology* 56 (1963): 362-65.

Krauss, H. H., et al. "Anxiety and Temporal Perspective Among Normals in a Stressful Life Situation." *Psychological Reports* 21.3 (1967): 721-24.

Krishnan, L., and N. K. Saxena. "Perceived Time: Its Relationship with Locus of Control, Filled Versus Unfilled Time Intervals, and Perceiver's Sex." *J of General Psychology* 110 (April 1984): 275-81.

Kristofferson, A. B. "Quantal and Deterministic Timing." *Annals of the New York Academy of Sciences* 423 (1984): 3-15.

——. "A Quantal Step Function in Duration Discrimination." *Perception and Psychophysics* 27 (1980): 300-06.

Kristofferson, A. B., and L. G. Allan. "Successiveness and Duration Discrimination." *Attention and Performance IV*. Ed. S. Kornblum. New York: Academic P, 1973. 737-49.

Kruup, K. "Influence of Method on Time Judgments." *Australian J of Psychology* 13 (1961): 44-53.

Langer, J., S. Wapner, and H. Werner. "The Effect of Danger upon the Experience of Time." *American J of Psychology* 74 (1961): 94-97.

Lavie, P., and W. B. Webb. "Time Estimates in a Long-Term Time-Free Environment." *American J of Psychology* 88 (1975): 177-86.

Lee, D. N., et al. "Visual Timing in Hitting an Accelerating Ball." *Quarterly J of Experimental Psychology: Human Experimental Psychology* 35 A.2 (1983): 333-46.

Legg, C. F. "Alpha Rhythm and Time Judgments." *J of Experimental Psychology* 78.1 (1968): 46-49.

Leonov, Aleksei Arkhipovich, and Vladimir Lebedev. *Space and Time Perception by the Cosmonaut*. Trans. B. Belitsky. Moscow: Mir Publishers, 1971.

Levine, R., L. West, and H. Reis. "Perceptions of Time and Punctuality in the United States and Brazil." *J of Personality and Social Psychology* 38.4 (1980): 541-50.

Levitt, Eugene E. "Estimating the Duration of Sexual Behaviour: A Laboratory Analog Study." *Archives of Sexual Behaviour* 12.4 (1983): 329-35.

Lockhart, J. M. "Ambient Temperature and Time Estimation." *J of Experimental Psychology* 73.2 (1967): 286-91.

Loehlin, J. C. "The Influence of Different Activities on the Apparent Length of Time." *Psychological Monographs: General and Applied* 73 (1959): 1-25.

Macar, F., and N. Vitton. "Contingent Negative Variation and Accuracy of Time Estimation: A Study on Cats." *EEG and Clinical Neurophysiology* 47 (1979): 213-28.

——. "Force and Movement Parameters in Timing Accuracy." *Acta Psychologica* 72 (1989): 247-61.

——. "Temporal Threshold and Response Bias in the Discrimination of Empty Intervals by Cats." *Behavioural Processes* 11 (1985): 337-48.

Macar, F., N. Vitton, and J. Requin. "Slow Cortical Potential Shifts Recorded in Cats Performing a Time Estimation Task: Preliminary Results." *Neuropsychologia* 14 (1976): 353-61.

——. "Time Discrimination Versus Time Regulation: A Study on Cats." *Timing and Time Perception*. Ed. J. Gibbon, and L. Allan. *Annals of the New York Academy of Sciences* 423 (1984): 622-24.

McGrath, J. J., and J. F. O'Hanlon. "Method for Measuring the Rate of Subjective Time." *Perceptual and Motor Skills* 24 (1967): 1235-40.

MacLeod, R. B., and M. F. Roff. "An Experiment in Temporal Disorientation." *Acta Psychologica* 1 (1935): 381-423.

Malhotra, Rekha, and K. D. Broota. "Perception of Brief Durations as a Function of Size." *Psychological Studies* 29.2 (1984): 125-30.

Marum, K. D. "Reproduction and Ratio-Production of Brief Duration Under Conditions of Sensory Isolation." *American J of Psychology* 81.1 (1968): 21-26.

Matsuda, F. "Development of Time Estimation: Effects of Frequency of Sounds Given During Standard Time." *Japanese J of Psychology* 36 (1966): 285-94.

May, J. G., et al. "Temporal Order Judgements in Good and Poor Readers." *Neuropsychologia* 26.6 (1988): 917-24.

Meade, R. D. "Time Estimates as Affected by Motivational Level, Goal Distance, and Rate of Progress." *J of Experimental Psychology* 58 (1959): 275-79.

——. "Time Estimates as Affected by Need Tension and Rate of Progress." *J of Psychology* 50 (1960): 173-77.

Meerloo, Joost A. M. *Along the Fourth Dimension: Man's Sense of Time and History*. New York: John Day, 1970.

Michon, John A. "Magnitude Scaling of Short Durations with Closely Spaced Stimuli." *Psychonomic Science* 9.6 (1967): 359-60.

——. "The Perception and Evaluation of Time from a Psychological Viewpoint." *Epistemologia* 1 (1978): 223-50.

——. "Studies in Subjective Duration I: Differential Sensitivity in the Perception of Repeated Temporal Intervals." *Acta Psychologica* 22 (1964): 441-50.

——. "Studies in Subjective Duration II: Subjective Time Measurement during Tasks with Different Information Content." *Acta Psychologica* 24 (1965): 205-19.

——. "Tapping Regularity as a Measure of Perceptual Motor Load." *Ergonomics* 9 (1966): 401-12.

Michon, John A., A. Bruinsma, and W. Riedel. "Concatenation of Short Time Intervals." *Limits in Perception*. Ed. A. J. van Doorn, W. A. van de Grind, and J. J. Koenderink. Utrecht: VNU Science P, 1984. 381-95.

Miller, G. W., R. E. Hicks, and M. Willette. "Effects of Concurrent Verbal Rehearsal and Temporal Set upon Judgments of Temporal Duration." *Acta Psychologica* 42 (1978): 173-79.

Mitchell, C. Thomas, and Roy Davis. "The Perception of Time in Scale Model Environments." *Perception* 16.1 (1987): 5-16.

Moiseeva, N. I. "Perception of Time by Human Consciousness." *Chronobiologia* 15.4 (October-December 1988): 301-17.

Müller, K. "Die phänomenale Dauer visueller Sukzessionen. Experimentelle Untersuchungen zum Problem der Abhängigkeit der Zeitwahrnehmung von der Geschehensstruktur." *Zeitschrift für experimentelle und angewandte Psychologie* 12 (1965): 98-123.

Murphy, L. E. "Absolute Judgments of Duration." *J of Experimental Psychology* 71.2 (1966): 260-63.

Nelson, T. M., S. H. Bartley, and J. F. Jordan. "Experimental Evidence for the Involvement of a Neurophysiological Mechanism in the Discrimination of Duration." *J of Psychology* 55 (1963): 371-85.

Ochberg, F. M., I. W. Pollack, and E. Meyer. "Reproduction and Estimation Methods of Time Judgments." *Perceptual and Motor Skills* 20 (1965): 653-56.

Otto, E. "Modifizierende und konstituierende Faktoren der Zeitwahrnehmung beim Menschen." *Biologische Rundschau* 1 (1963): 2-18.

——. "Unterschiedsempfindlichkeit für Zeitintervalle und automatisierte Motorik nach Injektion von Amytal-Na." *Zeitschrift für Psychologie* 168 (1963): 176-94.

Patterson, Robert. "Perceptual Moment Models Revisited." *Cognitive Models of Psychological Time*. Ed. Richard A. Block. Hillsdale: Erlbaum, 1990. 85-100.

Pfaff, D. "Effects of Temperature and Time of Day on Time Judgments." *J of Experimental Psychology* 76.3 Part 1 (1968): 419-22.

Philbin, T., and R. M. Seidenstadt. "Feedback and Time Perception." *Perceptual and Motor Skills* 57 (August 1983): 308.

Piéron, H. "Les problèmes psycho-physiologiques de la perception du temps." *Année Psychologique* 24 (1923): 1-25.

Polzella, D. J., S. M. Bower, and A. S. Gouse. "Manipulating Apparent Duration with Simultaneous Effects on Memory." *Bulletin of the Psychonomic Society* 20 (1982): 175-77.

Polzella, D. J., F. DaPolito, and M. C. Hinsman. "Cerebral Asymmetry in Time Perception." *Perception and Psychophysics* 21 (1977): 187-92.

Pöppel, E. "Time Perception." *Handbook of Sensory Physiology, Vol. 8*. Ed. R. Held, H. W. Leibowitz, and H.-L. Teuber. Berlin: Springer, 1978. 713-29.

Pöppel, E., and H. Giedke. "Diurnal Variation of Time Perception." *Psychologische Forschung* 34 (1970): 182-98.

Predebon, J. "Organization of Stimulus Events and Remembered Apparent Duration." *Australian J of Psychology* 36 (1984): 161-69.

———. "Remembered Duration and Perceptual Reversals." *Acta Psychologica* 69 (1988): 157-63.

———. "Retrospective Time Judgments and Clock Duration." *Perceptual and Motor Skills* 66 (February 1988): 19-24. Includes bibliography.

Quasebarth, K. "Zeitschätzung und Zeitauffassung optisch und akustisch ausgefüllter Intervalle." *Archiv für die gesamte Psychologie* 49 (1924): 379-432.

Rama Rao, Palamund. *Studies in Time Perception*. Delhi: Concept Publishing, 1978.

Rammsayer, T., et al. "Sex Differences in Time Perception." *Perceptual and Motor Skills* 68.1 (February 1989): 195-98.

Raslear, Thomas G. "Perceptual Bias and Response Bias in Temporal Bisection." *Perception and Psychophysics* 38.3 (1985): 261-68.

Richards, W. "Time Estimation Measured by Reproduction." *Perceptual and Motor Skills* 18 (1964): 929-43.

Richards, W. J., and P. V. Livingston. "Method, Standard Duration, and Inter-Stimulus Delay as Influences upon Judgment of Time." *American J of Psychology* 79 (1966): 560-67.

Richie, D. M., and M. H. Bickhard. "The Ability to Perceive Duration: Its Relation to the Development of the Logical Concept of Time."

Developmental Psychology 24 (May 1988): 318-23. Includes bibliography.

Robertson, L. C., and L. M. Gomez. "Figural Versus Configural Effects in the Filled Duration Illusion." *Perception and Psychophysics* 27 (1980): 111-16.

Roitblat, H. L., and K. N. J. Young. "Time and Order: A Comparative Perspective." *Cognitive Models of Psychological Time*. Ed. Richard A. Block. Hillsdale: Erlbaum, 1990. 119-52.

Ross, S., and L. Katchmar. "The Construction of a Magnitude Function for Short-Time Intervals." *American J of Psychology* 64 (1951): 397-401.

Rule, Stanley I. "Composition Rule for Perceived Duration of Simultaneous Events." *Perception and Psychophysics* 34.6 (1983): 569-72.

Rutschmann, J., and L. Rubinstein. "Time Estimation, Knowledge of Results and Drug Effects." *J of Psychiatric Research* 4.2 (1966): 107-14.

St. Jean, R. "Amnesia and Hypnotic Time Estimation." *International J of Clinical and Experimental Hypnosis* 30.2 (April 1982): 127-37.

———. "Attentional Versus Absorptive Processing in Hypnotic Time Estimation." *J of Abnormal Behavior* 95 (1986): 40-42.

St. Jean, R., and C. Macleod. "Hypnosis, Absorption, and Time Perception." *J of Abnormal Psychology* 92 (February 1983): 81-86.

Schaefer, V. G., and A. R. Gilliland. "The Relation of Time Estimation to Certain Physiological Changes." *J of Experimental Psychology* 23 (1938): 545-52.

Schiffman, H. R., and D. J. Bobko. "Effects of Stimulus Complexity on the Perception of Brief Temporal Intervals." *J of Experimental Psychology* 103 (1974): 156-59.

———. "The Role of Number and Familiarity of Stimuli in the Perception of Brief Temporal Intervals." *American J of Psychology* 90 (1977): 85-93.

Schoeffler, M. S., and D. M. Poole. "Accuracy and Variability in the Production of Short Durations." *Psychonomic Science* 7.12 (1967): 423-24.

Sebel, Aliza I., and W. E. Wilsoncroft. "Auditory and Visual Differences in Time Perception." *Perceptual and Motor Skills* 57 (August 1983): 295-300.

Siegman, A. W. "The Relationship between Future Time Perspective, Time Estimation, and Impulse Control in a Group of Young Offenders and in a Control Group." *J of Consulting Psychology* 25 (1961): 470-75.

Silverstein, C. I., and A. M. Owens. "Perception of Time During

Examination." *Perceptual and Motor Skills* 55 (1982): 298.

Sixtl, F. "Die Erfassung von Sukzessionen bei Ausschaltung der aktiven Vergegenwärtigung." *Archiv für die gesamte Psychologie* 114 (1962): 337-77.

Smirnov, A. G., and M. V. Polyakova. "Assessment of Time Interval by the Different Fingers." *Human Psychology* 9.3 (1984): 156-61.

Spencer, L. T. "An Experiment in Time Estimation Using Different Interpolations." *American J of Psychology* 32 (1921): 557-62.

Spreen, O. "The Position of Time Estimation in a Factor Analysis and Its Relation to Some Personality Variables." *Psychological Records* 13 (1963): 455-64.

Stalnaker, J. M., and M. W. Richardson. "Time Estimation in the Hypnotic Trance." *J of General Psychology* 4 (1930): 362-66.

Steinberg, H. "Changes in Time Perception Induced by an Anaesthetic Drug." *British J of Psychology* 46.4 (November 1955): 273-79.

Stephens, G. J., and F. Halberg. "Human Time Estimation." *Nursing Research* 14 (1965): 310-17.

Sterzinger, O. "Chemopsychologische Untersuchungen über den Zeitsinn." *Zeitschrift für Psychologie und Physiologie der Sinnesorgane* 134 (1935): 100-31.

——. "Neue chemopsychologische Untersuchungen über den menschlichen Zeitsinn (das Problem der 5-Minutenzeitstrecke)." *Zeitschrift für Psychologie und Physiologie der Sinnesorgane* 143 (1938): 391-406.

Stevens, L. T. "On the Time-Sense." *Mind* 11 (1886): 393-404.

Straus, E. W. *The Primary World of Senses*. New York: Free P of Glencoe, 1963.

Stroud, J. M. "The Fine Structure of Psychological Time." *Information Theory in Psychology*. Ed. H. Quastler. Glencoe: Free P, 1955. 174-207.

Stubbs, D. Alan, Leon R. Dryfus, and J. Gregor Fetterman. "The Perception of Temporal Events." *Annals of the New York Academy of Sciences* 423 (1984): 30-42.

Sturmer, G. von. "Stimulus Variation and Sequential Judgments of Duration." *Quarterly J of Experimental Psychology* 18 (1966): 354-57.

Sturmer, G. von, T. Wong, and M. Coltheart. "Distraction and Time Estimation." *Quarterly J of Experimental Psychology* 20 (1968): 380-84.

Summers, Jeffrey J., and Bruce D. Burns. "Timing in Human Movement Sequences." *Cognitive Models of Psychological Time*. Ed. Richard A. Block. Hillsdale: Erlbaum, 1990. 181-206.

Surwillo, W. W. "Time Perception in Relation to Pulse Rate in Healthy Males." *J of Psychology* 110, 1st Half (January 1982):

101-06.

Swartz, P. "Concerning the Continuous Present." *Perceptual and Motor Skills* 57 (August 1983): 159-71.

Swift, E. J., and J. A. McGeoch. "An Experimental Study of the Perception of Filled and Empty Time." *J of Experimental Psychology* 8 (1925): 240-49.

Thomas, E. A. C., and I. Brown, Jr. "Time Perception and the Filled-Duration Illusion." *Perception and Psychophysics* 16 (1974): 449-58.

Thomas, E. A. C., and N. E. Cantor. "On the Duality of Simultaneous Time and Size Perception." *Perception and Psychonomics* 18 (1975): 44-48.

——. "Simultaneous Time and Size Perception." *Perception and Psychophysics* 19 (1976): 353-60.

Thor, D. H. "Diurnal Variability in Time Estimation." *Perceptual and Motor Skills* 15 (1962): 451-54.

"Timing and Time Perception." *Annals of the New York Academy of Sciences* 423 (1984): 1-654. Contains a wide range of articles on this subject.

Treisman, M. "Temporal Discrimination and the Indifference Interval: Implications for a Model of the 'Internal Clock.'" *Psychological Monographs* 77 (1963): No. 576.

Underwood, G. "Attention and the Perception of Duration During Encoding and Retrieval." *Perception* 4 (1975): 291-96.

Underwood, G, and R. A. Swain. "Selectivity of Attention and the Perception of Duration." *Perception* 2 (1973): 101-05.

Vierordt, K. *Der Zeitsinn nach Versuchen*. Tübingen: Laupp, 1868.

Vroon, P. A. "Effects of Presented and Processed Information on Duration Experience." *Acta Physiologica* 34.1 (1970): 115-21.

——. "The Lengthening Effect in Sequential Estimations of a Short Interval." *Psychologische Forschung* 35 (1972): 263-76.

——. "Sequential Estimations of Time." *Acta Psychologica* 40 (1976): 475-87.

Vroon, P. A., and A. van Boxtel. "Testing Some Implications of the Sensory Physiological Model." *Psychologische Forschung* 35 (1972): 81-92.

Vroon, P. A., and A. G. Vroon. "Tapping Rate and Expectancy in Simple Reaction Time Tasks." *J of Experimental Psychology* 98 (1973): 85-90.

Walker, J. T., A. L. Irion, and D. G. Gordon. "Simple and Contingent Aftereffects of Perceived Duration in Vision and Audition." *Perception and Psychophysics* 29 (1981): 475-86.

Wallace, M., and A. I. Rabin. "Temporal Experience." *Psychological*

Bulletin 57 (1960): 213-36.

Warm, J. S., L. F. Greenberg, and H. C. S. Dube. "Stimulus and Motivational Determinants in Temporal Perception." *J of Psychology* 58 (1964): 243-48.

Warm, J. S., J. R. Morris, and J. K. Kew. "Temporal Judgment as a Function of Nosological Classification and Experimental Method." *J of Psychology* 55 (1963): 287-97.

Warm, J. S., R. P. Smith, and L. S. Caldwell. "Effects of Induced Muscle Tension on Judgment of Time." *Perceptual and Motor Skills* 25.1 (1967): 153-60.

Waterworth, John A. "Memory Mechanisms and the Psychophysical Scaling of Duration." *Perception* 14.1 (1985): 81-92.

Watts, F. N., and R. Sharrock. "Fear and Time Estimation." *Perceptual and Motor Skills* 59 (October 1984): 597-98.

Weber, A. O. "Estimation of Time." *Psychological Bulletin* 30 (1933): 233-52.

Weber, D. S. "A Time Perception Task." *Perceptual and Motor Skills* 21.3 (1965): 863-66.

White, C. T. "Temporal Numerosity and the Psychological Unit of Duration." *Psychological Monographs* 77 (1963): No. 575.

Whyman, A. D., and R. H. Moos. "Time Perception and Anxiety." *Perceptual and Motor Skills* 24 (1967): 567-70.

Woodrow, H. "The Effect of Practice upon Time-Order Errors in the Comparison of Temporal Intervals." *Psychological Review* 42.2 (March 1935): 127-52.

——. "Time Perception." *Handbook of Experimental Psychology*. Ed. S. S. Stevens. New York: Wiley, 1951. 1224-36.

Zakay, D. "The Evasive Art of Subjective Time Measurement: Some Methodological Dilemmas." *Cognitive Models of Psychological Time*. Ed. Richard A. Block. Hillsdale: Erlbaum, 1990. 59-84.

——. "An Integrated Model of Time Estimation." *Time and Human Cognition: A Life-Span Perspective*. Ed. I. Levin, and D. Zakay. Amsterdam: North-Holland, 1989. 365-97.

Zakay, D., D. Nitzan, and J. Glicksohn. "The Influence of Task Difficulty and External Tempo on Subjective Time Estimation." *Perception and Psychophysics* 34 (1983): 451-56.

Zelkind, I., and B. Spilka. "Some Time Perspective Time Perception Relationships." *Psychological Record* 15 (1965): 417-21.

Zimbardo, P. G., G. Marshall, and C. Maslach. "Liberating Behavior from Time-Bound Control: Expanding the Present through Hypnosis." *J of Applied Social Psychology* 1.4 (1971): 305-23.

Zimbardo, P. G., G. Marshall, and G. White. "Objective Assessment of Hypnotically Induced Time Distortion." *Science* 181 (1973): 282-84.

Zipf, George K. "The Repetition of Words, Time-Perspective, and Semantic Balance." *J of General Psychology* 32 (1945): 127-48.

Zoltobrocki, J. *Über systematische Fehler bei wiederholter Schätzung von Zeitintervallen*. Frankfurt: Kramer, 1965.

22. RELIGION

A. GENERAL

Balthasar, Hans Urs von. *A Theology of History*. London: Sheed and Ward, 1964.

Barr, James. *Biblical Words for Time*. Naperville: A. R. Allenson, 1962.

Braine, David. *The Reality of Time and the Existence of God: The Project of Proving God's Existence*. New York: Oxford UP, 1988.

Brandon, S. G. F. "The Deification of Time." *The Study of Time I*. Ed. J. T. Fraser, F. C. Haber, and G. H. Müller. New York: Springer, 1972. 370-82.

——. *History, Time, and Deity*. New York: Barnes and Noble, 1965.

——. *Man and His Destiny in the Great Religions*. Manchester: Manchester UP, 1962.

——. "The Personification of Death in Some Ancient Religions." *Bulletin of the John Rylands Library, Manchester* 43 (1961): 318-22, 333-35.

Brown, G. "Praying about the Past." *Philosophical Quarterly* 35 (January 1985): 83-86.

Cook, R. R. "God, Time and Freedom." *Religious Studies* 23 (March 1987): 81-94.

Corbin, Henry. "Cyclical Time in Mazdaism and Ismailism." *Man and Time: Papers from the Eranos Yearbooks*. Trans. Ralph Manheim. Ed. Joseph Campbell. New York: Pantheon, 1957. 115-72.

Corish, D. "Time, Space, and Freewill: The Leibniz-Clarke Correspondence." *The Study of Time III*. Ed. J. T. Fraser, N. Lawrence, and D. Park. New York: Springer, 1978. 634-55.

Cornford, F. M. *Plato's Cosmology*. London: Routledge and Kegan Paul, 1948.

Davis, Richard H., ed. *Religion and Aging: The Behavioral and Social Sciences Look at Religion and Aging*. Los Angeles: U of Southern California P, 1967.

Eliade, M. *The Myth of the Eternal Return*. New York: Pantheon, 1954.

Farrer, Austin. *Finite and Infinite*. 2nd ed. London: Dacre P, 1956.

Granville, William Anthony. *The Fourth Dimension and the Bible*. Boston: R. G. Badger, 1922.

Guitton, J. *Justification du temps*. Paris: Presses Universitaires de France, 1941.

Heninger, S. K. *The Cosmological Glass: Renaissance Diagrams of the Universe*. San Marino: Huntington Library, 1977.

Heschel, Abraham Joshua. *The Sabbath: Its Meaning for Modern Man.* New York: Farrar, Straus and Young, 1951.

Hesiod. *Theogony.* In *The Works and Days, Theogony, The Shield of Herakles.* Trans. Richmond Lattimore. Ann Arbor: U of Michigan P, 1959.

King, J. R. "The Moment as a Theological Category." *Studies in Formative Spirituality* 9 (February 1988): 79-95.

Leftow, B. "The Roots of Eternity." *Religious Studies* 24 (June 1988): 189-212.

Lewis, D. "Timelessness and Divine Agency." *International J for Philosophy of Religion* 21.3 (1987): 143-59.

Lucas, J. R. *The Future: An Essay on God, Temporality and Truth.* Cambridge: Blackwell, 1989.

Lynch, W. F. "Imagining Past, Present, and Future in One Piece." *Studies in Formative Spirituality* 6 (February 1985): 65-72.

Macey, Samuel L. *Patriarchs of Time: Dualism in Saturn-Cronus, Father Time, The Watchmaker God, and Father Christmas.* Athens: U of Georgia P, 1987.

Meerloo, J. A. M. *Dance and Sacred Dance.* London: Peter Owen, 1961.

Moberg, David O. "Religion and the Aging Family." *The Family Coordinator* 21.1 (1972): 47-60.

——. "Religion and the Later Years." *The Daily Needs and Interests of Older People.* Springfield: Charles C. Thomas, 1970. 175-91.

——. "Religiosity in Old Age." *The Gerontologist* 5.2 (1965): 78-87.

Moore, Ralph J., and Brooke N. Moore. *The Cosmos, God and Philosophy.* New York: P. Lang, 1989.

Mouroux, J. *Le mystère du temps, approche théologique.* Paris: Aubier, 1962.

Norris, Bill, and Judy Norris. *What the Bible Says about Growing Old.* Joplin: College P Publishing, 1988.

Orbach, Harold L. "Aging and Religion." *Geriatrics* 16.10 (1961): 530-40.

Panikkar, R. "Time and Sacrifice—The Sacrifice of Time and the Ritual of Modernity." *The Study of Time III.* Ed. J. T. Fraser, N. Lawrence, and D. Park. New York: Springer, 1978. 683-725.

Pike, Nelson. *God and Timelessness.* New York: Schocken Books, 1970.

Plessner, Helmuth. "On the Relation of Time to Death." *Man and Time: Papers from the Eranos Yearbooks.* Trans. Ralph Manheim. Ed. Joseph Campbell. New York: Pantheon, 1957. 233-63.

Pruyser, P. W. "The Idea of Destiny." *Hibbert J* 57 (1958-59): 380-85.

Puech, Henri-Charles. "Gnosis and Time." *Man and Time: Papers from the Eranos Yearbooks*. Trans. Ralph Manheim. Ed. Joseph Campbell. New York: Pantheon, 1957. 38-84.

Richardson, A. *History, Sacred and Profane*. London: SCM P, 1963.

Santillana, G. de, and H. von Dechend. *Hamlet's Mill: An Essay on Myth and the Frame of Time*. London: Macmillan, 1970.

Schragg, O. O. "An Existentialist Challenge to Philosophy of Religion." *International J for Philosophy of Religion* 4 (1973): 151-67.

Schwanitz, D. "Time Is Out of Joint, but Life Goes on: Hamlet and the Ephesian Matron." *J of the American Academy of Religion* 51 (March 1983): 37-53. Includes bibliography.

Sorabji, Richard. *Time, Creation and the Continuum: Theories in Antiquity and the Early Middle Ages*. London: Duckworth, 1983.

Stace, Walter Terence. *Time and Eternity: An Essay in the Philosophy of Religion*. Princeton: Princeton UP, 1952.

Stagg, Frank. *The Bible Speaks on Aging*. Nashville: Broadman P, 1981.

Terrien, S. "The Pantomime Cat: T. S. Eliot and Hebrew-Christian Dynamics." *Theology Today* 44 (January 1988): 450-61.

Van der Leeuw, G. "Primordial Time and Final Time." *Man and Time: Papers from the Eranos Yearbooks*. Trans. Ralph Manheim. Ed. Joseph Campbell. New York: Pantheon, 1957. 324-50.

Werfel, Franz V. *Star of the Unborn*. Trans. Gustave O. Arlt. New York: Viking, 1946.

B. CHRISTIAN

Baker, J. R. "Omniscience and Divine Synchronization." *Process Studies* 2 (1972): 201-08.

Barry, L. E. "Eternity and Critical Insight." *New Scholasticism* 48 (1974): 351-59.

Bowman, C. R. "Time and the Religious Consciousness." *Southern J of Philosophy* 11 (1973): 73-82.

Brabant, F. H. *Time and Eternity in Christian Thought*. London: Longmans, Green, 1937.

Case, S. J. *The Christian Philosophy of History*. Chicago: U of Chicago P, 1943.

Cullmann, Oscar. *Christ and Time: The Primitive Christian Conception of Time and History*. Trans. Floyd V. Filson. London: SCM P, 1967.

Dubé, Arthur Joseph. *The General Principles for the Reckoning of Time in Canon Law: An Historical Synopsis and Commentary*. Washington: Catholic U of America P, 1941.

Filson, F. V. *Primitive Christian Conception of Time and History*.

London: SCM P, 1951.

Gentile, Giovanni. *Giordano Bruno e il pensiero del rinascimento.* 2nd ed. Florence: Vallecchi, 1925.

Hick, J. "Resurrection Worlds and Bodies." *Mind* 82.7 (1973): 409-12.

Kuzniar, A. "Philosophic Chiliasm: Generating the Future or Delaying the End?" *Eighteenth-Century Studies* 19 (Fall 1985): 1-20.

Lamorte, André. *Le problème du temps dans le prophétisme biblique.* Paris: Librarie protestante, 1960.

Martin, John. *Mechanicus ana Flaven; or The Watch Spiritualized.* London, 1763.

Niebuhr, R. *Faith and History; a Comparison of Christian and Modern Views of History.* London: Nisbet, 1949.

Panikkar, R. "La misa como 'consecratio temporis': la tempiternidad." *Sanctum Sacrificium, Proceedings of the V Congreso Eucaristico de Zaragoza.* Zaragoza, 1960. 75-93.

——. "El presente tempiterno." *Teologia y mundo contemporáneo.* Ed. A. Vargas-Machuca. Madrid: Cristiandad, 1975. 133-75.

Rissi, Mathias. *Time and History: A Study on the Revelation.* Trans. Gordon C. Winsor. Richmond: John Knox P, 1966.

Russell, J. L. "Time in Christian Thought." *The Voices of Time.* Ed. J. T. Fraser. 2nd ed. Amherst: U of Massachusetts P, 1981. 59-76.

Shapiro, Herman. *Motion, Time and Space According to William of Ockham.* St. Bonaventura: Franciscan Institute, 1957.

Smalley, Beryl. "The Bible and Eternity: John Wycliff's Dilemma." *J of the Warburg and Courtauld Institutes* 27 (1964): 73-89.

Stapleton, Laurence. "Milton's Conception of Time in *The Christian Doctrine.*" *Harvard Theological Review* 57 (1964): 9-21.

Sturch, R. L. "The Problem of the Divine Eternity." *Religious Studies* 10.12 (1974): 487-93.

Teilhard de Chardin, Pierre. *The Phenomenon of Man.* Trans. Bernard Well. New York: Harper, 1959.

——. *Réflexions et prières dans l'espace-temps.* Ed. Édouard Bret, and Suzane Bret. Paris: Éditions du Seuil, 1972.

Tillich, Paul. *The Interpretation of History.* Trans. N. A. Rasetzki, and E. L. Talmey. New York: Scribner, 1936.

——. *Systematic Theology.* 3 vols. Chicago: U of Chicago P, 1951-57.

Torrance, Thomas Forsyth. *Space, Time and Incarnation.* London: Oxford UP, 1969.

Wieck, R. S., ed. *Time Sanctified: The Book of Hours in Medieval Art and Life.* New York: Braziller, 1988.

Wilson, M. L. "Time and the Religious Dialogue in Mid-Nineteenth-Century America." *Midwest Quarterly* 21 (Winter 1979): 175-95.

Includes bibliography.
Wingrove, C. Ray, and John P. Alston. "Age, Aging, and Church Attendance." *The Gerontologist* 11.4 Part 1 (Winter 1971): 356-58.
——. "Cohort Analysis of Church Attendance, 1939-69." *Social Forces* 53.2 (1974): 324-31.

C. CHRISTIAN ARGUMENT FROM DESIGN

Clark, F. Le Gros, ed. *Paley's Natural Theology. Revised to Harmonize with Modern Science*. London: The Christian Evidence Committee of the Society for Promoting Christian Knowledge, 1885.
Derham, W. *Astro-Theology: Or a Demonstration of the Being and Attributes of God, from a Survey of the Heavens*. London: W. Innys, 1715.
Mornay, Philip of. *A Woorke Concerning the Trewnesse of the Christian Religion*. Trans. Sir Philip Sidney and Arthur Golding. London: Thomas Cadman, 1587. For an early reference in English see 97-100.
Nieuwentyt. *Religious Philosopher*. Trans. John Chamberlayne. London: J. Senex and W. Taylor, 1718.
Paley, William. *Natural Theology: Or, Evidences of the Existence and Attributes of the Deity, Collected from the Appearances of Nature*. 4th ed. London: R. Faulder, 1803. Twenty editions during 1802-20.
Patrick, Simon. *A Brief Account of the New Sect of Latitude-Men*. Introduction by T. A. Birrell. 1662; reprint, Los Angeles: Augustan Reprint, 1963.
Ray, John. *The Wisdom of God Manifested in the Works of the Creation*. 9th ed. London: William and John Innys, 1727.
Search, Edward [Abraham Tucker]. *The Light of Nature Pursued*. London: T. Payne, 1768.
Smith, John. "Of the Existence and Nature of God." In his *Select Discourses*. London: W. Morden, 1660. 131.

D. CREATION

Anderson, Bernard W., ed. *Creation in the Old Testament*. Minneapolis: Augsburg Fortress Publications, 1984.
Arthur, R. T. W. "Continuous Creation, Continuous Time: A Refutation of the Alleged Discontinuity of Cartesian Time." *J of the History of Philosophy* 26 (July 1988): 349-75.

Asimov, Isaac. *Beginnings: The Story of Origins, of Mankind, Life, the Earth, the Universe*. New York: Berkley Publishing, 1989.

Barrow, John D., and Joseph Silk. *The Left Hand of Creation: The Origin and Evolution of the Expanding Universe*. New York: Basic Books, 1986.

Berry, R. J. *God and Evolution: Creation, Evolution, and the Bible*. Nashville: Abingdon P, 1989.

Bird, Wendell R. *The Origin of Species Revisited*. 2 vols. New York: Philosophical Library, 1989.

Bixler, R. Russell. *Earth, Fire and Sea: The Untold Drama of Creation*. Pittsburgh: Baldwin Manor P, 1986.

Blavatsky, Helena P. *The Secret Doctrine*. Ed. Boris de Zirkoff. 3 vols. 7th ed. Wheaton: Theosophical Publishing House, 1980.

Bliss, Richard. *Origins: Creation or Evolution*. El Cajon: Master Books, 1988. 64 pages.

Brandon, S. G. F. *Creation Legends of the Ancient Near East*. London: Hodder and Stoughton, 1963.

Brunner, Emil. *The Christian Doctrine of Creation and Redemption*. Trans. Olive Wyon. Louisville: Westminster John Knox, 1979.

Cole-Turner, R. S. "Genetic Engineering Co-Creation?" *Theology Today* 44 (October 1987): 338-49.

Dewey, Barbara. *The Creating Cosmos*. Inverness: Bartholomew Books, 1985.

Eaton, J. C. "Berkeley's Immaterialism and the Scientific Promise of the Christian Doctrine of Creation." *Harvard Theological Review* 80 (October 1987): 431-47.

Eilberg-Schwartz, H. "Creation and Classification in Judaism: From Priestly to Rabbinic Conceptions." *History of Religions* 26 (May 1987): 357-81.

Eldridge, Niles. *The Monkey Business: A Scientist Looks at Creationism*. New York: Washington Square P, 1982.

Follette, Marcel la. *Creationism, Science, and the Law: Arkansas Case Documents and Commentaries*. Cambridge: MIT P, 1983.

French, W. C. "Returning to Creation: Moltmann's Eschatology Naturalized." *J of Religion* 68 (January 1988): 78-86.

Fritzsch, Harald. *The Creation of Matter: The Universe from Beginning to End*. New York: Basic Books, 1988.

Godfrey, Laurie R., ed. *Scientists Confront Creationism*. New York: Norton, 1984.

Hamilton, Virginia. *In the Beginning: Creation Stories from Around the World*. San Diego: Harcourt Brace, 1988.

Handelman, D. "Myths of Murugan: Asymmetry and Hierarchy in a

South Indian Puranic Cosmology." *History of Religions* 27 (November 1987): 133-70.

Hanson, Robert W., ed. *Science and Creation: Geological, Theological and Educational Perspectives*. New York: Macmillan, 1985.

Harrold, Francis B., and Raymond A. Eve, eds. *Cult Archaeology and Creationism: Understanding Pseudoscientific Beliefs about the Past*. Iowa City: U of Iowa P, 1987.

Heidel, A. *The Babylonian Genesis*. Chicago: U of Chicago P, 1967.

Heller-Wilensky, Sarah. "Isaac Arama on the Creation and Structure of the World." *Essays in Medieval Jewish and Islamic Philosophy*. Ed. Arthur Hyman. New York: Ktav Publishing House, 1977.

Horigan, James E. *Chance or Design?* New York: Philosophical Library, 1979.

Jastrow, Robert. *God and the Astronomers*. New York: Norton, 1978.

Kass, L. R. "Evolution and the Bible: Genesis 1 Revisited." *Commentary* 86 (November 1988): 29-39.

Leslie, J. "Mackie on Neoplatonism's 'Replacement for God.'" *Religious Studies* 22 (September-December 1986): 325-42.

Lincoln, Bruce. *Myth, Cosmos and Society: Indo-European Themes of Creation and Destruction*. Cambridge: Harvard UP, 1986.

Long, C. H., ed. *Alpha, The Myths of Creation*. New York: Braziller, 1963.

Longsworth, R. "Two Medieval Cornish Versions of the Creation of the World." *Comparative Drama* 21 (Fall 1987): 249-58.

Murphy, R. E. "Wisdom and Creation." *J of Biblical Literature* 104 (March 1985): 3-11.

Niditch, Susan. *Chaos to Cosmos: Studies in Biblical Patterns of Creation*. Atlanta: Scholars P, 1985.

O'Brien, Joan, and Wilfred Major. *In the Beginning: Creation Myths from Ancient Mesopotamia, Israel, and Greece*. Atlanta: Scholars P, 1985.

Och, B. "The Garden of Eden: From Creation to Covenant." *Judaism* 37 (Spring 1988): 143-56; and 37 (Summer 1988): 340-51.

Pannenberg, W. "The Doctrine of Creation and Modern Science [with discussion]." *Zygon* 23 (March 1988): 3-21, 23-43.

Rudavsky, T. M. "Creation, Time and Infinity in Gersonides." *J of the History of Philosophy* 26 (January 1988): 25-44.

Silk, Joseph. *The Big Bang: The Creation and Evolution of the Universe*. 2nd ed. San Francisco: W. H. Freeman, 1988.

Staub, Jacob. *The Creation of the World According to Gersonides*. Atlanta: Scholars P, 1982.

Steiner, Rudolf. *The Evolution of the Earth and the Influence of the*

Stars. Hudson: Anthroposophic P, 1987.

Trefil, James. *The Moment of Creation: Big Bang Physics from Before the First Millisecond to the Present Universe*. New York: Macmillan, 1984.

Van Over, Raymond. *Sun Songs: Creation Myths from Around the World*. New York: New American Library, 1980.

Weigle, M. "Creation and Procreation, Cosmogony and Childbirth: Reflections on Ex Nihilo, Earth Diver, and Emergence Mythology." *J of American Folklore* 100 (October-December 1987): 426-35. Includes bibliography.

Zahan, D. "The Boat of the World as a Pendant." *African Arts* 21 (August 1988): 56-57.

Zolbrod, Paul G. *Dine Bahane: The Navajo Creation Story*. Albuquerque: U of New Mexico P, 1988.

E. EASTERN

Boyd, J. W., and D. A. Crosby. "Is Zoroastrianism Dualistic or Monotheistic?" *American Academy of Religion. J* 47 (December 1979): 557-88. Includes bibliography.

Brown, D. "Buddhism and Historical Thought in Japan before 1221." *Philosophy East and West* 24.4 (1974): 215-25.

Eliade, Mircea. "Time and Eternity in Indian Thought." *Man and Time: Papers from the Eranos Yearbooks*. Trans. Ralph Manheim. Ed. Joseph Campbell. New York: Pantheon, 1957. 173-200.

Govinda, A. "The Mystery of Time." *Main Currents in Modern Thought* 27.1 (1970): 18-23.

Kalupahana, D. J. "The Buddhist Conception of Time and Temporality." *Philosophy East and West* 24.4 (1974): 181-91.

Koller, J. M. "On Buddhist Views of Devouring Time." *Philosophy East and West* 24.4 (1974): 201-08.

Lancaster, L. R. "Discussion of Time in Mahayana Texts." *Philosophy East and West* 24.4 (1974): 209-14.

Nakamura, Hajime. "Time in Indian and Japanese Thought." *The Voices of Time*. Ed. J. T. Fraser. 2nd ed. Amherst: U of Massachusetts P, 1981. 77-91.

Ogawa, H. "The Concept of Time in Mithraic Mysteries." *The Study of Time III*. Ed. J. T. Fraser, N. Lawrence, and D. Park. New York: Springer, 1978. 658-79.

Panikkar, R. "Time and History in the Tradition of India: Kala and Karma." Translated in *Cultures and Time*. Paris: Payot-UNESCO,

1976. 63-68.

Puligandla, R. "Time and History in the Indian Tradition." *Philosophy East and West* 24.4 (1974): 165-70.

Schipper, Kristofer, and Wang Hsiu-Huei. "Progressive and Regressive Time Cycles in Taoist Ritual." *Time, Science, and Society in China and the West: The Study of Time V*. Ed. J. T. Fraser, N. Lawrence, and F. C. Haber. Amherst: U of Massachusetts P, 1986. 185-205.

Vermaseren, M. J. "A Magical Time God." *Mithraic Studies II*. Ed. J. Hinnels. Manchester: Manchester UP, 1975. 446-56.

Zaehner, R. C. *The Dawn and Twilight of Zoroastrianism*. London: Weidenfeld and Nicolson, 1961.

Zimmer, H. *Philosophies of India*. New York: Pantheon, 1953.

F. ESCHATOLOGY

Berdiaev, Nikolai A. *The Beginning and the End*. Trans. R. M. French. 1952; reprint, Westport: Greenwood, 1976.

Berthold-Bond, D. "Hegel's Eschatological Vision: Does History Have a Future?" *History and Theory* 27.1 (1988): 14-29.

Bratt, John H. *Final Curtain—Studies in Eschatology*. Grand Rapids: Baker Book House, 1978.

Brown, T. *The Fate of the Dead: A Study in Folk Eschatology in the West Country after the Reformation*. Wolfeboro: Longwood Publishing, 1979.

Brunt, John C. *Now and Not Yet: Adventist Eschatology and Ethics*. Ed. Richard W. Coffen. Hagerstown: Review and Herald Publishing, 1987.

Bultmann, R. *History and Eschatology*. Edinburgh: Edinburgh UP, 1957.

Cederroth, S., et al., eds. *On the Meaning of Death: Essays on Mortuary Rituals and Eschatalogical Beliefs*. Philadelphia: Coronet Books, 1988.

Chittick, W. C. "Death and the World of Imagination: Ibn al-'Arabi's Eschatology." *Muslim World* 78 (January 1988): 51-82.

Davies, P. R. "Eschatology at Qumran." *J of Biblical Literature* 104 (March 1985): 39-55.

Dienstag, Jacob I. *Eschatology in Maimonidean Thought, Messianism, Resurrection, and the World to Come*. New York: Ktav Publishing House, 1982.

Ewert, David. *And Then Comes the End*. Scottdale: Herald P, 1980.

Fischer, Ulrich. *Eschatologic und Jenseitserwartung in hellenistischen Diasporajudentum*. Hawthorne: De Gruyter, 1978.

Gowan, Donald E. *Eschatology in the Old Testament*. Minneapolis: Augsburg Fortress Publications, 1985.

Hassel, R. C. "Last Words and Last Things: St. John, Apocalypse, and Eschatology in *Richard III*." *Shakespeare Studies* 18 (1986): 25-40.

Konig, Adrio. *The Eclipse of Christ in Eschatology: Toward a Christ-Centered Approach*. Grand Rapids: Eerdmans, 1989.

MacDonald, Charles. *Church and World in the Plan of God: Aspects of History and Eschatology in the Thought of Père Yves Congar*. New York: Peter Lang, 1981.

Moessner, D. P. "Paul in Acts: Preacher of Eschatological Repentance to Israel." *New Testament Studies* 34 (January 1988): 96-104.

Oberman, H. "Teufelsdreck: Eschatology and Scatology in the 'Old' Luther." *Sixteenth Century J* 19 (Fall 1988): 435-50.

Pelikan, Jaroslav J. *The Shape of Death: Life, Death, and Immortality in the Early Fathers*. 1961; reprint, Westport: Greenwood, 1978.

Pentecost, J. Dwight. *Things to Come*. Grand Rapids: Zondervan Publishing, 1958.

Phan, Peter C. *Culture and Eschatology: The Iconographical Vision of Paul Evdokimov*. New York: Peter Lang, 1984.

Slaatte, Howard A. *Time and Its End: A Comparative Existential Interpretation of Time and Eschatology*. Lanham: UP of America, 1980.

——. *Time, Existence and Destiny*. New York: Peter Lang, 1989.

Smith, Jane I., ed. *The Precious Pearl: A Translation from the Arabic*. Atlanta: Scholars P, 1979.

Smith, Jane I., and Yvonne Y. Haddad. *Islamic Understanding of Death and Resurrection*. Albany: State U of New York P, 1981.

Suchocki, Marjorie H. *The End of Evil: Process Eschatology in Historical Context*. Albany: State U of New York P, 1988.

Sullivan, Clayton. *Rethinking Realized Eschatology*. Macon: Mercer UP, 1988.

Te Selle, E. "Toward an Augustinian Politics." *J of Religious Ethics* 16 (Spring 1988): 87-108. Includes bibliography.

Wainwright, Geoffrey. *Eucharist and Eschatology*. New York: Oxford UP, 1981.

Weigert, A. J. "Christian Eschatological Identities and the Nuclear Context." *J for the Scientific Study of Religion* 27 (June 1988): 175-91. Includes bibliography.

Wilder, Amos N. *Eschatology and Ethics in the Teaching of Jesus*. 1950; reprint, Westport: Greenwood, 1978.

Zandee, Jan. *Death as an Enemy According to Ancient Egyptian Conceptions*. Ed. Robert Kastenbaum. 1960; reprint, Salem: Ayer, 1977.

G. JUDAIC AND ISLAMIC

Beaucamp, E. *The Bible and the Universe: Israel and the Theology of History*. London: Burns and Oates, 1963.

Efros, Israel Isaac. *The Problem of Space in Jewish Mediaeval Philosophy*. New York: AMS P, 1966.

Fackenheim, Emile. "The Possibility of the Universe in Al-Farabi, Ibn Sina, and Maimonides." *Essays in Medieval Jewish and Islamic Philosophy*. Ed. Arthur Hyman. New York: Ktav Publishing House, 1977. 303-34.

Goldman, Steven L. "On the Beginnings and Endings of Time in Medieval Judaism and Islam." *The Study of Time IV*. Ed. J. T. Fraser, N. Lawrence, and D. Park. New York: Springer, 1981. 59-72.

Massignon, Louis. "Time in Islamic Thought." *Man and Time: Papers from the Eranos Yearbooks*. Trans. Ralph Manheim. Ed. Joseph Campbell. New York: Pantheon, 1957. 108-14.

Nasr, S. H. *An Introduction to Muslim Cosmological Doctrines*. Cambridge: Harvard UP, 1964.

Neusner, J. "Innovation Through Repetition: The Role of Scripture in the Mishnah's Division of Appointed Times." *History of Religions* 21 (August 1981): 48-76.

Salvaneschi, Enrico. *Il sistemo lessico-semantico del tempo in ebraico biblico*. Bologna: Patron, 1979.

Sutcliffe, E. F. *The Old Testament and the Future of Life*. 2nd ed. London: Burns and Oates, 1947.

23. SOCIOLOGY AND ANTHROPOLOGY

A. GENERAL

Adam, B. E. "Feminist Social Theory Needs Time: Reflections on the Relation Between Feminist Thought, Social Theory and Time as an Important Parameter in Social Analysis." *The Sociological Review* 37 (1989): 458-73.

——. "Social Versus Natural Time: A Traditional Distinction Re-examined." *The Rhythms of Society*. Ed. M. Young, and T. Schuller. London: Routledge, 1988.

——. *Time and Social Theory*. Cambridge: Polity P, 1990.

Allan, G. *The Importance of the Past: A Meditation on the Authority of Tradition*. New York: State U of New York P, 1985.

Archer, M. S. "Morphogenesis Versus Structuration: On Combining Structure and Action." *British J of Sociology* 33 (1982): 455-83.

Aubert, V., and H. White. "Sleep: A Sociological Interpretation." *Acta Sociologica* 4.2 (1960): 46-54.

Belloni, C., and E. Rampazi, eds. *Il tempo, lo spazio e l'attore sociale*. Milan: Angeli, 1988.

Bergmann, Werner. "Bewusstsein oder Handlung. Ansatzpunkte einer soziologischen Zeittheorie." *Ökonomie und Zeit. Beiträge zur interdisciplinären Zeitökonomie*. Ed. Eberhard Seifert. Frankfurt: Haag and Herchen, 1988. 70-101.

——. "Das Problem der Zeit in der Soziologie." *Kölner Zeitschrift für Soziologie und Sozialpsychologie* 35 (1983): 462-504.

——. "Die Temporalstrukturen des Wirtschaftssystems." *Zeit als Strukturelement von Lebenswelt und Gesellschaft*. Ed. F. Fürstenberg and T. Mörth. Linz: Trauner, 1986. 307-26.

Braudy, L. "Popular Culture and Personal Time." *Yale Review* 71(Summer 1982): 481-98.

Bruneau, T. J. "Silencing and Stilling Process: The Creative and Temporal Bases of Signs." *Semiotica* 56.3-4 (1985): 279-90.

Deutsch, Eliot. *Personhood, Creativity and Freedom*. Honolulu: U of Hawaii P, 1982.

Eames, E. Ramsden. "Mead's Concept of Time." *The Philosophy of George Herbert Mead*. Ed. W. R. Corti. Winterthur: Armiswiler Bücherei, 1973. 59-82.

Elchardus, M. "The Rediscovery of Chronos: The New Role of Time in Sociological Theory." *International Sociology* 3.1 (August 1986): 35-59.

Elias, Norbert. *Über die Zeit*. Trans. H. Fliessbach and M. Schröter. Frankfurt: Suhrkamp, 1984.

Fabian, J. *Time and the Other*. New York: Columbia UP, 1983.

Fraser, J. T. "The Individual and Society." *The Study of Time III*. Ed. J. T. Fraser, N. Lawrence, and D. Park. New York: Springer, 1978. 419-42.

Fürstenberg, F., and T. Mörth, eds. *Zeit als Strukturelement von Lebenswelt und Gesellschaft*. Linz: Trauner, 1986.

Gore, P. M., and J. B. Rotter. "A Personality Correlate of Social Action." *J of Personality* 31 (1963): 58-64.

Hall, Edward Titchell. *The Hidden Dimension*. Garden City: Doubleday, 1966.

Helson, R., et al. "Personality and Patterns of Adherence and Non-Adherence to the Social Clock." *J of Personality and Social Psychology* 46.5 (1984): 1079-96.

Hoffman, P. *Doubt, Time, Violence*. Chicago: U of Chicago P, 1987.

Hohn, H.-W. *Die Zerstörung der Zeit. Wie aus einem göttlichen Gut eine Handelsware wurde*. Frankfurt: Fischer Alternativ, 1984.

Imhof, A. E. *Die Lebenszeit. Vom aufgeschobenen Tod und von der Kunst des Lebens*. Munich: C. H. Beck, 1988.

Ingold, T. *Evolution and Social Life*. Cambridge: Cambridge UP, 1986.

Inkeles, A., and D. H. Smith. *Becoming Modern*. London: Heinemann, 1974.

Jacques, Elliot. *The Form of Time*. New York: Crane, Russak, 1982.

Kelly, J. R., and J. E. McGrath. *On Time and Method*. Newbury Park: Sage, 1988.

Kinget, G. M. *On Being Human: A Systematic View*. New York: Harcourt, Brace, 1975.

Lauer, Robert H. *Temporal Man: The Meaning and Uses of Social Time*. New York: Praeger, 1981.

Lévinar, Émmanuel. *Time and the Other and Additional Essays*. Trans. Richard A. Cohen. Pittsburgh: Duquesne UP, 1987.

Lowenthal, David. *The Past Is a Foreign Country*. Cambridge: Cambridge UP, 1985.

Luhmann, N. "World-Time and System History." *The Differentiation of Society*. Trans. S. Holmes and C. Larmore. New York: Columbia UP, 1982. 289-324.

Lüscher, K. K. "Time: A Much Neglected Dimension in Social Theory and Research." *Sociological Analysis and Theory* 4 (1974): 104-17.

Lynch, Kevin. *What Time is This Place?* Cambridge: MIT P, 1972.

Maines, D. R., ed. "Conceptions of Temporality in Sociological Theory [symposium]." *Sociological Quarterly* 28 (Fall 1987): 303-85.

Marschack, Alexander. *The Roots of Civilization*. New York: McGraw-Hill, 1972.
Mayr, Albert. "Social Time in Experimental Music and Art." *Time and Mind: The Study of Time VI*. Ed. J. T. Fraser. Madison: International Universities P, 1989. 217-28.
Mead, G. H. *The Philosophy of the Present (1932)*. Preface by J. Dewey. La Salle: Open Court, 1959.
Mongardini, Carlo. *Epistemologia e sociologia: temi e tendenzi della sociologia contemporanea*. Milan: Angeli, 1985.
Movahedi, S. "Cultural Preconceptions of Time: Can We Use Operational Time to Meddle in God's Time?" *Comparative Studies in Society and History* 27 (July 1985): 385-400. Includes bibliography.
Nowotny, Helga. *Eigenzeit: Entstehung und Strukturierung eines Zeitgefühls*. Frankfurt: Suhrkamp, 1989.
Pastore, N. "The Nature-Nurture Controversy: A Sociological Approach." *School and Society* 57 (1943): 373-77.
Pawelczynska, Anna. *The Time of Man* (in Polish). Wroclaw: Ossolineaum, 1986.
Reale, Paola, ed. *Tempo e identità*. Milan: Franco Angeli, 1988.
Schmied, G. *Soziale Zeit. Umfang, "Geschindigkeit" und Evolution*. Berlin: Duncker and Humblot, 1985.
Schwartz, Barry. *Queuing and Waiting: Studies in the Social Organization of Access and Delay*. Chicago: U of Chicago P, 1975.
Seligman, M. E. P. *Helplessness: On Depression, Development, and Death*. San Francisco: W. H. Freeman, 1975.
Sorokin, Pitirim Aleksandrovich. *Sociocultural Causality, Space, Time: A Study of Referential Principles of Sociology and Social Science*. Durham: Duke UP, 1943.
Sorokin, Pitirim Aleksandrovich, and R. K. Merton. "Social Time: A Methodological and Functional Analysis." *American J of Sociology* 42 (1937): 615-29.
Thrift, N. "A. Giddens: The Constitution of Society." *Sociology* 19.4 (November 1985): 609-24.
Wendorff, R. *Der Mensch und die Zeit*. Opladen: Westdeutscher Verlag, 1988.

B. ANTHROPOLOGY

Atkins, K. E. "'Kafir Time': Preindustrial Temporal Concepts and Labour Discipline in Nineteenth-Century Colonial Natal." *J of African History* 29.2 (1988): 229-44.

Bailey, G. N. "Concepts of Time in Quaternary Prehistory." *Annual Review of Anthropology* 12 (1983): 165-92.

——, ed. *Hunter-Gatherer Economy in Pre-history: A European Perspective*. Cambridge: Cambridge UP, 1983.

Barnes, J. A. "Time Flies Like an Arrow." *Man* 6 (1971): 537-52.

Bauschatz, Paul C. *The Well and The Tree: World and Time in Early Germanic Culture*. Amherst: U of Massachusetts P, 1982.

Bloch, M. E. F. *From Blessing to Violence: History and Ideology in the Circumcision Ritual of the Merita of Madagascar*. Cambridge: Cambridge UP, 1986.

——. "The Past and the Present in the Present." *Man* 12 (1977): 278-82.

Bohannan, Paul. "Concepts of Time among the Tiv of Nigeria." *Southwestern J of Anthropology* 9.3 (1953): 251-62.

Bollnow, Otto Friedrich. *Das Verhältnis zur Zeit. Ein Beitrag zur pädagogischen Anthropologie*. Heidelberg: Quelle and Meyer, 1972.

Bordieu, Pierre. "The Attitude of the Algerian Peasant toward Time." *Mediterranean Countrymen*. Ed. J. Pitt-Rivers. The Hague: Mouton, 1963. 55-72.

Clutton-Brock, J., and C. Grigson, eds. *Animals and Archaeology, III: Early Herders and Their Flocks*. Oxford: British Archaeological Reports, 1984.

Daniel, Glyn Edmund, and Colin Renfrew. *The Idea of Prehistory*. Edinburgh: Edinburgh UP, 1988. Includes bibliography.

Darlington, C. D. *The Evolution of Man and Society*. New York: Simon and Schuster, 1971.

D'Azevedo, Warren L. "Uses of the Past in Gola Discourse." *J of African History* 3.1 (1962): 11-34.

Evans-Pritchard, E. E. *The Nuer (1940)*. Oxford: Oxford UP, 1969.

Farriss, N. M. "Remembering the Future, Anticipating the Past: History, Time, and Cosmology among the Maya of Yucatan." *Comparative Studies in Society and History* 29 (July 1987): 566-93. Includes bibliography.

Frankenberg, R. "Your Time or Mine? An Anthropological View of the Tragic Temporal Contradictions of Biomedical Practice." *The Rhythms of Society*. Ed. M. Young, and T. Schuller. London: Routledge, 1988. 118-53.

Geertz, C. "Person and Conduct in Bali: An Essay in Cultural Analysis." Cultural Report Series 14, *Southeast Asia Studies*. New Haven: Yale UP, 1966.

Givens, Douglas R. *An Analysis of Navajo Temporality*. Washington: UP of America, 1977.

Gurevich, A. J. "Time as a Problem of Cultural History." *Cultures and Time*. Paris: Unesco P, 1976. 229-45.

Hall, Edward Twitchell. *The Dance of Life: The Other Dimension of Time*. Garden City: Doubleday, 1983.

Harkin, M. "History, Narrative, and Temporality: Examples from the Northwest Coast." *Ethnohistory* 35 (Spring 1988): 99-130. Includes bibliography.

Kubik, Gerhard. "African Space/Time Concepts and the Tusonia Ideographs in Luchazi Culture with a Discussion of Possible Cross Parallels in Music." *African Music* 6.4 (1987): 53-89. Includes bibliography.

——. "Oral Notation of Some West and Central African Time-Line Patterns." *Review of Ethnology* 3.22 (1972): 169-76.

Leach, Edmund. "Two Essays Concerning the Symbolic Representation of Time." *Rethinking Anthropology*. London: Athlone P, 1961. 124-63.

Leon-Portilla, Miguel. *Time and Reality in the Thought of the Maya*. 2nd ed. Norman: U of Oklahoma P, 1988.

Lyle, E. "The Dark Days and the Light Month." *Folklore* 95 (1984): 221-23.

Maxwell, R. J. "Anthropological Perspectives." *The Future of Time*. Ed. H. Yaker, H. Osmond, and F. Cheek. London: Hogarth P, 1972.

Mbiti, J. S. *African Religions and Philosophy* (1969). London: Heinemann, 1985.

McElwain, T. "Seneca Iroquois Concepts of Time." *Cosmos 4. Amerindian Cosmology*. Edinburgh: Traditional Cosmology Society, 1988.

Ortiz, Alfonso. *The Tewa World: Space, Time, Being, and Becoming in a Pueblo Society*. Chicago: U of Chicago P, 1972.

Partington, G. "The Australian Aborigines and the Human Past." *Mankind (Australia)* 15 (April 1985): 26-40. Includes bibliography.

Pocock, D. "The Anthropology of Time-Reckoning." *Contributions to Indian Sociology* 7 (1964): 18-29.

Rigby, P. "Time, Historical Consciousness and the Masai." *Comparative Studies in Society and History* 25 (1983): 428-56.

Rouse, Irving. *Migrations in Prehistory: Inferring Population Movements from Cultural Remains*. New Haven: Yale UP, 1986.

Rutz, H., ed. *The Politics of Time*. New York: American Ethnological Society, 1991.

Stone, R. M. *Dried Millet Breaking: Time, Words, and Song in the Woi Epic of the Kpelle*. Bloomington: Indiana UP, 1988.

Theunissen, Bert. *Eugène Dubois and the Ape-Man from Java: The History of the First "Missing Link" and Its Discoverer*. Dordrecht:

Kluwer Academic, 1989.

Thornton, Robert J. *Space, Time, and Culture Among the Iraqw of Tanzania*. New York: Academic P, 1980.

Urton, Gary. *At the Crossroads of the Earth and the Sky: An Andean Cosmology*. Austin: U of Texas P, 1988.

Wenner-Gren Foundation for Anthropological Research: 45th Anniversary Report. 3 vols. New York: The Foundation, 1986. Vols. 2 and 3 are bibliography.

Young, David E., and Robson Bonnischen. *Understanding Stone Tools: A Cognitive Approach*. Orono: Centre for the Study of Early Man, U of Maine at Orono, 1984.

C. COMMUNICATIONS AND COMPARATIVE STUDIES

Bruneau, Tom. "Chronemics: The Study of Time in Human Interaction." *Communication: J of the Communications Association of the Pacific* 6.1 (1977): 1-30.

——. "Theoretical Perspectives on the Temporal Bases of Intercultural Communication." *Communication: J of the Communications Association of the Pacific* 9.1 (1980): 77-115.

Gerbner, G., and M. Siefert, eds. *World Communications*. New York: Longmans, 1984.

Green, H. B. "Temporal Attitudes in Four Negro Subcultures." *The Study of Time I*. Ed. J. T. Fraser, F. C. Haber, and G. H. Müller. New York: Springer, 1972. 402-17.

Hamelink, C. J. *Cultural Autonomy in Global Communications*. New York: Longmans, 1983.

Migliore, S. "Punctuality, Pain and Time-Orientation among Sicilian-Canadians." *Social Science and Medicine* 28.8 (1989): 851-59.

Monge, P. R., et al. "The Process of Studying Process in Organizational Communication." *J of Communication* 34.1 (1984): 22-43.

Munroe, R. H., et al. "Time Allocation in Four Societies." *Ethnology* 22 (October 1983): 355-70. Includes bibliography.

Ornauer, H., ed. *Images of the World in the Year 2000: A Comparative Ten-Nation Study*. Hawthorne: Mouton, 1976.

Robinson, John P., et al. *The Rhythm of Everyday Life: How Soviet and American Citizens Use Their Time*. Boulder: Westview P, 1988.

Spangler, Michael. "Time Proverbs and Social Change in Belgrade, Yugoslavia." *The Study of Time IV*. Ed. J. T. Fraser, N. Lawrence, and D. Park. New York: Springer, 1981. 223-33.

Szalai, Alexander, et al., eds. *The Use of Time: Daily Activities of*

Urban and Suburban Populations in Twelve Countries. Hawthorne: Mouton, 1972.

D. FAMILY

Bawin-Legros, B., ed. *La dynamique familiale et les constructions sociales du temps*. Liège: Université de Liège, 1988.

Beach, B. A. "Time Use in Rural Home-Working Families." *Family Relations* 36 (October 1987): 412-16. Includes bibliography.

Hardesty, C., and J. Bokemeier. "Finding Time and Making Do: Distribution of Household Labor in Nonmetropolitan Marriages." *J of Marriage and the Family* 51 (February 1989): 253-67. Includes bibliography.

Khandker, S. R. "Determinants of Women's Time Allocation in Rural Bangladesh." *Economic Development and Cultural Change* 37 (October 1988): 111-26.

LaRossa, R. "Transition to Parenthood and the Social Reality of Time." *J of Marriage and Family* 45 (August 1983): 579-89. Includes bibliography.

Lawrence, F. C., and P. H. Wozniak. "Rural Children's Time in Household Activities." *Psychological Reports* 61 (December 1987): 927-37. Includes bibliography.

Nock, S. L., and P. W. Kingston. "Time with Children: The Impact of Couples' Work-Time Commitments." *Social Forces* 67.1 (September 1988): 59-85. Includes bibliography.

Shelton, B. A., and J. Firestone. "Time Constraints on Men and Women: Linking Household Labor to Paid Labor." *Sociology and Social Research* 72 (January 1988): 102-05. Includes bibliography.

Wallman, S. *Eight London Households*. London: Tavistock, 1984.

E. FUTURE ORIENTATION, LIFE CYCLE, AND SOCIAL DEVELOPMENT

Barabasz, A. F. "Temporal Orientation and Academic Achievement in College." *J of Social Psychology* 80 (1970): 231-32.

Barndt, R., and D. Johnson. "Time Orientation in Delinquents." *J of Abnormal and Social Psychology* 51 (1955): 343-45.

Bell, W., and J. Mau, eds. *The Sociology of the Future*. New York: Russell Sage Foundation, 1971.

Cantril, H. *The Pattern of Human Concerns*. New Brunswick: Rutgers

UP, 1965.

Cottle, T. J., and S. L. Klineberg. *The Present of Things Future: Explorations of Time in Human Experience*. New York: Free P, 1974.

Crandall, V. J., D. Solomon, and R. Kellaway. "Expectancy Statements and Decision Times as Functions of Objective Probabilities and Reinforcement Values." *J of Personality* 24 (1955): 192-203.

Doob, Leonard W. *Patterning of Time*. New Haven: Yale UP, 1971.

Dunne, J. S. *Time and Myth: A Meditation on Storytelling as an Exploration of Life and Death*. London: SCM P, 1973.

Elchardus, M. "Life Cycle and Life Course, the Scheduling and Temporal Integration of Life." *Population and Societal Outlook*. Ed. S. Feld, and R. Lesthaeghe. Brussels: Koning Boudewijnstichting, 1984. 251-67.

Galtung, J. *Images of the World in the Year 2000: A Synthesis of the Marginals of a Ten Nations Study*. Vienna: European Coordination Centre for Research and Documentation in Social Science, 1970.

Geissler, K. A. *Zeit leben—Vom Hasten und Rasten, Arbeiten und Lernen, Leben und Sterben*. Weinheim: Beltz Verlag, 1985.

Heckhausen, H. *The Anatomy of Achievement Motivation*. New York: Academic P, 1967.

Klineberg, S. L. "Future Time Perspective and the Preference for Delayed Reward." *J of Personality and Social Psychology* 8 (1968): 253-57.

Kolaja, Jiris Thomas. *Social System and Time and Space: Introduction to the Theory of Recurrent Behavior*. Pittsburgh: Duquesne UP, 1969.

LeShan, L. L. "Time Orientation and Social Class." *J of Abnormal and Social Psychology* 47 (1952): 589-92.

Luhmann, N. "The Future Cannot Begin: Temporal Structures in Modern Society." *The Differentiation of Society*. Trans. S. Holmes and C. Larmore. New York: Columbia UP, 1982. 271-89.

Mallmann, Carlos A., and Oscar Nudler, eds. *Time, Culture and Development*. Bariloche, Argentina: Fundacion Bariloche, 1986.

——, eds. *Time, Quality of Life and Social Development*. Bariloche, Argentina: Fundacion Bariloche, 1982.

Nowotny, Helga. "Mind, Technologies, and Collective Time Consciousness: From the Future to an Extended Present." *Time and Mind: The Study of Time VI*. Ed. J. T. Fraser. Madison: International Universities P, 1989. 197-213.

Parelius, Ann P. "Lifelong Education and Age Stratification: Some Unexplored Relationships." *American Behavioral Scientist* 19.2 (1975): 206-23.

Rezsohazy, R. *Temps social et développement*. Brussels: La renaissance du livre, 1970.
Robertson, J. *Future Work*. London: Gower, 1985.
Schneider, L., and S. Lysgaard. "The Deferred Gratification Pattern." *American Sociological Review* 18 (1953): 142-49.
Spangler, Michael, and Olivera Petrovich. "Future Time Perspective and Feeling Tone: A Study in the Perception of the Days of the Week by Yugoslav and American Students." *J of Social Psychology* 105.2 (1978): 189-93.
Talmon, Y. "Millenarian Movements." *Archives Européens de Sociologie* 7 (1966): 159-200.
Trevico, R. G. *L'immagine del futuro*. Milan: F. Angeli, 1986.
Trommsdorff, G., and H. Lamm. "An Analysis of Future Orientation and Some of Its Social Determinants." *The Study of Time II*. Ed. J. T. Fraser, and N. Lawrence. New York: Springer, 1975. 343-62.
Turnbull, C. *The Human Cycle*. London: Jonathan Cape, 1984.
Zern, D. "The Influence of Certain Developmental Factors in Fostering the Ability to Differentiate the Passage of Time." *J of Social Psychology* 72.1 (1967): 9-17.

F. SOCIETY AND ORGANIZATION

Belloni, C. *La tempo della citta*. Milan: Angeli, 1984.
Bergmann, Werner. "Zeit, Handlung und Sozialität bei G. H. Mead." *Zeitschrift für Soziologie* 10.4 (1981): 351-63.
——. "Zeitgrenzen. Die zeitliche Dimension der Grenzerhaltung sozialer Systeme." *Schweizerische Zeitschrift für Soziologie* 1 (1989): 1-15.
——. *Die Zeitstrukturen sozialer Systeme: Eine systemtheoretische Analyse*. Berlin: Duncker and Humblot, 1981.
Bourdieu, Pierre. "La société traditionelle, attitude à l'égard du temps et conduite économique." *Sociologie du Travail* 1 (1963): 24-44.
Braudel, Fernand. *The Structures of Everyday Life: The Limits of the Possible*. Vol. 1 of *Civilization and Capitalism: 15th-18th Century*. Trans. and revised by Sian Reynolds. New York: Harper and Row, 1981.
Chapin, F. S., Jr. *Human Activity Patterns in the City: Things People Do in Time and Space*. New York: Wiley, 1974.
Daheim, H. "Soziale Herkunft, Schule und Rekrutierung der Berufe." *Kölner Zeitschrift für Soziologie und Sozialpsychologie* (special

volume) 5 (1970): 200-17.

Ellis, L. M., et al. "Time Orientation and Social Class: An Experimental Supplement." *J of Abnormal and Social Psychology* 51 (1955): 146-47.

Feather, N. T., and M. J. Bond. "Time Structure and Purposeful Activity among Employed and Unemployed University Graduates." *J of Occupational Psychology* 56 (1983): 241-54.

Fineman, F., ed. *Unemployment: Personal and Social Consequences*. London: Tavistock, 1987.

Fortes, Meyer. *Time and Social Structure and Other Essays*. Atlantic Highlands: Humanities P International, 1970.

Giddens, A. *The Constitution of Society*. Cambridge: Polity P, 1984.

———. "Time and Social Organisation." In his *Social Theory and Modern Sociology*. Cambridge: Polity P, 1987.

Gurvitch, Georges. *The Spectrum of Social Time*. Trans. Myrtle Korenbaum. Dordrecht: Reidel, 1964.

Hägerstrand, T. "Time and Culture." *Time Preferences: An Interdisciplinary Theoretical and Empirical Approach*. Ed. G. Kirsch, P. Nijkamp, and K. Zimmermann. Berlin: Wissenschaftszentrum, 1985.

Jacobs, J. *The Death and Life of Great American Cities*. New York: Vintage, 1961.

Janelle, D., and M. Goodchild. "Diurnal Patterns of Social Group Distributions in a Canadian City." *Economic Geography* 59 (October 1983): 403-25.

Küng, Emile. *Arbeit und Freizeit in der nachindustriellen Gesellschaft*. Tübingen: Mohr, 1971.

Levine, R., and E. Wolff. "Social Time: The Heartbeat of Culture." *Psychology Today* (March 1985): 28-35.

Lewis, J. David, and Andrew J. Weigert. "The Structures and Meanings of Social Time." *Social Forces* 60.2 (December 1981): 432-62.

McGrath, Joseph Edward, ed. *The Social Psychology of Time: New Perspectives*. Newbury Park: Sage, 1988.

McGrath, Joseph Edward, and Janice R. Kelly. *Time and Human Interaction: Toward a Social Psychology of Time*. New York: Guilford P, 1986.

Maines, D. R. "Sociological Import of G. H. Mead's Theory of the Past." *American Sociological Review* 48 (April 1983): 161-73.

Melbin, Murray. "City Rhythms." *The Study of Time III*. Ed. J. T. Fraser, N. Lawrence, and D. Park. New York: Springer, 1978. 444-65.

———. *Night as Frontier: Colonizing the World after Dark*. New York: Free

P, 1987.

Mercure, D., and A. Wallemacq, eds. *Temps sociaux*. Brussels: De Boeck, 1988.

Moore, W. E. *Man, Time and Society*. New York: Wiley, 1963.

Nock, S. L., and P. W. Kingston. "The Division of Leisure and Work." *Social Science Quarterly* 70 (March 1989): 24-39. Includes bibliography.

Nowotny, Helga. "From the Future to the Extended Present—Time in Social Systems." *Time Preferences: An Interdisciplinary Theoretical and Empirical Approach*. Ed. G. Kirsch, P. Nijkamp, and K. Zimmerman. Berlin: Wissenschaftszentrum, 1985.

——. "Time Structuring and Time Measurement: On the Interrelation Between Timekeepers and Social Time." *The Study of Time II*. Ed. J. T. Fraser, and N. Lawrence. New York: Springer, 1975. 325-42.

Rinderspacher, Jürgen P. *Gesellschaft ohne Zeit. Individuelle Zeitverwendung und Soziale Organisation der Arbeit*. Frankfurt: Campus, 1985.

Roth, H. *Time Tables: Structuring the Passage of Time in Hospital Treatment and Other Careers*. Indianapolis: Bobbs-Merrill, 1963.

Schöps, Martina. *Zeit und Gesellschaft*. Stuttgart: Ferdinand Enke, 1980.

Sirianni, Carmen J. *Work, Time and Equality*. Oxford: Polity P, 1987.

Starkey, K. "Time and Work Organisation: A Theoretical and Empirical Analysis." *The Rhythms of Society*. Ed. M. Young, and T. Schuller. London: Routledge, 1988.

Tabboni, S. *La rappresentazione sociale del tempo*. Milan: F. Angeli, 1984.

——. *Tempo e società*. Milan: F. Angeli, 1985.

Tarkowska, Elzbieta. *Time in Society: Problems, Traditions, Directions of Research* (in Polish). Wroclaw: Ossolineum, 1987. Contains a bibliography of some 400 items.

Thornton, Robert J. "Time Scales and Social Thought." *Time and Mind: The Study of Time VI*. Ed. J. T. Fraser. Madison: International Universities P, 1989. 181-93.

Walker, J. "Social Problems of Shift Work." *Hours of Work*. Ed. S. Folkard, and T. H. Monk. Chichester: Wiley, 1985.

Williams, R. "Public Time for Public Good." *Social Policy* 14 (Summer 1983): 41-44.

Young, Michael D. *The Metronomic Society: Natural Rhythms and Human Timetables*. Cambridge: Harvard UP, 1988.

Young, Michael D., and Tom Schuller, eds. *The Rhythms of Society*. London: Routledge and Kegan Paul, 1988.

24. TIME MANAGEMENT

A. GENERAL

Adam, E. E., Jr., and R. J. Ebert. "A Comparison of Human and Statistical Forecasting." *AIIE Transactions* 8.1 (March 1976): 120-27.

Ahl, D. H. "New Product Forecasting Using Consumer Panels." *J of Marketing* 7.2 (May 1970): 159-67.

Ammons, David N., and Charldean Newell. *City Executives: Leadership Roles, Work Characteristics and Time Management.* Albany: State U of New York P, 1989.

Bell, Daniel. "The Clock Watchers: Americans at Work." *Time* September 8, 1975: 55-57.

——. *The Coming of the Post-Industrial Society.* New York: Basic Books, 1973.

Bennis, W., K. Benne, and R. Chin. *The Planning of Change.* New York: Holt, Rinehart, 1961.

Bental, B., et al. "Management Contribution and the Allocation of Time in a Labor-Managed Firm." *J of Comparative Economics* 6 (December 1982): 353-62.

Benton, W. K. *Forecasting for Management.* Reading: Addison-Wesley, 1972.

Berk, Richard A., and Sarah Fenstermaker Berk. *Labor and Leisure at Home: Content and Organization of the Household Day.* Beverly Hills: Sage, 1979.

Best, Fred. *Flexible Life Scheduling: Breaking the Education-Work-Retirement Lockstep.* New York: Praeger, 1980.

Blyton, P. *Changes in Working Time: An International Review.* London: Croom Helm, 1985.

Brouha, L. *Physiology in Industry.* New York: Pergamon, 1960.

Brown, G. T. "Cybernetics and Work Study." *Work Study and Management Services* 17 (April 1973): 228-32.

Bryman, A., et al., eds. *New Perspectives on the Life Cycle.* London: Macmillan, 1987.

Carlstein, Tommy. *Time Resources, Society, and Ecology: On the Capacity for Human Interaction in Space and Time.* London: Allen and Unwin, 1982.

Chambers, J. S., S. K. Mullick, and D. D. Smith. "How to Choose the Right Forecasting Technique." *Harvard Business Review* 49.4 (July-August 1971): 55-64.

Clark, P. "Temporal Inventories and Time Structuring in Large Organizations." *The Study of Time III*. Ed. J. T. Fraser, N. Lawrence, and D. Park. New York: Springer, 1978. 391-416.

Deacon, R. T., and J. Sonstelie. "Rationing by Waiting and the Value of Time: Results from a Natural Experiment." *J of Political Economy* 93 (August 1985): 627-47.

Denhardt, Robert B. *In the Shadow of Organization*. Lawrence: The Regents P of Kansas, 1981.

Drawbaugh, Charles C. *Time and Its Use: A Self Management Guide for Teachers*. New York: Teachers College P, 1984.

Eberts, R. W. "Union Effects on Teacher Productivity." *Industrial and Labor Relations Review* 37 (April 1984): 346-58.

Engstrom, Ted W., and Alec R. MacKenzie. *Managing Your Time: Practical Guidelines on the Effective Use of Time*. Revised ed. Grand Rapids: Zondervan Publishing, 1988.

Ferner, Jack D. *Successful Time Management*. New York: Wiley, 1980.

Fisher, Charles W., and David C. Berliner. *Perspectives on Instructional Time*. New York: Longmans, 1985.

Folkard, S., and T. Monk. *Hours of Work: Temporal Factors in Work Scheduling*. Chichester: Wiley, 1985.

Goodloe, Alfred, et al. *Managing Yourself: How to Control Emotion, Stress, and Time*. New York: Watts, Franklin, 1984.

Gorz, Andre, ed. *The Division of Labour, the Labour Process and Class Struggle in Modern Capitalism*. London: Harvester, 1976.

Gramm, W. S. "Labor, Work, and Leisure: Human Well-Being and the Optimal Allocation of Time." *J of Economic Issues* 21 (March 1987): 167-88. Includes bibliography.

Grandjean, E. *Fitting the Task to the Man*. London: Taylor and Francis, 1969.

Grossbard-Shechtman, A. "A Theory of Allocation of Time in Markets for Labour and Marriage." *Economic J* 94 (December 1984): 863-82. Includes bibliography.

Grudin, Robert. *Time and the Art of Living*. San Francisco: Harper and Row, 1982.

Gulick, L. "Time and Public Administration." *Public Administration Review* 47 (January-February 1987): 115-19.

Hareven, T. K. *Family Time and Industrial Time: The Relationship between the Family and Work in a New England Industrial Community*. Cambridge: Cambridge UP, 1982.

Harvey, Andrew S., and Susan Clark. *Descriptive Analysis of Halifax Time-Budget Data*. Halifax: Institute of Public Affairs, Dalhousie U, 1975.

Harvey, Andrew S., et al. *Time Budget Research*. Berlin: Campus Verlag, 1984.

Herzberg, F. *Work and the Nature of Man*. Cleveland: World Publishing, 1966.

Herzberg, F., B. Mausner, and B. Snyderman. *The Motivation to Work*. New York: Wiley, 1959.

Heward, J. H., and P. M. Steele. *Business Control through Multiple Regression Analysis*. London: Gower P, 1972.

Hobbs, Charles. *Time Power: The Revolutionary Time Management System That Can Change Your Professional and Personal Life*. New York: Harper and Row, 1987.

Holmes, Ivory H. *The Allocation of Time by Women Without Family Responsibilities*. Lanham: UP of America, 1983.

Hoyt, John S., Jr. *Personal Time Management Manual*. Saint Paul: Telstar, 1981.

Jay-Rayon, Jean Claude. *Pour prendre le temps d'être mieux*. Sillery: Québec Science, 1983.

Juster, Thomas F., and Frank P. Stafford, eds. *Time, Goods and Well-Being*. Ann Arbor: Survey Research Center, Institute for Social Research, U of Michigan, 1985.

Lee, Mary Dean, and Rabindra N. Kanungo, eds. *Management of Work and Personal Life: Problems and Opportunities*. New York: Praeger, 1984.

McClelland, D. C. *The Achieving Society*. Princeton: Van Nostrand, 1961.

Macey, Samuel L. "British Columbia's Open Learning Institute." *Management Services* 27 (August 1983): 20-25.

——. "Work Study Methods Applied to Learning." *Work Study and Management Services* 13 (February 1969): 84-87.

MacKenzie, R. Alec. *The Time Trap*. New York: McGraw-Hill, 1975.

MacKenzie, R. Alec, and Kay C. Waldo. *About Time! A Woman's Guide to Time Management*. New York: McGraw-Hill, 1981.

Makridakis, S., and S. C. Wheelwright. *Forecasting: Methods and Applications*. Santa Barbara: Wiley, 1978.

——. *Interactive Forecasting*. 2nd ed. San Francisco: Holden-Day, 1978.

Malcolm, D. G., and A. J. Rowe. *Management Control Systems*. New York: Wiley, 1960.

Medrich, Elliott A., et al. *The Serious Business of Growing Up: A Study of Children's Lives Outside School*. Berkeley: U of California P, 1982. Deals with time management surveys.

Miles, L. D. *Techniques of Value Analysis and Engineering*. New York: McGraw-Hill, 1961.

Mueller, E. "The Value and Allocation of Time in Rural Botswana." *J of Development Economics* 15 (May-August 1984): 329-60.

Mullins, Laurie J. *Management and Organizational Behaviour*. 2nd ed. London: Pitman, 1989.

Mumford, Lewis. "The Monastery and the Clock." *Technics and Civilization*. New York: Harcourt, Brace and World, 1962.

Murrell, K. F. H. *Ergonomics: Man in His Working Environment*. London: Chapman and Hall, 1965.

Needham, Joseph. *Clerks and Craftsmen in China and the West*. Cambridge: Cambridge UP, 1970.

Pappas, P. *Time Allocation Study*. Athens: Athens Center of Ekistics, 1968.

Parker, G. C., and E. L. Segura. "How to Get a Better Forecast." *Harvard Business Review* 49.2 (March-April 1971): 99-109.

Patrushev, V. D. "Past and Future Changes in Soviet Workers' Time-Budget." *International Social Science J* 38.1 (1986): 77-88.

Presser, Harriet B. "Shift Work among American Women and Child Care." *J of Marriage and the Family* 48 (August 1986): 551-63.

——. "Work Shifts of Full-Time Dual-Earner Couples: Patterns and Contrasts by Sex of Spouse." *Demography* 24.1 (February 1987): 99-112.

Pütz, Karl. *Zeitbudgetforschung in der Sowjetunion: zur empirischen Sozialforschung in der USSR*. Meisenheim am Glan: A. Hain, 1970.

Rézsoházy, R. "The Methodological Aspects of a Study about the Social Notion of Time in Relation to Economic Development." *The Use of Time*. Ed. A. Szalai. The Hague: Mouton, 1972. 449-60.

Roberts, S. D., and R. Reed. "The Development of a Self-Adaptive Forecasting Technique." *AIIE Transactions* (December 1969): 314-22.

Robinson, John P. *How Americans Use Time: A Social-Psychological Analysis of Everyday Behavior*. New York: Praeger, 1977.

Sacks, Michael Paul. *Women's Work in Soviet Russia: Continuity in the Midst of Change*. New York: Praeger, 1976.

Scaglion, R. "The Importance of Nighttime Observations in Time Allocation Studies." *American Ethnologist* 13 (August 1986): 537-45. Includes bibliography.

Schmidt, R. M. "Who Maximizes What? A Study in Student Time Allocation." *American Economic Review. Papers and Proceedings* 73 (May 1983): 23-28.

Scribner, J. B., and B. A. Gutek. "Some Time Dimensions of Work: Measurement of an Underlying Aspect of Organization Culture." *J of Applied Psychology* 72 (November 1987): 642-50. Includes bibliography.

Seiwert, Lothar J. *Time is Money: Save It*. Trans. Edward J. Zajac. Homewood: Dow Jones-Irwin, 1989.

Shipman, Neil J., and Jack B. Martin. *Effective Time-Management Techniques for School Administrators*. Englewood Cliffs: Prentice-Hall, 1983.

Sullivan, W. G., and W. W. Claycombe. *Fundamentals of Forecasting*. Reston: Reston Publishing, 1977.

Toffler, Alvin. *Future Shock*. New York: Bantam Books, 1970.

Tonn, B. E. "A Sociopsychological Contribution to the Theory of Individual Time-Allocation." *Environment and Housing* 16.2 (1984): 201-23.

Unger, S. L. "Effect of Actual and Perceived Availability of Time on Voluntarism." *Perceptual and Motor Skills* 65 (October 1987): 524-26. Includes bibliography.

Walker, Kathryn E., and Margaret E. Woods. *Time Use: A Measure of Household Production of Family Goods and Services*. Washington: Center for the Family of the American Home Economics Association, 1976.

Webber, Ross A. *Time is Money! Tested Tactics That Conserve Time for Top Executives*. New York: Free P, 1980.

Wheelwright, S. C., and S. Makridakis. *Forecasting Methods for Management*. 2nd ed. New York: Wiley, 1977.

Whybark, D. Clay. "A Comparison of Adaptive Forecasting Techniques." *The Logistics and Transportation Review* 8.3 (1972): 13-26.

Wiener, N. *Cybernetics*. 2nd ed. Cambridge: MIT P, 1963.

——. *Kybernetik, Regelung und Nachrichtenübertragung im Lebewesen und in der Maschine*. Düsseldorf: Econ, 1963.

Zuzanek, Jiri. *Work and Leisure in the Soviet Union*. New York: Praeger, 1980.

B. LEISURE

Brooks, C. M. "Leisure Time Physical Activity Assessment of American Adults through Analysis of Time Diaries Collected in 1981." *American J of Public Health* 77 (April 1987): 455-60.

Cowgill, Donald O., and Norma Baulch. "The Use of Leisure Time by Older People." *The Gerontologist* 2.1 (Spring 1962): 47-50.

DeSalvo, J. S. "A Model of Urban Household Behavior with Leisure Choice." *J of Regional Science* 25 (May 1985): 159-74. Includes bibliography.

Friedsam, Hiram J., and Cora A. Martin. "Travel by Older People as a Use of Leisure." *The Gerontologist* 13.2 (Summer 1973): 204-07.

Gihring, T. A. "Leisure-Time Activities in an Urban Nigerian Setting: Attitudes and Experience." *J of Leisure Research* 15.2 (1983): 108-24.

Grazia, S. de. *Of Time, Work and Leisure*. New York: Twentieth Century Fund, 1962.

Hantrais, L. "L'évolution des pratiques de loisir et le changement social." *Loisir et Société* 9.2 (1986): 363-76.

——. "Leisure Lifestyles and the Synchronization of Family Schedules." *World Leisure and Recreation* 27.2 (1985): 18-24.

——. "Time, the Family and Leisure within the Franco-British Perspective: A Report on an International Research Seminar." *World Leisure and Recreation* 29.2 (1987): 16-22.

Hantrais, L., P. Clark, and N. Samuel. "Time-Space Dimensions of Work, Family and Leisure in France and Great Britain." *Leisure Studies* 3.3 (1984): 301-18.

Kaplan, Max. *Leisure: Lifestyle and Lifespan: Perspectives for Gerontology*. Philadelphia: W. B. Saunders, 1979.

Kaplan, Max, and P. Besserman. *Technology, Human Values and Leisure*. Nashville: Abingdon P, 1971.

Kleemeier, Robert W., ed. *Aging and Leisure*. New York: Oxford UP, 1961.

——. "Leisure and Disengagement in Retirement." *The Gerontologist* 4.4 (Winter 1964): 180-84.

Linder, Staffan. *The Harried Leisure Class*. New York: Columbia UP, 1970.

Samuel, N. "Free Time in France: A Historical and Sociological Survey." *International Social Science J* 38.1 (1986): 49-63.

Shaw, S. M. "Leisure, Recreation or Free Time? Measuring Time Usage." *J of Leisure Research* 18.3 (1986): 177-89. Includes bibliography.

Simpura, J. "Drinking: An Ignored Leisure Activity." *J of Leisure Research* 17.3 (1985): 200-11. Includes bibliography.

C. TECHNOLOGY

Bose, C. E., et al. "Household Technology and the Social Construction of Housework." *Technology and Culture* 25 (January 1984): 53-82.

Burnes, Bernard. *New Technology in Context: The Selection, Introduction and Use of Computer Numerically Controlled Machine Tools*. Aldershot, Hampshire: Gower, 1989.

De Keyser, Veronique, et al., eds. *The Meaning of Work and Technological Options*. New York: Wiley, 1988.

Dubinskas, Frank A., ed. *Making Time: Ethnographies of High-Technology Organizations*. Philadelphia: Temple UP, 1988.

Durfee, W. F. "The First Systematic Attempt at Interchangeability in Firearms." *Cassier's Magazine* 5 (April 1904): 469-77.

Fitch, Charles H. "Manufactures of Interchangeable Mechanisms, V: The Manufacture of Watches." *Report on the Manufactures of the United States at the Tenth Census (June 1, 1880)*. Produced by the U.S. Department of the Interior, Census Office. Washington: Government Printing Office, 1883. 60-67.

Habakkuk, H. J. *American and British Technology in the Nineteenth Century*. Cambridge: Cambridge UP, 1962.

Hinton, Robert. *Information Technology, and How to Use It: A Handbook of Effective Practice*. Cambridge: ICSA Publishing, 1988.

James, C. F., Jr. "Incentives for Machine-Paced Operations." *Industrial Engineering* (September 1975): 52-55.

Kaganovich, M. "Efficiency of Sliding Plans in a Linear Model with Time-Dependent Technology." *Review of Economic Studies* 52 (October 1985): 691-702. Includes bibliography.

Kingham, E. G., et al., eds. *Real Time Data Handling and Process Control*. New York: Elsevier, 1984.

Knave, B. G., et al., eds. *Ergonomic Principles in Office Automation: State of the Art Reports and Guidelines on Human Factors in the Office Environment*. Philadelphia: Coronet Books, 1984.

Lock, Dennis. *Handbook of Engineering Management*. London: Heinemann, 1989.

Marsh, E. A. *The Evolution of Automatic Machinery as Applied to the Manufacture of Watches at Waltham, Mass., by The American Waltham Watch Company*. 1896; reprint, Exeter: Adams Brown, 1968.

Mayr, Otto, and Robert C. Post, eds. *Yankee Enterprise: The Rise of the American System of Manufactures*. Washington: Smithsonian Institution, 1981.

Mirsky, Jeanette, and Allan Nevins. *The World of Eli Whitney*. New York: Collier Books, 1962.

Needham, Joseph. *The Development of Iron and Steel Technology in China*. London: Newcomen Society, 1958.

Panikkar, R. "Technique et temps: la technochronie." *Technica e casistics*. Ed. E. Castelli. Padua: CEDAM, 1964. 195-229.

Rolt, Lionel T. C. *A Short History of Machine Tools*. Cambridge: MIT P, 1965.

Rosenberg, Nathan, ed. *The American System of Manufactures*. Edinburgh: Edinburgh UP, 1969.

——. *Perspectives on Technology*. Cambridge: Cambridge UP, 1976.

Serazzi, G., ed. *Workload Characterization of Computer Systems and Computer Networks*. New York: Elsevier, 1986.

Smith, Merritt Roe. "Eli Whitney and the American System of Manufacturing." *Technology in America: A History of Individuals and Ideas*. Ed. Carroll W. Pursell. Cambridge: MIT P, 1981. 45-61.

Strassmann, W. Paul. *Risk and Technological Innovation: American Manufacturing Methods during the Nineteenth Century*. Ithaca: Cornell UP, 1959.

Weiss, Marvin D. *Microprocessors in Industrial Measurement and Control*. Blue Ridge Summit: TAB Books, 1987.

Wight, O. W. *Production and Inventory Management in the Computer Age*. Boston: Cahhers Brooks, 1974.

D. TIME AND METHOD STUDY

Elements of this discipline have also been known by such names as Scientific Management, Industrial Engineering, Work Study, Organization and Methods, Time and Motion Study, Management Services, and Management Consulting.

Advanced Methods and Work Measurement. Flint: General Motors Institute, 1973.

Antis, W., J. M. Honeycutt, Jr., and G. N. Koch. *The Basic Motions of M.T.M* (Methods-Time-Measurement). Pittsburgh: The Maynard Foundation, 1968.

Ascher, Kate. *The Politics of Privatization: Contracting Out Public Services*. London: Macmillan Education, 1987.

Atkinson, Henry. "Scientific Management." *Engineering and Industrial Management* July 31, 1919: 137.

Atkinson, Philip E. "What Is the Best Use of Your Time Right Now?" *Management Services* 32.4 (1988): 12-17.

Avery, M. *Time Study, Incentives and Budgetary Control*. London: Business Publications, 1964.

Bailey, G. E., and R. Presgrave. *Basic Motion Time-Study*. New York: McGraw-Hill, 1958.

Barnes, Ralph M. *Motion and Time Study: Design and Measurement of*

Work. 7th ed. New York: Wiley, 1980. Includes bibliography.

——. *Work Sampling*. New York: Wiley, 1957.

Bayhylle, J. E. *Productivity Improvements in the Office*. Bristol: Engineering Employers Federation, 1968.

Birn, Serge A., and Eric G. Brightford. "Industrial Engineering in Europe Today." *Work Study and Management Services* 14 (April 1970): 311-15 (reprinted from the *J of Industrial Engineering*).

Birn, Serge A., R. M. Crossan, and R. W. Eastwood. *Measurements and Control of Office Costs*. New York: McGraw-Hill, 1966.

Blyton, Paul, et al. *Time, Work and Organization*. New York: Routledge, Chapman, and Hall, 1989.

Brisley, Chester L. "Comparison of Predetermined Time Systems (PTS)." *Proceedings of AIIE Spring Annual Conference*. Norcross: American Institute of Industrial Engineers, 1978.

——. "Work Measurement in the 1980's." *43rd Annual IMS Clinic Proceedings*. Des Plaines: Industrial Management Society, 1979.

British Standards Institution. *Glossary of Terms in Work Study* (B.S. 3138: 1959). London: British Standards Institution, 1959.

"The British Work-Measurement Data Foundation." *Work Study and Management Services* 14 (May 1970): 376-78.

"BWD [Basic Work Data] Goes Metric." *Work Study and Management Services* 14 (June 1970): 484-86.

Carroll, P. *Time Study for Cost Control*. New York: McGraw-Hill, 1954.

——. *Time Study Fundamentals for Foremen*. New York: McGraw-Hill, 1951.

Cemach, H. P. *Work Study in the Office*. London: *Office Magazine*, 1958.

Crossan, Richard M., and Harold W. Nance. *Master Standard Data: The Economic Approach to Work Measurement*. Revised ed. Melbourne: Krieger, 1980.

Currie, R. M. *Financial Incentives Based on Work Measurement*. London: British Institute of Management, 1963.

——. *The Measurement of Work*. London: British Institute of Management, 1967.

——. *Simplified P.M.T.S.* (Predetermined Motion Time Systems). London: British Institute of Management, 1964.

——. *Work Study*. 2nd ed. London: Pitman, 1963.

Davis, P. R., ed. *Performance under Sub-Optimal Conditions*. New York: Taylor and Francis, 1971.

Dow, R. *Marketing and Work Study*. Elmsford: Pergamon, 1969.

Duckworth, E. *A Guide to Operational Research*. London: University Paperbacks, 1965.

Dudley, N. A. *Work Measurement—Some Research Studies*. London: Macmillan, 1963.

Duncan, Acheson J. *Quality Control and Industrial Statistics*. 5th ed. Homewood: Irwin, 1986.

Eady, Karl. "Today's International MTM [Methods Time Measurement] Systems—Decision Criteria for Their Use." *Proceedings, AIIE Spring Annual Conference*. Norcross: American Institute of Industrial Engineers, 1977.

Edosomwan, Johnson Aimie. *Integrating Productivity and Quality Management*. New York: Marcel Dekker, 1987.

Enrick, Norbert L. *Time Study Manual for the Textile Industry*. 2nd ed. Melbourne: Krieger, 1982.

Evans, F. "Tape Data Analysis [MTM or Methods Time Measurement]." *Work Study and Management Services* 13 (December 1969): 835-37.

Explorations in Time Use. Ottawa: Employment and Immigration Canada, 1983.

Force Limits in Manual Work. Stoneham: Butterworth Publications, 1980. 24 pages.

Gantt, H. L. "A New System of Rewarding Machine Shop Labour." *Cassier's Magazine* November, 1902.

Gershoni, H. "An Analysis of Time Study Based on Studies Made in the United Kingdom and Israel." *AIIE Transactions* 1.3 (September 1969): 244-51.

Gilbreth, Frank B. *Bricklaying System*. Chicago: Myron C. Clark, 1909.

——. *Motion Study*. 1911; reprint, Easton: Hive Publishing, 1972.

Gilbreth, Frank B., and Lillian M. Gilbreth. *Applied Motion Study*. 1919; reprint, Easton: Hive Publishing, 1973.

——. *Fatigue Study*. 1916; reprint, Easton: Hive Publishing, 1973.

——. *The Writings of the Gilbreths*. Ed. William R. Spriegel, and Clark E. Myers. Homewood: Irwin, 1953.

Gitlow, Howard S., and Shelly J. Gitlow. *The Deming Guide to Quality and Competitive Position*. Hemel Hempstead, Hertfordshire: Prentice-Hall, 1987.

Glassey, W. C. *The Theory and Practice of Time Study*. London: Business Books, 1966.

Gomberg, W. *A Trade Union View of Time Study*. New York: Prentice-Hall, 1955.

Heiland, R. E., and W. J. Richardson. *Work Sampling*. New York: McGraw-Hill, 1957.

Hockey, G. R. J., ed. *Stress and Fatigue in Human Performance*. Chichester: Wiley, 1983.

Hopeman, R. J. *Production and Operations Management*. 4th ed. Columbus: Charles E. Merrill, 1980.

Horn, C. A. "The Origins of Work Study in the UK." *Work Study and Management Services* 15 (April 1971): 260-67.

Hoxie, R. F. *Scientific Management and Labour*. New York: Appleton, 1935.

Hutchins, David. *Just in Time*. Aldershot, Hampshire: Gower Technical P, 1988.

International Labor Office. *Introduction to Work Study*. Revised ed. Geneva, 1974.

Introduction to Work Study. 3rd ed. Washington: International Labour Office, 1981.

Ishikawa, Kaoru. *What Is Total Quality Control? The Japanese Way*. Hemel Hempstead, Hertfordshire: Prentice-Hall, 1985.

Karger, Delmar W., and Franklin H. Bayha. *Engineered Work Measurement*. 3rd ed. New York: Industrial P, 1977.

Karger, Delmar W., and Walton Hancock. *Advanced Work Measurement*. New York: Industrial P, 1982.

Kearsey, H. E. "The Bedaux Work Unit Method." *Work Study and Management Services* 14 (September 1970): 725.

Kelley, James E., Jr. "CPM [Critical Path Method]: Present and Future." *New Horizons in Industrial Engineering*. Ed. Seymour M. Selig, and Morton Ettelstein. Baltimore: Spartan, 1963. 1-6.

Knight, Frank B., ed. *Essentials of Brownian Motion and Diffusion*. Providence: American Mathematical Society, 1981.

Kourim, G. *Wertanalyse*. Munich: R. Oldenbourg, 1968.

Lockyer, K. G. *Critical Path Analysis*. London: Pitman, 1964.

Louden, J., J. Keith, and J. Wayne Deegan. *Wage Incentives*. New York: Wiley, 1959.

Lowry, S. M., H. B. Maynard, and G. J. Stegemerten. *Time and Motion Study and Formulas for Wage Incentives*. 3rd ed. New York: McGraw-Hill, 1940.

McConnell, Charles. *Cost Analysis Through Work Sampling*. Chicago: American Medical Research Association, 1976. 24 pages.

Macey, Samuel L. *The History and Practice of Work Study*. Victoria: School of Public Administration, U of Victoria, 1976. 42 pages.

——. "Work Study before Taylor: An Examination of Certain Preconditions for Time and Motion Study that Began in the Seventeenth Century." *Work Study and Management Services* 18 (October 1974): 530-36.

Maynard, H. B. *Industrial Engineering Handbook*. New York: McGraw-Hill, 1963.

Maynard, H. B., G. J. Stegemerten, and J. L. Schwab. *Methods-Time Measurement*. New York: McGraw-Hill, 1962.

Menipaz, Ehud. *Essentials of Production and Operations Management*. Englewood Cliffs: Prentice-Hall, 1984. Includes bibliographies.

Merrick, Dwight V. *Time Studies As a Basis for Rate Setting*. 1919; reprint, Easton: Hive Publishing, 1980.

Methods Analysis and Work Measurement. Flint: General Motors Institute, 1963.

Milward, G. E., and P. H. S. Wroe. *Office Economy by O & M* (Organization and Methods). London: Macmillan, 1964.

Minors, D. S., and J. M. Waterhouse. "Circadian Rhythms in Deep Body Temperature, Urinary Excretion and Alertness in Nurses on Night Work." *Ergonomics* 28 (1985): 1523-30.

Monks, J. G. *Operations Management: Theory and Problems*. New York: McGraw-Hill, 1977.

Moore, F., and T. Hendrick. *Production Operations Management*. 8th ed. Homewood: Irwin, 1980.

Morrow, R. L. *Motion Economy and Work Measurement*. New York: Ronald P, 1957.

——. *Time Study and Motion Economy*. New York: Ronald P, 1946.

Mundel, Marvin Everett. *Motion and Time Study: Improving Productivity*. 6th ed. Englewood Cliffs: Prentice-Hall, 1985.

Musgrave, P. "Why Everything Takes 2.71828...Times as Long as Expected." *American Economic Review* 75 (March 1985): 250-52.

Nance, H. W., and R. E. Nolan. *Office Work Management*. New York: McGraw-Hill, 1971.

Niebel, Benjamin W. *Motion and Time Study*. 8th ed. Homewood: Irwin, 1987.

Noel-Brown, S. J. "The History of Work Study." *Work Study and Management Services* 13 (December 1969): 816-19, 822.

Oliver, S. "An Outline of Ergonomics." *Work Study and Management Services* 14 (February 1970): 117-21.

Outline of Work Study, Part 2: Method Study. London: British Institute of Management, 1955.

Outline of Work Study, Part 3: Work Measurement. London: British Institute of Management, 1957.

Perham, J. C. "Pay Off in Performance Bonuses." *Dunn's Review* 103 (May 1974): 51-55.

Polk, Edward J. *Methods Analysis and Work Measurement*. New York: McGraw-Hill, 1984. Includes bibliography.

Presgrave, R. *Dynamics of Time Study*. New York: McGraw-Hill, 1945.

Quick, J. H., J. H. Dundan, and J. A. Malcolm, Jr. *Work-Factor Time*

Standards. New York: McGraw-Hill, 1962.

Reamur, Ferchault de. *Art de l'épingler...avec des additions de M. Duhamel du Monceau, et des remarques extraites de M. Perronet*. Paris, 1762.

Rice, R. S. "Survey of Work Measurement and Wage Incentives." *Industrial Engineering* 9.7 (July 1977): 18-31.

Richmond, S. B. *Operational Research for Management Decisions*. New York: Ronald P, 1963.

Roll, Sir Eric. *An Early Experiment in Industrial Organization*. London: Frank Cass, 1968.

Salvendi, G. "Hand Size and Assembly Performance." *AIIE Transactions* 3.1 (March 1971): 32-36.

Schafer, G. *Functional Analysis of Office Requirements*. Chichester, Sussex: Wiley, 1988.

Schaffer, L. H. "Rhythm and Timing in Skill." *Psychological Review* 89.2 (March 1982): 109-22.

Schutt, W. H. *Time Study Engineering*. New York: McGraw-Hill, 1943.

Shaw, Anne G. *The Purpose and Practice of Motion Study*. London: Harlequin P, 1952.

Shell, Richard L., ed. *Work Measurement: Principles and Practice*. Norcross: Institute of Industrial Engineers, 1986.

Shumard, F. W. *A Primer in Time Study*. New York: McGraw-Hill, 1940.

Singleton, W. T., and P. Spurgeon, eds. *Measurement of Human Resources*. Melbourne: Krieger, 1975.

Sitnek, L. N. "Performance Rating." *Industrial Management* 19.1 (January-February 1977): 11-16.

Swanson, Richard A., and Deane Gradous. *Performance at Work: A Systematic Program for Analyzing Work*. New York: Wiley, 1986.

Swanston, M. T., and R. H. Crowder. "Ratings of the Speed and Efficiency of Work by Untrained Observers." *Perceptual and Motor Skills* 58 (February 1984): 275-78.

Taylor, Frederick Winslow. "A Piece-Rate System." *Cassier's Magazine* October, 1895.

——. *Scientific Management*. Comprising *Shop Management*, *The Principles of Scientific Management*, and *Testimony Before the Special House Committee*. New York: Harper and Brothers, 1947.

Thurman, J. E., et al. *Higher Productivity and a Better Place to Work: Practical Ideas for Owners and Managers of Small and Medium Sized Industrial Enterprises. Action Manual*. London: International Labour Office, 1988.

Webber, Ross A. *Time and Management*. Nutley: Moffat Publishing,

1981.
Whitmore, Dennis A. *Measurement and Control of Indirect Work.* London: Heinemann, 1970.
——. *Work Measurement.* London: Heinemann for I.W.S.O.M., 1976. Includes bibliography.
——. *Work Study and Related Management Services.* 3rd ed. London: Heinemann, 1976. Includes bibliography.
Wrege, Charles D., and Armedeo G. Perroni. "Taylor's Pig-Tale." *Work Study and Management Services* 18 (November 1974): 564-75.
Zandin, K., ed. *Most Work Measurement Systems.* New York: Dekker, 1980.

25. TIME'S MEASUREMENTS AND DIVISIONS

A. GENERAL

Asimov, Isaac. *The Clock We Live On*. New York: Collier-Macmillan, 1963.

Aveni, Anthony F. *Empires of Time: Calendars, Clocks, and Cultures*. New York: Basic Books, 1989.

——, ed. *World Archaeoastronomy*. Cambridge: Cambridge UP, 1989.

Bibliographie générale de la mesure du temps, suivie d'un essai de classification technique et géographique. Paris: Tardy, 1947.

Bilfinger, Gustav. *Zeitrechnung*. Stuttgart: C. Leibich, 1883-1901.

Blair, Byron Emerson. *Time and Frequency: A Bibliography of NBS Literature Published July 1955-December 1970*. Washington: U.S. National Bureau of Standards, 1971. 50 pages.

Carrasco, D. *Quetzalcoatl and the Irony of Empire: Myths and Prophecies in Meso-America*. Chicago: U of Chicago P, 1982.

Carrasco, D., and Jane Marie Swanberg, eds. *Waiting for the Dawn: Mircea Eliade in Perspective*. Boulder: Westview P, 1985.

Chevallier, R. *Aiôn: le temps chez les Romains*. Paris: A. and J. Picard, 1976.

Cordoliani, A. "Contribution à la littérature du comput au Moyen Age." *Studi Medievali* Ser. 3, Vol. 1 (1960):107-38; Vol. 2 (1961): 169-208. Includes bibliography.

Grotefend, Hermann. *Taschenbuch der Zeitrechnung des deutschen Mittelalters und der Neuzeit*. Hanover: Verlag Hahnsche Buchhandlung, 1971.

Gumpach, Johannes von. *Die Zeitrechnung der Babylonier und Assyrer*. 1852; reprint, Walluf bei Wiesbaden: Sändig, 1972.

Jones, C. W. *Bedae opera de temporibus*. Ithaca: Cornell UP, 1943.

——. *Bedae pseudepigrapha: Scientific Writings Falsely Attributed to Bede*. Ithaca: Cornell UP, 1939.

——. "The Lost Sirmond Manuscripts of Bede's Comptus." *English Historical Review* 52 (1937): 204-19.

Kubitschek, Wilhelm. *Grundriss der antiken Zeitrechnung*. Munich: Beck, 1928.

McCarthy, Dennis D., and John D. Pilkington, eds. *Time and the Earth's Rotation*. Norwell: Kluwer Academic, 1979.

O'Daly, G. J. P. "Augustine on the Measurement of Time: Some Comparisons with Aristotelian and Stoic Texts." *Neoplatonism and*

Early Christian Thought. Ed. H. J. Blumenthal, and R. A. Markus. London: Variorum, 1981. 171-79.

Poole, Reginald L. *Medieval Reckonings of Time*. New York: Gordon P, 1977.

Richard, M. "Comput et chronographie chez Saint Hippolyte." *Mélanges de science religieuse* 7 (1950): 237-57.

Samuel, Alan E. *Greek and Roman Chronology: Calendars and Years in Classical Antiquity*. Munich: Beck, 1972.

Stevens, Wesley M. *Bede's Scientific Achievement*. Jarrow, Durham: St Paul's Church, 1985. 44 pages.

——. "Sidereal Time in Anglo-Saxon England." *Sutton Hoo: Voyage to the Other World*. Ed. C. B. Kendall, and P. S. Wells. Minneapolis: U of Minnesota P, 1990.

Tannenbaum, Beulah, and Myra Stillman. *Understanding Time: The Science of Clocks and Calendars*. New York: Whittlesey House, 1958.

Tedlock, Barbara. *Time and the Highland Maya*. Albuquerque: U of New Mexico P, 1982.

Tedlock, Dennis. *Days from a Dream Almanac*. Urbana: U of Illinois P, 1990.

Thompson, J. E. S. *The Rise and Fall of Maya Civilization*. Norman: U of Oklahoma P, 1966.

B. CALENDARS

Achelis, Elisabeth. *The World Calendar: Addresses and Occasional Papers Chronologically Arranged on the Progress of Calendar Reform since 1930*. Ann Arbor: Gryphon Books, 1971.

Alavi, Is'haqunnabi. *The Arab Calendar Prevalent during the Lifetime of Muhammad: An Indian Discovery*. Compiled and trans. by Abid Raza Bedar. Delhi: Rampur Institute of Oriental Studies, 1968. 48 pages.

Alkon, Paul. "Changing the Calendar." *Eighteenth-Century Life* 7.2 (January 1982): 1-18.

Bergmann, Werner. "Der römische Kalender: Zur sozialen Konstruktion der Zeitrechnung." *Saeculum* 35 (1984): 1-16.

Boll, Franz J. *Griechische Kalender*. 5 vols. Heidelberg: C. Winter, 1910-20.

Borst, Arno. *Computus: Zeit und Zahl in Mittelalter*. Cologne: Bohlau, 1988.

Bredon, Juliet, and Igor Mitrophanow. *Das Mondjahr: chinesische Sitten, Bräuche und Feste*. Berlin: P. Zsolnay, 1937.

Casas, Johanna Broda de. *The Mexican Calendar as Compared to Other Mesoamerican Systems*. Vienna: Engelbert Stiglmayr, 1969.

Coler, Johann. *Calendarium perpetuum et libri oeconomici*. Leipzig: F. Schnelboltz, 1598.

Colson, Francis Henry. *The Week: An Essay on the Origin and Development of the Seven-Day Cycle*. Westport: Greenwood, 1974.

Harvey, O. L. *The Chinese Calendar and the Julian Day Number*. Silver Spring: Harvey, 1977. 30 pages.

——. *Miscellaneous Calendars*. Silver Spring: Harvey, 1977. 31 pages.

Hawkins, G. C. *Stonehenge Decoded*. New York: Doubleday, 1965.

Herberger, Charles F. *The Thread of Ariadne: The Labyrinth of the Calendar of Minos*. New York: Philosophical Library, 1972.

Jones, C. W. "The Victorian and Dionysiac Paschal Tables in the West." *Speculum* 9 (1934): 408-21.

Kaletsch, Hans. *Tag und Jahr: die Geschichte unserers Kalenders*. Zurich: Artemis, 1970.

Kaltenbrunner, Ferdinand. *Die Vorgeschichte der gregorianischen Kalenderreform*. Vienna: K. Gerold, 1876.

Kiyosi, Yabuuti. "The Calendar Reforms in the Han Dynasties and Ideas in their Background." *Archives internationales d'histoire des sciences* 24 (1974): 51-65.

Kubitschek, Wilhelm. *Die Kalenderbücher von Florenz, Rom und Leyden*. Vienna: A. Hölder, 1915.

Landsberger, Benno. *Der kultische Kalender der Babylonier und Assyrer, I. Die altbabylonischen Lokalkalender*. Leipzig: A. Preis, 1914.

Lechner, Anton. *Mittelalterliche Kirchenfeste und Kalendarien in Bayern*. Freiburg: Herder, 1891.

Lunardi, Ernesto. *Il calendario maya: acienza, religione, cadenze esistenziali di una grande civiltà*. Genoa: AISA, 1984.

Lyle, E. "Archaic Calendar Structure Approached through the Principle of Isomorphism." *Semiotica* 61 (1986): 243-57.

Malmstrom, V. H. "Origin of Mesoamerican 260 Day Calendar." *Science* 181 (1973): 939-41.

Moyer, Gordon. "The Gregorian Calendar." *Scientific American* 246.5 (May 1982): 144-53.

O'Neil, W. M. *Time and the Calendars*. Sydney, Australia: Sydney UP, 1975.

Parise, Frank, ed. *The Book of Calendars*. New York: Facts on File, 1982.

Patrick, J. "Midwinter Sunrise at Newgrange." *Nature* 249 (1974): 517-19.

Piper, Ferdinand. *Karls des Grossen Kalendarium und Ostertafel*. 1858;

reprint, Walluf bei Wiesbaden: Sändig, 1974.

Price, Derek de Solla. "Gears from Greeks: The Antikythera Mechanism—a Calendar Computer from ca. 80 B.C." *Transactions of the American Philosophical Society* n.s., 64 (1974): 1-70.

Rohner, Ludwig. *Kalendergeschichte und Kalender*. Wiesbaden: Akademische Verlagsgesellschaft Athenaion, 1978.

Schram, Robert Gustav. *Kalendariographische und chronologische Tafeln*. Leipzig: J. C. Hinrichs, 1908.

Spier, Arthur. *The Comprehensive Hebrew Calendar: Its Structure, History, and One Hundred Years of Corresponding Dates, 5660-5760/-1900-2000*. New York: Behrman House, 1952.

Tedlock, Barbara. *Time and the Highland Maya*. 2nd ed. Albuquerque: U of New Mexico P, 1987.

Wilson, H. A., ed. *The Calendar of St. Willibrord*. London: Henry Bradshaw Society, 1918.

Zerubavel, Eviatar. *Hidden Rhythms: Schedules and Calendars in Social Life*. Chicago: U of Chicago P, 1981.

——. *The Seven-Day Circle: The History and Meaning of the Week*. New York: Free P, 1985.

Ziolkowski, M. S., and R. M. Sadowski, eds. *Time and Calendars in the Inca Empire*. Oxford: British Archaeological Reports, 1989.

C. CHRONOLOGY

Adler, William. *Immemorial: Archaic History and Its Sources in Christian Chronography from Julius Africanus to George Syncellus*. Washington: Dumbarton Oaks, 1989.

al-Biruni. *The Chronology of Ancient Nations*. Trans. and ed. C. Edward Sachau. 1879; reprint, Frankfurt: Minerva, 1969.

Alkon, Paul K. "Johnson and Chronology." *Greene Centennial Studies*. Ed. Paul J. Korshin and Robert R. Allen. Charlottesville: UP of Virginia, 1984. 143-71.

Ball, R. "Generational Dating in Herodotus." *Classical Quarterly* n.s., 29 (1979): 276-81.

Bickerman, E. J. *Chronology of the Ancient World*. 2nd ed. London: Thames and Hudson, 1980.

Bond, John James. *Handy-Book of Rules and Tables for Verifying Dates with the Christian Era. Giving an Account of the Chief Eras, and Systems Used by Various Nations*. 1889; reprint, New York: Russell and Russell, 1966.

Borst, A. "Computus: Zeit und Zahl im Mittelalter." *Deutsches Archiv*

44.1 (1988): 1-82.

Burkhardt, Johannes. *Die Entstehung der modernen Jahrhundertrechnung.* Groppingen: A. Kümmerle, 1971.

Chatelain, J. C. Vital. *Le Père Denis Petau d'Orléans, Jésuite. Sa vie et ses oeuvres.* Paris, 1884.

Frank, Edgar. *Talmudic and Rabbinical Chronology.* Spring Valley: Feldheim, 1978.

Gahlot, Sukhvir Singh. *Historians' Calendar: Giving Equivalent Dates of Christian, Vikram, Shaka, Hijari, Bengali, and Kollam Eras. Vol. I: 1544 A.D.-1643 A.D.* Jodhpur: Hindi Sahitya Mandir, 1979.

Ginzel, F. K. *Handbuch der Chronologie.* 2 vols. Leipzig, 1911.

——. *Handbuch der mathematischen und technischen Chronologie.* 3 vols. 1906-14; reprint, Leipzig: J. C. Heinrichs, 1958.

Grafton, Anthony. "Joseph Scaliger and Historical Chronology: The Rise and Fall of a Discipline." *History and Theory* 14.2 (1975): 156-85.

Haber, Francis C. *The Age of the World.* 2nd ed. Westport: Greenwood, 1978.

Hale, Sir Matthew. *The Primitive Origination of Mankind.* London: William Shrowsbury, 1677.

Hardouin, Jean. "Le fondement de la chronologie de M. Newton, Anglois, imprimé à Londres en 1726, sappé le P.J.H." *Mémoires de Trèvoux* 29 (1729): 1570-77. Reprint, Geneva: Slatkine Reprints, 1968.

Hewitt, James Francis Katherinus. *Primitive Traditional History: The Primitive History and Chronology of India, South-eastern and South-western Asia, Egypt, and Europe, and the Colonies Thence Sent Forth.* 2 vols. London: J. Parker, 1907.

Jones, C. W. *Saints' Lives and Chronicles in Early England.* Ithaca: Cornell UP, 1947.

Ketkar, Venkatesh Bapuji. *Indian and Foreign Chronology, with Theory, Practice and Tables. B.C. 3102 to 2100 A.D. and Notices of the Vedic, the Ancient Indian, the Chinese, the Jewish, the Ecclesiastical and the Coptic Calendars.* Bombay: British Indian P, 1923.

Krusch, Bruno. *Studien zur christlich-mittelalterlichen Chronologie: die Entstehung unserer heutigen Zeitrechnung.* Berlin: Akademie der Wissenschaften, 1938.

La Harpe, Jean de. *Genèse et mesure du temps.* Neuchâtel: Université, 1941.

Lietmann, Hans. *Zeitrechnung der römischen Kaiserzeit, des Mittelalters und der Neuzeit für die Jahre 1-2000 nach Christus.* Berlin: W. de Gruyter, 1934.

Mahler, Eduard. *Handbuch der jüdischen Chronologie*. 1916; reprint, Hildesheim: G. Olms, 1967.
Maurice, Thomas. *The History of Hindostan*. Ed. Burton Feldman, and Robert D. Richardson. New York: Garland, 1984.
Musset, L. *The Germanic Invasions: The Making of Europe, A.D. 400-600*. Trans. E. James and C. James. London: Elek, 1975.
Neugebauer, Paul V. *Astronomische Chronologie*. Berlin: W. de Gruyter, 1929.
Patrides, C. A. "Renaissance Estimates of the Year of Creation." *Huntington Library Quarterly* 26 (1963): 315-22.
Petavius, Domenicus. *Opus de doctrina temporum*. Ed. J. Hardouin. 3 vols. Venice, 1705.
——. *Rationarium temporum*. Paris, 1633.
Richardson, C. C. "The Quartodecimans and the Synoptic Chronology." *Harvard Theological Review* 33 (1940): 177-90.
Rühl, F. *Chronologie des Mittelalters und der Neuzeit*. Berlin, 1897.
Rosa, Pietro di. "Denis Petau et la cronologia." *Archivium historicum Societatis Jesu* 29 (1960): 3-54.
St. Allais, M., ed. *L'art de verifier les dates*. 41 vols. Paris, 1819-44.
Scaliger, Joseph. *Isagogici canones*. In vol. 2 of *Thesaurus temporum*. Lyons, 1606. Reprint, Osnabrück: Otto Zeller, 1968.
——. *Opus novum de emendatione temporum*. Paris, 1583.
Steinberg, Sigfrid Henry. *Historical Tables, 58 B.C.-A.D. 1972*. 9th ed. New York: St. Martin's P, 1973.
Teres, Gustav. "Time Computations and Dionysius Exiguus." *J for the History of Astronomy* 15 (1984): 177-88.
Turriano, Juanelo. *Breve discurso a su Majestad el Rey Catolico en torno a la reduccion del año y reforma del calendario*. Madrid: Fundacion Juanelo Turriano, 1990.
Ussher, James. *Annals of the World*. Oxford, 1658.
——. *Chronologia sacra*. Oxford, 1660.
Villani, Giovanni. *Cronica*. 4 vols. Florence, 1844-45.
Wall, C. Edward, ed. *Book of Days: An Encyclopedia of Information Sources on Historical Figures and Events, Keyed to Calendar Dates*. Ann Arbor: Pierian P, 1987.
Walton, Robert C. *Chronological and Background Charts of Church History*. Grand Rapids: Zondervan Publishing, 1986.
[Webb, Jane]. *Conversations Upon Comparative Chronology and General History from the Creation of the World to the Birth of Christ*. London, 1830.
Wilcox, Donald J. *The Measure of Times Past: Pre-Newtonian Chronologies and the Rhetoric of Relative Time*. Chicago: U of Chicago P,

1987.

D. DIVISIONS OF THE DAY

Bartky, I. R. "The Adoption of Standard Time." *Technology and Culture* 30 (January 1989): 25-56.

Bilfinger, Gustav. *Die antiken Stundenangaben.* Stuttgart: W. Kohlhammer, 1888.

——. *Die mittelalterlichen Horen und die modernen Stunden.* 1892; reprint, Wiesbaden: Sändig, 1969.

Bolton, Lyndon. *Time Measurement: An Introduction to Means and Ways of Reckoning Physical and Civil Time.* London: G. Bell, 1924.

Cowan, Harrison J. *Time and Its Measurement: From the Stone Age to the Nuclear Age.* Cleveland: World Publishing, 1958.

Dugmore, C. W. "Canonical Hours." *A Dictionary of Liturgy and Worship.* Ed. John G. Davies. New York: Macmillan, 1972.

Gasquet, Cardinal, trans. "Of Daily Manual Labour." *The Rule of Saint Benedict.* London: Medieval Library, 1925. Ch. 48.

Hellyer, Brian. *Man the Timekeeper.* London: Priory P, 1974.

Hood, Peter. *How Time Is Measured.* 2nd ed. London: Oxford UP, 1969. 64 pages.

Krauss, Joseph. *Vom Messen der Zeit im Wandel der Zeiten.* Wolfshagen-Scharbeutz: F. Westphal, 1950.

Leclercq, Jean. "Experience and Interpretation of Time in the Early Middle Ages." *Studies in Medieval Culture, Vol. 5.* Ed. John R. Sommerfeldt, et al. Kalamazoo: Medieval Institute, Western Michigan U, 1975.

Lunardi, Heinrich, et al. *Beim dritten Ton war es genau.... Die Ahnen unserer Uhr.* Basel: GS-Verlag, 1976.

Nilsson, Martin P. *Primitive Time Reckoning.* Lund, Sweden: C. W. K. Gleerup, 1920.

Pollak, O. B. "Efficiency, Preparedness and Conservation: The Daylight Savings Time Movement." *History Today* 31 (March 1981): 5-9.

Review of British Standard Time. London: Her Majesty's Stationery Office, 1970.

Risley, Allan S. *The National Measurement System for Time and Frequency.* Washington: U.S. Department of Commerce, National Bureau of Standards, 1976. 64 pages.

Sethe, Kurt H. *Die Zeitrechnung der alten Aegypter im Verhältnis zu der der andern Völker.* Göttingen, 1919-20.

Springer, Max. *Mensch, Zeit, Uhr: zur Geschichte der Zeitmessung*. Berlin: Ullstein, 1927.

Standard Time in the United States: A History of Standard and Daylight Saving Time in the United States and an Analysis of the Related Laws. Washington: U.S. Department of Transportation. Office of Assistant General Counsel for Regulation, 1970. 27 pages.

Welch, Kenneth F. *Time Measurement: An Introductory History*. Newton Abbot: David and Charles, 1972.

AUTHOR INDEX

When an author is not always listed in the text with the same number of first names, the Author Index generally restricts his or her initials to the names common to all entries.

SUBJECT INDEX

Page numbers in bold print refer to subjects that form all or part of the title of a section or subsection in the text.